UFO Attacks in Brazil

The True Story of the Most Terrifying Events in the History of Ufology

UFO ATTACKS IN BRAZIL

The True Story of the Most Terrifying Events
in the History of Ufology

Thiago Luiz Ticchetti

ISBN 978-1-969800-99-3

Publisher:

Un-X Media

PO Box 1166

Independence, MO 64051

www.unxmedia.com

UNXMEDIA

PUBLISHING

ACKNOWLEDGEMENTS & DEDICATION

First and foremost, I express my deepest gratitude to all those who walked alongside me during the research, writing, and publication of this book.

To my family - my wife Mariana, my children Luiz Guilherme and Marcela - and to my extended family and friends: your unwavering support, love, and understanding throughout this long journey gave me the strength to keep going.

To my colleagues, fellow researchers, and field investigators who generously shared their time, insights, and experiences - your contributions and dedication have enriched every page of this work.

I am grateful to George Knapp for our recent, yet already strong friendship, Phillip Mantle and to Jeremy Corbell, who has also been an invaluable partner on this path.

A Special Thanks to:

Fernanda Pires - Director of MUFON CAG International, Regional Director for Canada, Central & South America, MUFON ERT, Founder of (CIFE) Scientific Channel of UFOs Phenomena & Space Research, International Consultant, Revista UFO Columnist, Documentary Producer, Screenwriter, Encounters Latin America Film Producer in Brazil, Co-author & Producer of the Book "Incident in Varginha: Space Creatures in The South of Minas", and Margie Kay - President of KUNX Digital Broadcasting Network Corp., Un-X Media Publishing/Films, Founding Member of the Hollywood Disclosure Alliance, MUFON Central Region Director, and Assistant State Director of Missouri MUFON - thank you both for your essential roles in coordinating, reviewing, and supporting this project with vision and tireless commitment. Fernanda, your passion for ufology and your remarkable ability to connect people across borders made this book possible and Un-X Media Publishing/Films for believing in me and publishing Colares Book.

I also thank the late and unforgettable A. J. Gevaerd. Without him, I would never be where I am in ufology. My gratitude extends to the entire Revista UFO magazine team and to all the researchers who contributed to this book - whether through their invaluable bibliographies or through countless hours of conversations about the facts and mysteries of the UFO phenomenon.

Finally, to every reader who believes that looking up at the sky is still worth it - may this book inspire you to keep questioning, researching, and seeking the truth.

And above all, I dedicate this book to the witnesses, experiencers, and countless individuals who opened their lives and stories to me - especially the victims, witnesses, and researchers of the "Chupa-Chupa" phenomenon. You are all part of the history of ufology. Thank you for your courage and trust.

CONTENTS

My Introduction to UFOs

When I was 8 years old, I had my first and so far only sighting of a UFO. It was in 1982, in Rio de Janeiro. It was night when I saw a sphere of red light descend from the sky and "disappear" behind the Pedra de Gávea (Mountain in the Rio de Janeiro). Secondly, the object reappeared, began to rise and increase its brightness. In the blink of an eye, it sped away and disappeared into space.

Figure 1. Artistic representation of the object seen by Thiago Ticchetti in 1982. (Credit: Author files).

I already liked the subject, but this experience caused an interest in Ufology to awaken inside me. From that day on my reading was divided between ufological themes and comic books.

Much of this interest came from my father, because he really encouraged me. As a pilot of the (FAB) Brazilian Air Force he never had a sighting during his flights, but as a young man he twice saw a classic flying saucer along with my aunt, his sister, in São Paulo. As I grew older, my interest also increased.

When I was 17, in the early 1990's, my father worked at the 'A2', the Air Force Intelligence Center. He got divorced from my mother and almost every Friday my brother and I took a bus to go to his house, where we stayed until Sunday.

Knowing my interest in Ufology, my father took copies of reports that were published in the newspapers, which were filed in the A2 and gave them to me. However, that was not enough for me. Every time I asked, *"Dad, show me the documents that you have filed concerning flying saucers",* He always said he had nothing. I kept insisting, and my request was becoming automatic, until one day I finally got a result!.

My father called a soldier and asked him to fetch two boxes of files and bring them to his office. My bones froze and my heart quickened. *"I would be seeing documents that few people had seen",* I was sure.

With the two boxes in front of me I ran to open them. Inside there were many documents stamped, "Secret." They were transcripts of dialogues between pilots and flight controllers, documents with graphs, numbers and drawings of UFOs. But suddenly something caught my attention: a fantastic photo of a UFO taken from inside the cockpit of an airplane. The photo was perfectly clear and sharp. The object was a flying saucer of the classic type, but without windows. It seemed to be a single piece of silvery metal. The photo had been taken during day, and the sun reflected on the structure of the UFO.

I asked my father if I could have the picture. He said no. I asked if I could copy it. He denied it again, but gave me 10 minutes to copy, by hand, whatever I wanted. Well, that's what I did. Unfortunately, I took the smaller cases so I could have more of them, but I was wrong in choosing quantity rather than quality.

Years later, the Brazilian Air Force released thousands of pages of documents that were secret. But none of the ones I copied that day, much less the modern UFO picture that have been released to the public.

This is just one example of what we encounter here in Brazil. Brazilian Ufology ranks among the most important in the world, thanks to the work of pioneers such as Irene Granchi, Walter Bhüller, Silvio Pereira, Claudeir Covo, Marco Antônio Petit, and Ademar Gevaerd - researchers who have built, and continue to build, a lasting legacy for future generations. I count myself among those deeply influenced by their dedication and discoveries.

This book will unveil in a way never before presented - a phenomenon that once terrorized communities in the states of Pará and Maranhão, in northern and northeastern Brazil: the infamous Chupa-Chupa, also known as The Light or Vampire Lights. These strange events began in 1977 and persisted through 1978, with some cases reported as late as 1982. They ultimately led to the creation of two covert missions by the Brazilian Air Force, which became known as Operation Saucer.

The city of Colares, in the state of Pará, was one of the epicenters of the attacks carried out against the local population by these mysterious lights, craft, and beings. Fear for their lives gripped the residents. In these pages, the reader will encounter never-before-published accounts from that period, along with more recent testimonies - offering an unparalleled glimpse into another facet of the enigmatic UFO phenomenon.

This book also presents dozens of photographs, illustrations, images, and newspaper clippings from the time, documenting these UFOs, their assaults, and the widespread panic they caused. Nearly fifty years later, we are still left without answers as to why these lights, craft, and beings targeted us, draining the blood of residents in multiple cities.

The fact is: something real attacked people in the north and northeast of Brazil and UFO activities are still a reality in those regions.

Preface

This is the story of how a journalist from a colorful city in the Mojave Desert had the good fortune to travel to a rustic but beautiful community in the Amazon delta. And it explains how an American TV reporter met a trusted colleague living on another continent, thousands of miles away from the bright lights of Las Vegas. I knew of Thiago Ticchetti when this adventure began because of his excellent reputation, but I had no idea we would take such a deep dive into one of the most important but least understood UFO incidents in history, and I did not expect that we would become friends as well as colleagues. The journey that led me to finally visit Colares has a few twists and turns, and I hope the reader will patiently tolerate my rambling explanation.

Government agencies, mainstream science, major media outlets and an army of self-appointed experts and influencers have engaged in a relentless, widespread, and ongoing effort to dissuade, discourage, and intimidate millions of people around the world who claim to have witnessed genuine UFO. Despite the enormous power of those institutions and personalities, neither the public nor the UFO themselves, seem to have received the memo. Public interest in these unknown objects, craft, orbs, lights, whatever they are, has ebbed and flowed over the decades, but has never gone away. At this moment, toward the end of 2024, an aroused citizenry all over the planet is asking tough questions, pressuring governments and military leaders for answers, and is refusing to accept the feeble lack of clarity and the often-ridiculous attempts to debunk astounding, well-documented encounters that have characterized UFO discussions for the past 80 years. Vast amounts of solid data, derived from hundreds of thousands of credible UFO incidents, has been accumulated by secretive programs and investigations conducted by multiple governments over several decades, and by civilian organizations born out of the frustration that erupts when people know they are being fed lies and disinformation.

As an American journalist who has covered this complex subject for nearly four decades, my opinion is that the U.S. government, in particular, has used its enormous influence to keep its allies and trading partners in line with a longstanding but unwritten edict that the public either doesn't have a right to know the truth, or can't handle it. Mention the name "Colares" and anyone who is not a serious student of the subject might be puzzled. But anyone who has spent more than a few months of digging into the history of the flying saucers or rumored aliens should know the significance. Of all the encounters, all the cases and files, all the stove-piped investigations by shadowy governments, Colares is in a category of its own. It is an oddball, an outlier, a very weird chapter in a seemingly preposterous premise, namely, that an unknown intelligence, possibly non-human, unleashed a fleet of spectacular vehicles which exhibited technological capabilities far beyond what was known to exist in 1977. The craft, many of them matching the description of "flying saucers" depicted in 1950's sci fi movies, descended from the sky, or rose from underwater, or materialized from under the earth and then began to attack innocent residents of a peaceful community living on an isolated island in Brazil. That is the essence of the story, one that has been told and retold in many versions and on numerous media platforms over the past five decades. UFOs came out of nowhere and started attacking random people on Colares with what appeared to be some sort of energy beams. Hundreds, perhaps thousands of people reported serious injuries from the attacks. Frightened resident of Colares fled their homes in an attempt to escape the unknown menace. The Brazilian Air Force launched an official and very serious investigation that lasted for months.

The brief description above is the essence of what most people, even UFO devotees, know about the Colares incidents. It is probably how I might respond if someone asked me what I knew. But telling the story in such broad strokes does not come close to giving it the scrutiny it deserves. In all of the millions of eyewitness accounts and fuzzy photos and secret files collected since 1947, there is nothing like it. Whoever launched those unknown objects did, in fact, target and injure hundreds of humans. There are voluminous files which provide intricate details about the shape, size, and capabilities of these mysterious trans-medium machines, craft that displayed astounding speed and power in the sky and in the waters of the Amazon River delta. There is no credible argument to be made that any military or government or corporation or secret cabal on Earth possessed the technological sophistication to build and deploy a fleet of seemingly magical machines in 1977. It is unlikely any power on Earth has such technology even today. But the central question remains. What was the objective? Why did this happen? Why were these people, living in this remote place, relatively unconnected to and unknown by most of the outside world, targeted by whoever controlled the mystery machines? I have always been curious about Colares and have struggled for years to get a more comprehensive assessment of what really happened and how it can possibly make sense. There have been other incidents where humans were injured during UFO encounters. Most of those injuries seems to have been incidental, a result of being in the wrong place at the wrong time. There are few credible cases in which UFO deliberately attacked people, and even fewer where multiple people were attacked and injured. In my opinion, Colares is as significant as the 1947 Roswell crash, the 1952 Washington DC overflights, the 2004 Tic Tac encounter, and the well-documented incidents in which unidentified object of unknown origin exhibited their ability to control our nuclear weapons. Colares is enormously important, even if we do not fully understand it half a century later.

I am a journalist in Las Vegas Nevada, USA. My interest in the UFO topic started percolating in 1987 when I conducted an unusual interview that aired on KLAS TV, my employer since 1981. The interview was with a highly experienced pilot named John Lear who, for a time, had flown cargo planes for the CIA. In that interview, Lear alleged that an ongoing coverup about the reality of UFO was being carried out by U.S. Intelligence agencies, defense contractors, and the military, that media companies had been compromised, that government operatives actively spread disinformation, and that civilian allies spied upon UFO researchers and organizations. The big secret, Lear claimed, is that a few governments around the world have obtained and hidden UFO technology, including intact craft, as well as the dead bodies of odd beings found inside the craft or in the wreckage of crash sites. It seemed preposterous when I first heard this, but it aroused my curiosity enough that I started reading. The more I read, the more curious I became. And while it was apparent that nearly all UFO cases---90 to 95 percent—are likely explainable in prosaic ways, the remaining percent are difficult to dismiss.

My involvement in the UFO topic was a combination of luck and coincidence. I wrote and produced a 9-part series about what was clearly a complicated subject, and one I found to be monumentally significant. If the allegations proved to be true, it would be the biggest news story, maybe the biggest event in all of human history. My boss, news director Robert Stoldal, fully supported my interest in continuing to investigate UFO stories, including a shadowy military base that, for a time, supposedly did not exist. Few people knew of it back then, but today, nearly everyone is aware of Area 51. Those news reports in the late 80's and early 90's led to discreet collaboration –and friendships--with two men who became monumentally influential figures in the UFO arena, and whose actions decades ago are directly responsible for the tumultuous changes that are unfolding today. The stigma that has discouraged serious inquiries into UFO matters is finally, if reluctantly, melting away.

One of the men was Harry Reid. Few could have predicted back in 1989 that a freshman senator from a sparsely populated state would eventually rise to become the Majority Leader of the US Senate, one of the most powerful positions in the entire government. The second individual was businessman Robert Bigelow, who built a hotel chain and real estate holding into a multi-billion dollar empire. Bigelow had a lifelong interest in UFO and related subjects. Reid wasn't all that familiar with the history of UFO investigations, but he had a curious and open mind. In early 1996, I saw an opportunity arose to bring those two men together. I had no idea that their meeting that year would be a central reason why I would be on a plane to Brazil some 28 years later. Reid and Bigelow stayed in touch with each other, and I stayed in touch with both.

In 2007, senior intelligence analysts with the Defense Intelligence Agency (DIA) proposed the creation of a new UFO program, an investigation to be overseen by the DIA but conducted by a private entity. Senator Reid was the primary sponsor of the program. Robert Bigelow created a new business entity, with the acronym BAASS, and was awarded the government contract. As far as the public knows, the program that sprung from that agreement was the largest acknowledged government-funded UFO investigation in history. Within a few months, BAASS had hired fifty full time investigators, who were dispatched to investigate new UFO sighting reports as well as hotspots where UFO sightings sometimes overlapped with other unusual phenomena. (One of those locations was a property in Utah that had been purchased by Bigelow. Today, it is known by the name Skinwalker Ranch.) The public, the Pentagon, the CIA, and even most of the analysts inside the DIA had no idea this program existed. But in a mere 27 months, the effort produced an enormous body of work, including more than 100 highly detailed investigations of UFO cases and encounters, including the very first official investigation of the Tic Tac incident, in which a flotilla of US Navy warships, including the USS Nimitz aircraft carrier, detected dozens of brief intrusions by unknown objects. Another accomplishment by the BAASS team was the creation of a voluminous UFO data warehouse, which contained files on more than 260,000 cases and incidents from around the world. The man who designed the data warehouse was Dr. Jacques Vallee, who was not only an early pioneer in computer technology but was a distinguished astrophysicist who had written many of the best known and most respected books about the UFO mystery ever published.

Valle was among the first outside investigators to travel to Colares and speak to witnesses. Vallee's personal history with the Colares events was a significant factor in the decision by BAASS to send investigator to Brazil to make contact with the Brazilian government, in particular the Brazilian Air Force, which had conducted a lengthy investigation of the Colares attacks---as well as subsequent events. In 1977 when the attacks occurred, Brazil's government was controlled by the military. The investigation was conducted in secrecy. Courageous journalists who tried to report what was happening encountered serious opposition from the top of Brazil's power structure. Many years later, after the junta was replaced by a civilian government, UFO activists (including Thiago) lobbied for the release of files stashed inside military archives, and they succeeded. Hundreds previously hidden pages were released. But did the public get to see everything? That is one of the questions addressed in this book.

The team from BAASS spent months negotiating with Brazilian officials and were successful in obtaining copies of key materials including the Colares documents. I am reasonably confident in saying they may have obtained considerable information that had not been released to the public by the Air Force. The files went into the BASSS data banks and were analyzed then incorporated into thick reports submitted to DIA. Again, it must be noted that none of the actual documents have been made public in the US, but I am able to describe some of those contents, including the Colares material, and the reason, again, is luck. As a civilian without a security clearance, I was not allowed to see the work product produced for DIA. But in 2018, after BAASS had ended and the DIA investigation was finished, Senator Reid invited me to Washington for a meeting with the DIA scientist who was a driving force in the creation of the program, His name was Dr. James Lacatski. He was proud of the work that had been accomplished under his directorship of the program and he indicated a willingness to help get some of the information out to the public. That discussion with Lacatski and Reid later resulted in a partnership that peeled back the layers of secrecy surrounding Colares…and the BAASS investigation. Dr. Lacatski agreed to seek authorization to write about the DIA effort, which operated another acronym—AAWSAP. He asked a friend and colleague Dr, Colm Kelleher to be a partner in this undertaking. Kelleher, who earned his doctorate in Biochemistry, had worked for Bigelow as part of a previous program which investigated UFO and had been the hands-on program manager in Las Vegas for the BAASS/AAWSAP endeavor, which had been conducted in total secrecy in one of the most vibrant cities in the world. Colm Kelleher and I had collaborated years earlier on a book about the Utah Ranch, Hunt for the Skinwalker. That book is what caught the attention of Lacatski and another DIA intelligence official named Jay Stratton and it was the spark that led to the creation of the DIA program.

Lakatski, Kelleher and I wrote a detailed book about the BAASS/AAWSAP investigation. The manuscript had to be approved by Pentagon censors, a process that took a laborious 14 months to complete. Because of the research on the book, I was able to familiarize myself with some key files, including Colares. In our book (Skinwalkers at the Pentagon) the authors offered this description of the Colares events. "Several different types of unidentified airborne phenomena were reported—some big, some small, saucer-shaped cigar shaped, barrel shaped, luminous or not. Some objects were observed originating from under water, flying out of the water or into the water. A number of Colares inhabitants reported being targeted by strange lights, beamed from the objects, with many describing similar ill effects ad a few deaths." The DIA data warehouse revealed much more about the Colares files and the high ranking military officials who assisted in transferring the materials. Our book reports that a substantial number of photos, along with hour of film of UFO events, were collected by the Brazilian military, som of it is said to be crystal clear images of UFOs emerging form rivers and hovering directly in front of the film cameras. It is not clear where the film and photos are stored today. Perhaps Thiago can get to the bottom of this. Someone, somewhere, has a storehouse of data and documents about Colares, as well as other UFO matters investigated in Brazil. My interest in Colares has never wavered. I was able to revisit writings by Jacques Vallee and to meet with him face to face to discern what he knows. Dr, Vallee is, in my opinion, the deepest thinker to ever write about this strange subject. He cautioned me that some of the more sensational details about the attacks on Colares residents may not be as dark as they seem on the surface. I was not sure what he meant then, but I understand it now because I was able to visit Colares myself and to meet the people who saw these events with their own eyes. I was approached by a television producer who was working with Netflix, the huge platform which streams content, including movies and documentaries, to 190 countries in 124 languages. Netflix has 280 million subscribers, and a much larger number of total viewers it is huge. The producer Aengus James invited me to be part of a Netflix project looking at the broad questions about UFOs. We discussed possible cases and locations, and I was delighted to learn that a trip to Colares was on our schedule. And, as it turned out, the production team that traveled with us was comprised of top-notch talent. I have had the pleasure of visiting Brazil previously, but those other trips were nothing like this one. Colares is not an easy place to find. We spent days in Belem, then recorded interviews in Barcarena, then took two ferries across rivers to get to Colares. The people were friendly and curious. The accommodations were spartan by bit city standards but that was part of the sense of adventure for our journey. At every step of the way, we relied on the good sense of Thiago. I can't imagine how we would have been able to get what we wanted without him.

His easy-going manner and knowledge of UFO history is deep and impressive, and he was an excellent translator as we conducted sensitive interviews with people who had been attacked decades earlier, suffered still-visible injuries, and were still trying to understand what had happened to them and why they were chosen. Before this trip, I already knew about Thiago because I owned an encyclopedic book he wrote about the astonishing history of UFO incidents in his country (UFO Contacts in Brazil.) It has been many centuries since Copernicus proved that Earth is not the center of the universe. I must humbly admit that some of my fellow North Americans never quite embraced that concept. They still seem to think that the US is the only country that matters, that everything revolves around us, and that attitude sometimes seeps into the UFO topic too.

Cases that occur in some other country are somehow not as important as some light in the sky that appears over New York or LA. What I learned during our visit to Colares and conversations with witnesses who had never spoken on camera before was deeply moving and also a bit scary. We still have no idea why saucers and other craft shot beams of light at rural residents, caused injuries, and seemingly were intent on collecting blood and tissue samples rather than blasting frightened people with laser beams. By any standard, Colares stands out. We may never get solid answers. Whoever was in charge of those craft have not volunteered any explanations. If this was a nonhuman intelligence behind these incidents, why was their agenda so different form the brief, less harmful cases that are the norm around the world? Much of what the government investigators in Brazil and the US obtained from their investigations is still hidden from the public. If anyone can find the truth, Thiago might be the one.

George T. Knapp
Journalist, Investigative reporter for KLAS TV and host at Coast-to-Coast AM

Introduction

The "Chupa-Chupa" phenomenon is the popular name given to a series of extraordinary events that began in 1977 and spread across parts of northern and northeastern Brazil. Mysterious luminous flying objects - sometimes reported as manned - displayed what was clearly intelligent behavior, targeting residents of several cities with concentrated beams of light. Those struck were left physically weakened, injured, and deeply traumatized. The scale and severity of these occurrences triggered widespread panic in the affected communities, compelling the Brazilian Air Force to launch a dedicated operation to investigate and monitor the phenomenon.

From 1977 through 1979, multiple states, particularly Amazonas and Pará in the North, along with Maranhão and Piauí in the Northeast became the stage for these disturbing incidents. Night after night, strange, glowing craft roamed the skies, and in many cases, their beams of light struck unsuspecting residents. Women were disproportionately targeted, and many victims suffered both physical and psychological consequences that lingered long after the encounters.

The first case allegedly occurred on the night of April 25th to 26th, 1997, in the region of Ilha dos Caranguejos (Crabs Island), on the coast of Maranhão state, when four fishermen, Firmino, Apolinário, Aureliano and José, were fishing in the region aboard the boat 'Maria Bonita'. On April 26th, the boat was found adrift, with one of the crew members dead and the others alive, but quite weak. They stated that they had been struck by a light coming from the sky.

Cases continued to occur in the region, and in the last quarter of 1977, the region north of Belém was ravaged by a phenomenon, nicknamed the *"Chupa-Chupa"* (Suck-Suck) Phenomenon, due to the nature of its attacks. This situation created an atmosphere of collective terror and panic. Residents of the areas where the phenomenon was prevalent began to abandon the region. The situation reached critical levels. The city's health center was full of victims of the phenomenon. They were weakened and terrified.

The mayor of Vigia, one of the cities most affected by the phenomenon - sent a letter to the Air Force warning that UFOs (now referred to as UAPs) were frightening local fishermen and causing panic in the region.

In just a few months of activity, the team of Air Force officers obtained dozens of photographs from several video recordings of unidentified objects flying over the Colares region in Pará. The team collected testimonies from dozens of people from various cities affected by the phenomenon. All this material generated a technical-operational report, with hundreds of pages containing testimonies, personal accounts from the Brazilian Air Force (FAB) team, sketches and maps and drawings of the occurrences.

The Operation Saucer has always been present in the national and international UFO community. There were some copies of the photographs and the operational report of the operation in the possession of Ufologists, but there were no statements from the military personnel involved in the operation. In 1997, these military personnel began to report their experiences publicly. The first of them was the commander of the operation, then Colonel Uyrangê Bolivar Sores Nogueira de Hollanda Lima, who gave details of the operation in an interview with the TV show *Fantástico*, on Rede Globo TV. Later, in a long interview with Revista UFO Magazine Brazil, he presented to A.J. Gevaerd and Marco Antonio Petit, previously unpublished details about the *"Chupa-Chupa"* Phenomenon and the operation's experiences in dealing with the mysterious enigma.

The phenomena lasted until the end of 1977, with some cases continuing into 1978. A new FAB group was sent to the site in 1982 to investigate similar phenomena. The fear was that the attacks would return, but to everyone's relief, this never happened, at least not with the same intensity as before.

Almost 40 years later, residents of cities in northern Brazil continue to report the appearance of strange lights and unknown flying objects, not only in the sky, but also in the water.

Much like the modern 'Tic-Tac' UAPs reported by U.S. Air Force pilots - observed at high altitudes, hovering and maneuvering over the ocean, and at times diving into the water before re-emerging in apparent defiance of known laws of physics - residents of Colares described similar encounters. In this region, the vast Amazon rivers have become the stage for UAP activity, with objects seen entering and emerging from the water, traversing the city's skies, and disappearing into the dense forest.

Chapter 1

The Beginning of the "Chupa-Chupa" Phenomenon

The most intense period of the so-called "*Chupa-Chupa*" Phenomenon occurred between the last months of 1977 and the first months of 1978. However, the events were not restricted to this specific period. The areas affected by the events had long been the scene of conventional sightings, almost always occurring at night. It was from April 26th, 1977, that the phenomena became more aggressive, causing injuries and physiological effects on the people involved. Some cases, unfortunately, resulted in the death of some people and widespread panic among residents of the affected areas.

Figure 2. UFOs were attacking the population of several cities. (Credit: Author's files)

The states affected by the mysterious phenomenon included Amazonas and Pará, in the North region, and Maranhão and Piauí, in the Northeast. Today, with information gathered from civilians, researchers, journalists, and other professionals involved in the events - as well as from the military through Operation Saucer - we have a clearer understanding of what took place.

The first one that became well known occurred on the date of April 26th, 1977. On that date, four fishermen traveled by boat to the vicinity of Ilha dos Caranguejos (Crab Island), in Maranhão state, where they spent the night, and were due to return the following day. On that fateful night, an intense light positioned itself over the boat. The result was tragic. One of the fishermen was dead and the other two were injured and weakened. Only one of the fishermen was physically well, and only with some difficulty managed to bring the boat back to the city of São Luís, from where they had set off. This case, which became known as the Crab Island Case, was the first in a series of strange and frightening events that caught the attention of Brazilian authorities, who created a special operation to investigate these cases. Operation Saucer, as it became known, interviewed hundreds of witnesses and victims of the phenomenon, documenting both the injuries observed on their bodies and the various effects attributed to these encounters. The investigation also succeeded in capturing, on both film and in photographs, the strange lights associated with the phenomena.

In the weeks and months following the Crab Island Case, conflicting reports began to circulate about strange lights that appeared in remote places, paralyzing witnesses and somehow drawing blood. Due to the precariousness of communications and access to these villages, these reports went unnoticed by the rest of the country, isolating entire communities in nights of terror where not even their homes seemed safe.

The focus of the apparitions was in the Gurupi River region, on the border with the state of Pará. Cities such as São Vicente Ferrer, São Bento, Pinheiro and Bequimão accounted for most of the cases. However, it was not long before some cases began to occur on the Pará side as well, near the border with Maranhão. Several cases were documented in Vizeu, São José do Pintá, Augusto Correa, Bragança and Capanema.

Due to the seriousness of the reports and the panic that was emerging in these communities, the press in Maranhão and Pará began to report the occurrences more seriously. Between the months of June, July and August, many newspapers published reports on the phenomenon. The newspaper *O Estado do Maranhão*, dated July 20th, 1977, describes:

"The appearance of an unidentified flying object in the skies of Pinheiro is causing panic among the population, to the point where some claim that the unidentified device approaches people to daze them with a beam of light and drain their blood. The presence of a strange object in the skies of Baixada has been definitively confirmed, and the population of São Luís will have the opportunity to confirm this when they watch the film made here by Difusora TV. The UFO that has been seen by thousands of people in this region and more frequently between Pinheiro and São Bento, where a strange shape similar to a "Y" with a flame at the bottom has been seen. The atmosphere in the region is one of widespread fear, and people do not dare to go out at night because of rumors that, when approaching, the UFO emits an extremely hot beam of light that burns people's skin."

Figure 3. Manoel Paiva, mayor of Pinheiro (MA) at the time of the Chupa-Chupa cases in Maranhão. (Credit: www.fenomenon.com.br).

In view of this, the mayor of Pinheiro at the time, Maneco Paiva, sent a letter to the Brazilian Air Force Command in São Luís requesting that measures be taken to prevent widespread panic in the city. The newspapers reported the news and reported that the Command sent a note to the Belém Air Force Base, which forwarded it to the Air Force Ministry. Around this time, the second phase of the *"Chupa-Chupa" Phenomenon* began, this was concentrated in the north of Pará, in the Baía do Sol region, affecting the cities of Vigia, Colares, Santo Antônio do Tauá and even the outskirts of the state capital, Belém.

The news of these occurrences came to the capital, São Luís, not only through the press, but also directly from the municipal authorities of the various locations affected by the apparitions. For example, the São Vicente de Ferrer police officer, José Ribamar Mendes, reported the UFO sightings directly to the director of the Public Security Secretariat of Maranhão. He even got his gun to shoot one of those objects, which flew over his residence at low altitude and emitted an intense light.

The situation in these regions reached increasingly alarming levels. As cases continued to rise, residents altered their routines, avoiding going out at night to work. In some instances, they gathered in large groups after dark and lit bonfires in an attempt to drive away the mysterious phenomenon. When these measures failed, many chose to leave the area entirely, abandoning all they owned. This wave of fear led the mayor of Vigia to send a letter to the 1st Regional Air Command in Belém (1st COMAR), reporting the events and requesting intervention. The Air Force's response was a military investigation known as the aforementioned Operation Saucer.

Figure 4. "Flying saucer is seen over the Costa do Sol". (Credit: newspaper "Jornal Pequeno", July 27th, 1977)

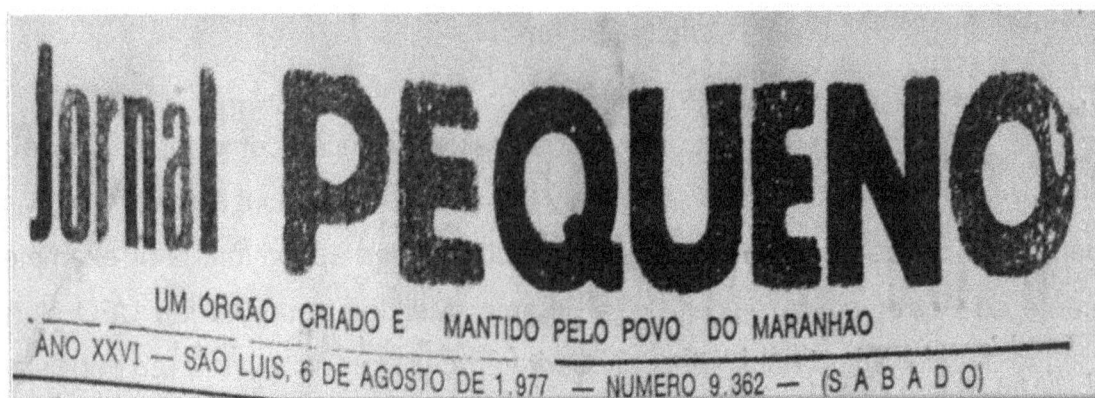

Integrantes da "GANG" vinda de ITAPECURU que agia em S. Luís, conforme nota publicada, ontem, por esta Folha.

Figure 5. "Fire torch follows car". (Credit: newspaper "Jornal Pequeno", August 6th, 1977)

Chapter 2

Death on Crab Island

The Crab Island case took place on April 25, 1977, and is regarded as the starting point of a wave of harmful UAP activity in the states of Pará and Maranhão. The incident involved three brothers—José Corrêa, 22, Apolinário Corrêa, 31, and Firmino Corrêa, 38—and their brother-in-law, Auleriano Bispo Alves, 36. All were residents of the municipality of Alcântara, Maranhão, and were aboard their boat, Maria Rosa.

Figure 6. Place where the boat Maria Rosa docked and where the fishermen were attacked on Crabs Island. (Credit: Pablo Villarrubia Mauso).

They arrived at the island early in the afternoon, anchored in a creek just inside its perimeter, and spent the remainder of the day cutting down thin trees and pruning branches. Their intention was to sell the wooden poles for use in construction. The island itself measures approximately 25 miles in length and 7 miles in width. Isolated, swampy, and uninhabited, it is infested with mosquitoes and covered with dense bushes and trees. The only reason people were going there was to collect wood or catch crabs.

They worked all afternoon cutting and stacking logs. They stopped at 6:00 p.m., just as the sun was beginning to set, and had a meal of meat and rice. The tide was low, and the boat was anchored in the mud of the empty creek. They chatted until 8:00 p.m. and then went to sleep inside the boat, covering the hatch with a piece of canvas to keep out mosquitoes. A small, closed window at the back of the cabin allowed air to circulate. A lantern with the wick low hung on one side of the cabin. The men planned to wake up around midnight when the tide came in, take the logs to the boat, and return to São Luís when the tide went out. José, Apolinário, and Auleriano had made this trip at least a hundred times before, and they never failed to wake up with the tide. The rocking of the boat as the creek filled up and the sound of the water lapping against the hull were a great wake-up call.

Firmino was the only newcomer. The fourth man who usually accompanied the group was sick, and Firmino, a farmer, asked to go in his place because he needed wooden poles for an addition he was building to his house in the rainforest. It was his first trip, and he would later regret taking it. Something terrible happened while they were sleeping. By midnight, José was dead and Firmino and Auleriano were seriously injured, but no one knew what had happened or why. No one knew then and we still do not know.

Frightening Discovery

Instead of waking at midnight with the tide, they did not wake until around 5:00 a.m., just as the sun was rising. Apolinário, who had slept on a rug on the floor of the cabin, heard Auleriano shouting for help at the front of the boat. He was puzzled, because Auleriano had gone to sleep in a hammock at the back of the boat, just over a meter behind Apolinário's rug. Apolinário staggered forward, crouched under the other hammock, where José had been, and removed the canvas covering the hatch.

With the cargo area suddenly visible in the first light of dawn, Apolinário looked down and saw Auleriano lying in several inches of water in the hold. He asked what was wrong, but Auleriano did not know. He was in pain and could not get up, and did not know how he had gotten there.

Apolinário helped Auleriano out through the hatch and onto the quarterdeck and noticed that he was burned on both shoulder blades. Auleriano pulled down his shorts and saw that he also had a burn on his buttocks, on the left side. Oddly, his shorts were not burned. Apolinário started to make Auleriano some tea but heard someone groaning in the back of the boat. He went down to the cabin, crouched down again under José's hammock, and saw Firmino lying on the floor under Auleriano's hammock. This was another surprise, because Firmino had gone to sleep at the front of the boat, where Auleriano had been found. But Apolinário's surprise turned to shock when he examined Firmino. *"He was burned and swollen, and his skin had fallen off,"* Apolinário said. *"I tried to talk to him, but he wouldn't respond. His eyes were closed, and I could not open them. I was terrified."*

Frightened, Apolinário ran to José's hammock to ask for help, but as soon as he touched him, he realized that he was dead. He tried to feel for a pulse, but there was no pulse. José's body was cold and rapidly stiffening, with one leg hanging out of the hammock. Overcome with grief, Apolinário thought he should put his leg back in the hammock, but it was a struggle to do so. Desperate, he wanted to cry, but he was the only healthy man on board, and he would have to take the others back to São Luis. There was no medicine or first-aid kit on the boat, and he could do nothing to treat the men's burns. Worse still, the tide was low, and the boat was stuck in the mud again.

He had to wait more than 8 hours for the tide to come in again. Around 2:00 p.m., he began to take the boat back to São Luis. It was a difficult journey because it usually takes at least three men to man the sails and rudder of the 12-meter boat, and Apolinário had to do it all alone. José was dead, Firmino was unconscious and Auleriano was in great pain. During the journey, Firmino rolled from side-to-side on the floor of the cabin as the boat hit the waves in the bay. *"God helped me. Without His help, we would all have died,"* said Apolinário, a thin man who was only 1.50m tall. The sun was setting when they reached the Port of Itaqui, near São Luis, but the young man's nightmare was not over. The only people in the small, deep-water port were two security guards, who were unable to help him. He had to walk 10km to São Luis, tell the police what had happened and go home to call his older brother, Pedrinho. The two returned to the port by car at 9:00 p.m. and took Firmino to a hospital. Although Auleriano was in great pain, he stayed by José's body.

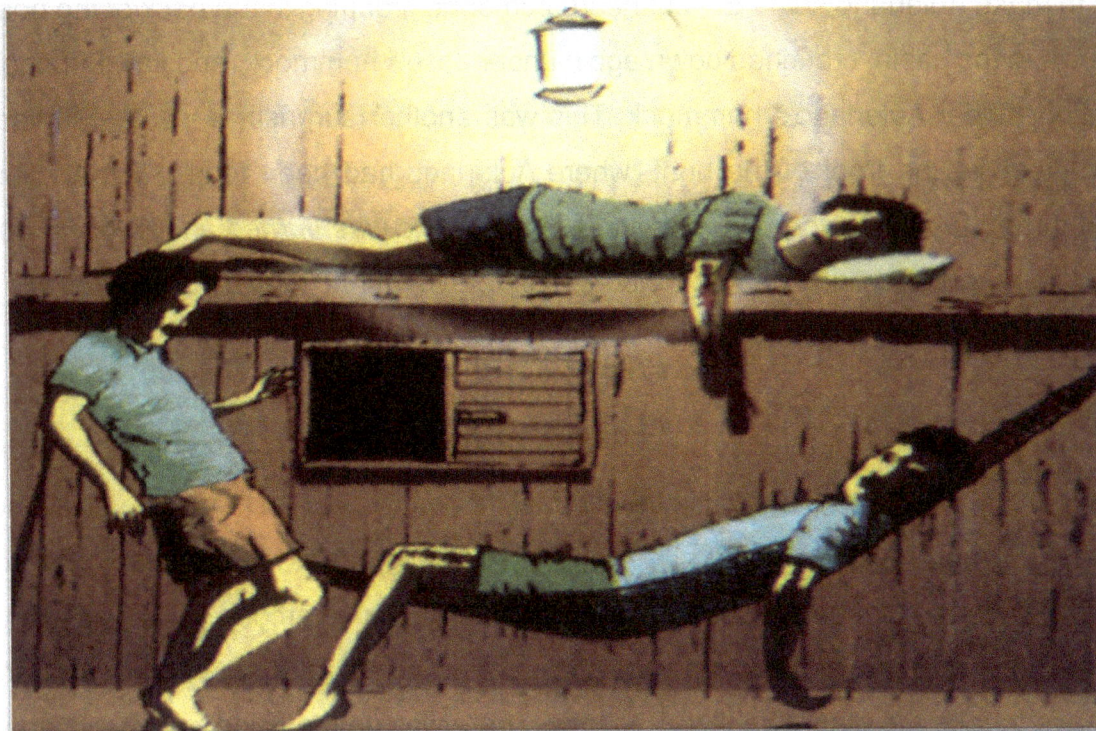

Figure 7. Apolinário runs to help his friends, but it was too late. (Credit: Jamil Villanova).

Figure 8. The Maria Rosa boat in an image produced by Ufologist and artist Jamil Vilanova. It was the boat that the victims of the attack on Crabs Island were on. (Credit: Jamil Villanova).

The police only arrived at the boat at 1:00 a.m. José's body was taken to the Forensic Medical Institute, and only then Auleriano went to a hospital for treatment. His burns would leave scars, but he was released that night. Firmino remained in a coma for a week and had to spend more than a month in the hospital. Much of his body had suffered second-degree burns. The most serious were on the left side of his ribs, on the inside of his left arm and on his forehead.

The muscles in his arm were so damaged that the fingers on his left hand were permanently twisted inward, with almost no mobility. An autopsy was not performed on José's body. São Luis is close to the equator and after 24 hours in the heat, his body was already in an advanced state of decomposition. The doctor who examined him at the Forensic Medical Institute said in his report that there were no cuts or bruises on his body. The death certificate stated that José had suffered a "…*stroke, caused by high blood pressure, as a consequence of an emotional shock*." The cause of death was attributed to 'emotional shock.'

Emotional Shock

There was no explanation for what this emotional shock could have been. The police were unable to determine what had happened on Crab Island. Investigators went there, examined the area where the boat had been anchored, inspected the boat itself, and spoke to the survivors and people who knew them. There was no evidence that the men had been drinking or taking drugs, had suffered from food poisoning or been exposed to toxic gases, or had even been in a physical fight. The police found no sign of fire on the boat or on the island. The only conclusion was that the three survivors really did not know what had happened.

None of the three men remembers the slightest detail of that night, not even under deep hypnosis. A burn must be one of the most excruciating pains anyone can suffer, yet the two men were severely burned before midnight and neither knew anything about the accident, one the following morning and the other when he came out of a coma a week later. How could such things have happened without the victims having the slightest memory of how they were burned? What or who could inflict such injuries and completely block the painful experience in the victims' minds? Why did a healthy young man like Joseph simply die in his sleep, without any apparent cause?

These are some of the questions that so intrigued the Maranhão police and have never been answered. There is no direct evidence that a UAP or UFO was involved in the incident, as the men reported seeing nothing unusual. The event took place on the night of April 25, 1977, during a period marked by numerous sightings of unidentified objects across the region. Newspapers, along with radio and television stations in São Luís, quickly picked up the story and attributed it to a UAP, citing the mystery surrounding the case and the high number of recent sightings. Despite the media attention, the Crab Island case was not reported beyond São Luís. Why?

Media Attention

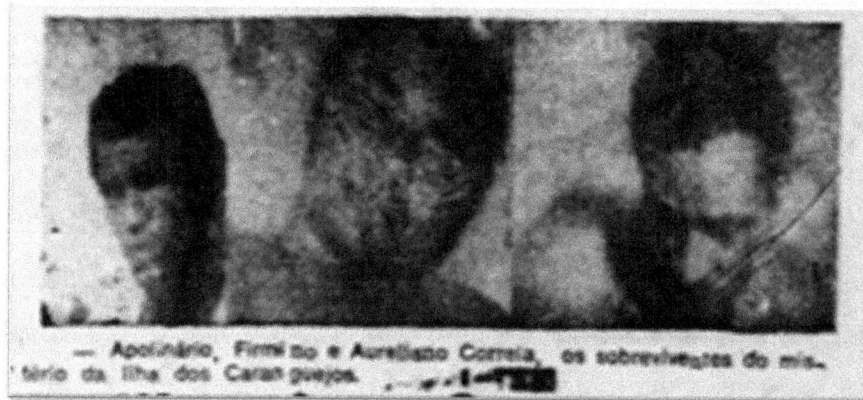

Figure 9. For many, the Crab Island Case marked the beginning of the "Chupa-Chupa" attacks. (Credit: newspaper "Jornal Pequeno", April 29th, 1977).

However, if it was not a UAP, or the "Chupa-Chupa" Phenomenon, what could have caused José Luiz's death? There is no shortage of possibilities.

He saw a fire

This last claim—that Firmino, in his delirium, had mumbled something about a "fire"—was the only discernible link to a UFO. Fire is probably the most common term for UFO in Brazil. The other doctor who defended the lightning theory was José Oliveira, a member of the Forensic Medical Institute (IML) team at the time.

Clésio Muniz, head of criminal investigations for the Maranhão police, recalled:
"I saw the men with those strange burns, and I don't think they were caused by a common fire". I don't believe in UFOs, but this is a strange phenomenon that I can't explain. I had heard reports of 'fireballs' being seen in the towns around Crab Island and to the west of there. Many people had seen these 'fireballs,' both before and after the incident. Moreover, from the testimonies I heard, the fireballs did not look like shooting stars - they went up and down, moved left and right, horizontally and vertically, sometimes slowly, sometimes quickly, and sometimes shifting from very slow to very fast. It is an unusual phenomenon, and I do not know what it is.

Another possibility considered was that lightning had caused the deaths and burns. According to this theory, lightning struck the sand or mud near the boat, ricocheted upward, and traveled horizontally into the cabin, striking three of the four men who were sleeping.

Two doctors from the Forensic Medical Institute (IML), who examined Firmino at the hospital, also believed lightning was the likely cause. One of them, Dr. Carneiro Belfort - then director of the institute and later a professor of medicine at a university in São Luís - stated:

"I wanted to see Firmino because the newspapers were saying that the injuries had been caused by UFOs, and I needed to verify them for myself. I have never seen a UFO and I do not believe they exist. The burns were characteristic of lightning, but I cannot say what caused them. If it was not lightning, I do not know what could have happened. The man told me that he saw 'a fire' before he passed out."

Figure 10. Firmino days after the incident, in hospital. (Credit: book "UFO Danger Zone". PRATT, Bob).

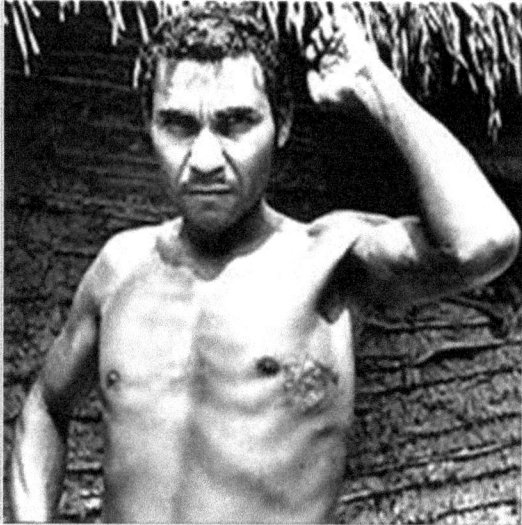

Figure 11. Firmino, already recovered, showing the scar acquired in the experience. (Credit: book "UFO Danger Zone". PRATT, Bob)

"Firmino had many second-degree burns and could have died. In my opinion, it was lightning. But, on the other hand, lightning would have caused some damage or burns to the boat, and the man who died would also have been burned." None of the doctors saw the boat or José's body, but the death certificate stated that there were no marks or injuries on the body.

Doctor Oliveira examined the institute's records on the injured men. Regarding the burn on Auleriano's buttocks, he said that *"…probably, if he had been struck by lightning, his clothes would have been burned too."* Auleriano's and Apolinário's shorts were intact. Clésio Muniz, the chief criminal investigator, vehemently disagreed with the lightning theory, as did Sergeant Antenor Costa, a meteorologist with the Brazilian Air Force (FAB) at São Luis Airport.

The airport is 4 km northeast of Crabs Island. At the time four national airlines, two regional airlines, and several air-taxi companies were using the airport. Weather station records indicate that there was no storm or lightning between 5:00 p.m. on April 25th and 6:00 a.m. on the 26th. There was some light rain at 11:00 p.m. and more rain at midnight, but otherwise the night was clear and calm.

"It would be impossible for lightning to strike, hit the sand, bounce upward, and deflect to the side, hitting the boat. That does not happen. If that were the case, the lightning would have burned through the canvas as well and would not have struck two or three men at the same time, because their positions in the boat were very different. To do that, lightning would have to be like a winding path. Furthermore, it is unlikely that it would have killed a man without burning him. It is simply not possible for lightning to burn two men and kill a third without leaving a mark on his body," said Sergeant Costa.

Natalino Filho, director of the meteorological station, said that lightning could have struck the water and passed through it to the boat, since water is a good conductor of electricity. *"But if that had happened, Apolinário would have died because he was lying on the ground at the point closest to the water,"* added Natalino. There were no burns on the boat.

American journalist Bob Pratt interviewed all the witnesses on three occasions: 1978, 1981 and 1992.

Figure 12. Bob Pratt aboard the small boat in which the fishermen were on that fateful night.
(Credit: book "UFO Danger Zone". PRATT, Bob.

"When I interviewed the three survivors, I hoped that the mental block had disappeared and that perhaps their memories would begin to be reactivated. But perhaps that will never happen. I went back in 1981 and spoke to Auleriano and Apolinário, and again in 1992, when I spoke to all three. None of them remembered anything. Interestingly, the two who suffered burns, Firmino and Auleriano, are now in excellent health, but Apolinário, who apparently

was not injured, is currently experiencing health problems. A year and a half after the incident, he began to feel weakness in his left arm. In 1981, the year he turned 36, he could not hold anything with his left hand without dropping it. In 1992, at the age of 46, he had little strength in his left hand and arm, suffered from severe headaches, and walked with difficulty, with a somewhat stiff gait. He does not know why. He has never had any accidents or debilitating illnesses. When he can work, he makes charcoal.

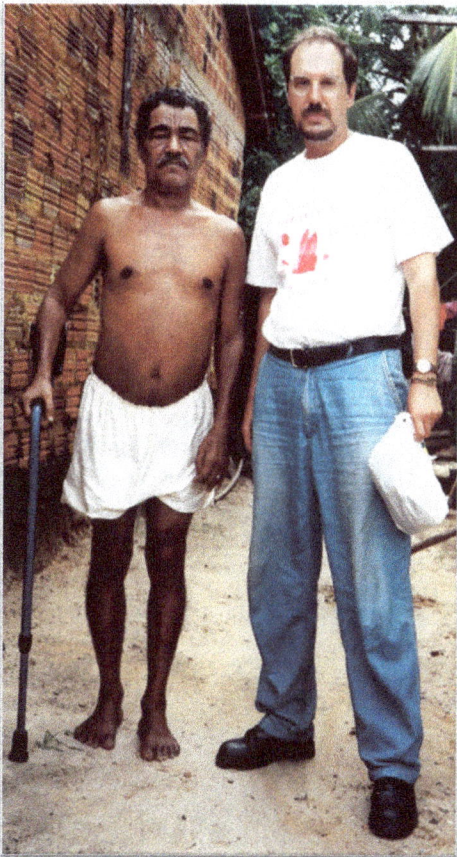

Figure 13. Apolinário Correia, one of the survivors of the shocking and tragic attack on Crabs Island, with the author Pablo Villarrubia Mauso. (Credit: book "Luzes do Medo**", MAUSO, Pablo Villarrubia).**

Firmino, who had lost weight, could hardly do anything for several years after the incident, and sometimes even seemed a bit dazed, according to his wife, but he is now robust and mentally agile again. He can do light manual labor, despite having a crooked left hand. He and Maria also own and operate a small grocery store in one of the poorest neighborhoods in São Luis.

Auleriano's scars have practically disappeared. Two years after the incident, he began going to Crabs Island to collect wood again and continued doing so until 1991 without any further unusual incidents. But he gave up that job and went to work as a security guard for a construction company. Neither Apolinário nor Firmino ever returned to Crab Island".

Another Death on Crab Island

That is not the end of the Crab Island story. Almost the same thing happened nine years later to another group of men, leaving one dead, one burned, and two mysteriously ill. On April 28th, 1986, the four men went to the island in a similar boat to collect wood. They worked for two days cutting down more than 300 logs and piling them on the riverbank near the boat. On April 30th, they stopped work at 6:00 p.m. and one of them, Juvêncio, 22, started cooking.

33

Veríssimo, 21, said he did not feel well and asked Juvêncio for garlic to rub on his arms, as it would make him feel better, but Juvêncio suddenly became dizzy and fell to the deck, unconscious. In quick succession, the other two men, Anselmo and Lázaro, both in their 40s, also passed out.

No one knows what happened to Veríssimo. Lázaro regained consciousness at noon the following day and found Veríssimo dead, lying on the deck. There were no marks on him, but some blood was running from his mouth. Anselmo woke up two hours later and Juvêncio regained consciousness at 5:00 p.m., almost 24 hours after he had passed out. The right side of his head was burned and swollen. Anselmo and Lázaro tried to load the wood onto the boat but gave up after loading no more than thirty logs. They began to sail the boat back to São Luís, but it was difficult because all three felt seasick and nauseous.

The second death on Crab Island was also not reported outside of São Luis. As in the first case, none of the three survivors knew what happened that night, except that they all felt dizzy and fainted. The port authorities questioned them and told me that it seemed as if the men were telling the truth. All three were certain that the problem was not food poisoning. They had not yet eaten and were feeling fine until they became dizzy. The authorities ruled out the possibility of some kind of poisonous gas from the swamp. Juvêncio said that no one had smelled anything strange before the dizziness.

An autopsy was not performed on Veríssimo. As in the first case, when the boat arrived at the port, his body was already in an advanced state of decomposition. Veríssimo's death certificate simply lists the cause of death as "undetermined." The connection to a UFO in this case is also tenuous. Something strange happened shortly before the men passed out. They heard a loud bang in the woods somewhere near their boat. In the darkness they could not see what it was, and they do not know what might have caused the noise. The island can only be reached by boat or helicopter, and the men were unaware of the presence of other people there with them. Some Ufologists might interpret the bang as a clear indication that a UAP/UFO had landed, crushing trees in its path, while debunkers will claim that the noise must have been caused by a falling tree. There is no way to prove who is right, but the men would have recognized the sound of a falling tree.

Another case was described by other fishermen. Marcos Rogério said that he had had a similar UFO encounter on a boat not far from Crab Island one night in 1983. His boat was anchored in a creek on the west side of the bay, when a large, bright object came down and hovered over him, shooting a light over the boat. The man and his companions jumped out of the boat and hid in the bushes until the UFO had moved away. He said several people on boats in the area had also had UFO encounters that year.

The three cases are remarkably similar, except that neither of the men in the first incident felt dizzy. It is quite possible that there was no UFO in two of the three cases, since the victims do not remember seeing anything strange and there were no other witnesses. But, if the villains in these cases are not UFOs, then some equally strange phenomenon was responsible. Either way, it is all part of a strange mystery that hurts and sometimes kills people.

Chapter 3

The Gurupi Phase of the "Chupa_Chupa" Phenomenon

The Ufologist Daniel Rebisso Giese, one of the most prominent *"Chupa-Chupa" Phenomenon* researchers, divided the phenomenon into two phases: the First Phase, also known as the 'Gurupi Phase', in Maranhão state and the Second Phase, which occurred in Baía do Sol (Sun Bay), in Pará state.

In the Gurupi Phase, the focus of the sightings was in the Gurupi River region, on the border with the State of Pará. Cities such as São Vicente Ferrer, São Bento, Pinheiro and Bequimão accounted for most of the cases. The region around the city of Pinheiro was the worst affected and practically the entire local community witnessed the strange events between April and July 1977. Most of the time the phenomena occurred after dark, when workers were returning home after a day of work. The reports were similar. An intensely illuminated object would suddenly appear over the deserted roads of the region, frightening the witnesses. It was generally described as a ball of fire, silent or emitting small noises, which approached witnesses who later presented visual sensitivity, fevers, chills, dizziness and in some cases burns.

A common characteristic in all the reports was the astonishing speed of the objects. One moment they were just a point of light that could be confused with the stars and a few seconds later they were objects very close by, or even above the witnesses' heads, illuminating the place.

Baixada Maranhense

The UFO wave of the "*Chupa-Chupa*" phenomenon evolved in the direction of the São Marcos Bay, in the state of Maranhão, towards the delta-estuary of the Amazon River.

During the second week of July 1977, the focus of UFO/UAP observations was not only Viseu, but also other municipalities in Maranhão, such as Pinheiros, São Bento, São Vicente de Ferrer and Bequimão, all in the Baixada Maranhense.

Figure 14. Baixada Maranhense. Main areas of UFO sightings, July/1977. (Google Maps).

The UFO contacts in Baixada Maranhense clearly revealed the appearance of luminous objects emitting rays, which, according to popular belief, would be capable of sucking the blood of their victims. From the Brazilian Air Force's own documents, it was possible to get an idea of the injuries caused by the "Vampire Lights" among the residents of Baixada Maranhense. Most of the victims of the "device" said that when they were hit by the light, they felt that their movements were neutralized, followed by a sensation of intense heat and weakness, causing some of them to faint. According to the Maranhão press, the UFOs chose

38

small rural communities, hitting isolated people, or people in small groups. In addition to psychological effects, the injuries, in their entirety, were represented by small, superficial burns. The newspaper 'O Liberal', on July 16th, 1977, published the following reports:

"The one who actually saw the mysterious object was the farmer Vicente Gomes. According to him, it was approximately 3:00 a.m. on July 14th, 1977, when he was riding his horse along a road in the village of Guarapiranga, in the municipality of São Bento. While on his horse he raised his head and suddenly saw a mysterious light in the shape of a kite appear coming in his direction. 'The light was so strong that it enveloped me and I fainted,' said Vicente.

At 12:00 p.m. on July 8th, 1977, an employee of the Ariquipa farm, Raimundo 'Socó' Corrêa, was burned by a mysterious torch, causing injuries to his body. In his opinion, the object was in the shape of a large ball. He was burned while traveling from the municipality of Boquimão.

In that municipality, São Vicente de Ferrer, the versions are somewhat different, but they were the same in one respect, that is, in terms of the fact that the devilish thing (or flying saucer) had come down in search of human blood.

On the morning of Sunday, July 24th, 1977, at around 8:00 a.m., in the municipality of Bom Jardim, a woman was attacked by a strange light.

Mrs. Coucima Gonçalves da Silva, a resident of the town of Boa Vista, in Bom Jardim, was taking care of her household chores when she had to go to the back of her property. It was then that she saw a strange ball of fire, from which a ray of lightning appeared and struck her and threw her to the ground. She said that from that moment on, she could not remember anything else and it was her relatives who said they found her unconscious and took her inside the house. When she woke up, she was talking incoherently."

According to the newspaper "O Estado do Maranhão", Mrs. Coucima was admitted to the Santa Casa de Saúde Santo Antônio hospital in Santa Inês; and thanks to the care of doctor Pedro Guimarães, she fully recovered. There were no reports of burns, only psychological impairment and amnesia.

Among the adults affected by the "*Chupa-Chupa*" are farmers, housewives, hunters and fishermen; as was the case on the night of July 21th, 1977, when a group of fishermen from the municipality of Pindará were forced to abandon their boat in the face of the lights from the "device".

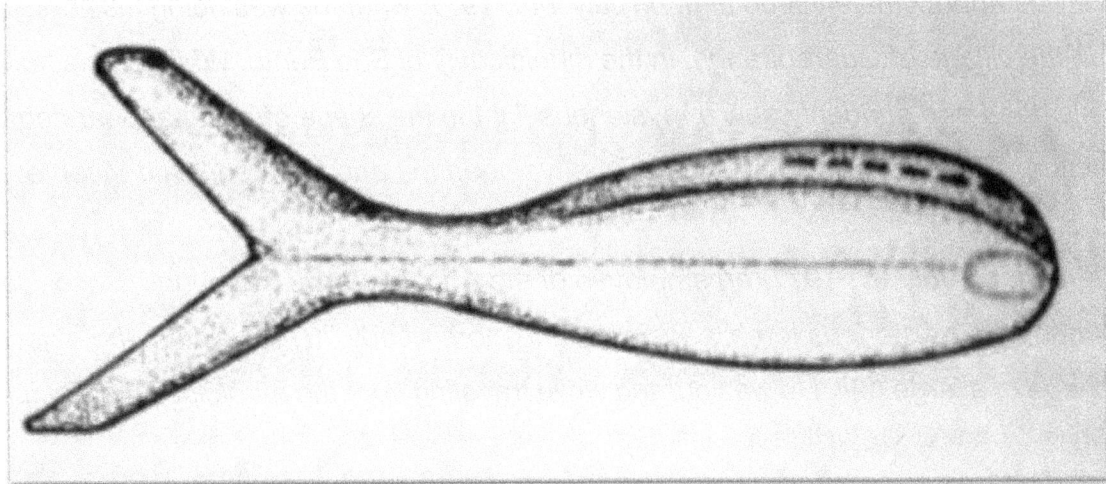

Figure 15. The flying object seen by the residents of São Bento. Drawing published in the newspaper "O Estado do Maranhão", on July 20th, 1977 (Credit: book "Vampiros Extraterrestresna Amazônia", GIESE, Daniel Rebisso).

Figure 16. "Ships" flew over São Bento, striking people like farmer Vicente Gomes with paralyzing rays on July 14th, 1977. (Credit: book "Vampiros Extraterrestresna Amazônia", GIESE, Daniel Rebisso).

Authorities on Alert

The fear and panic caused by the increasingly intense appearance of "extraterrestrial ships" in Maranhão airspace led to the mobilization of the 3rd Air Zone of the Brazilian Air Force and municipal authorities, mainly mayors and police chiefs. Unlike the official statements from Pará, there were many politicians and police chiefs who confirmed the existence of the flying devices. The press in São Luís was clear on several occasions, stating in the pages of several newspapers that "the strange luminous object really exists and moves intelligently in space".

During the penultimate week of July 1977, the police themselves were accompanied by a UFO at night while traveling between the village of Paca and the city center of Pinheiro. Several military police officers were in the car, such as a soldier, Mário Pontes Filho, who witnessed the exchange of light signals between the UFO and the police car. This fact led Lieutenant Amujacy Araújo Silva, sheriff of Pinheiro, to send a team to the location in the hope of seeing the UFO again. The mayor of Pinheiro, Manoel Paiva, also confirmed the existence of the device, to which he was an eyewitness.

The UFO sightings over the city of Pinheiro were so frequent, that the mayor had no doubts about requesting help from the Brazilian Air Force. The newspaper "*O Liberal*" published, "*The communication was received and the Air Force authorities arranged for a letter to be sent to the Air Force Base in Recife and to the Ministry of Aeronautics in Brasília, so that measures could be taken*".

However, before the measures were taken, another authority in Baixada Maranhense witnessed an unknown aircraft. The commissioner of the 3rd Police District, Edgar Sales, reported that, in the back of his farm located in the city of São Vicente de Ferrer, he saw a strange craft on the shores of the Barreira do Pascoal lake. The city's sheriff, José de Ribamar Mendes, informed the Director of the Maranhão Public Security Department about the occurrences in the region. The sheriff himself saw a UFO flying over his residence, and at the time, he tried to shoot the object with his rifle but was unable to do so due to the intense light that blinded him.

Mysterious Deaths

At the time (July 1977), strange rumors circulated that some dead farmers had been found in the municipality of Bequimão. Near the corpses were five hundred cruzeiro bills (the currency of Brazil at the time). These deaths were allegedly associated with the presence of three foreign-looking men who, for most of the day, were confined to a hotel in the city. Almost every night, they would go out in a vehicle of unknown make to carry out their secret activities.

Figure 17. Strange creatures wearing tight clothing, like reptile skin (shiny scales), have been seen carrying devices such as flashlights in their hands. Some have helmets or caps, like divers. (Credit: book " Vampiros Extraterrestres na Amazônia ", GIESE, Daniel Rebisso).

Humanoid Sightings

Sightings of extraterrestrial beings during the UFO wave of the "Chupa-Chupa" phenomenon are rare and imprecise.

The newspaper "O Estado do Maranhão" published the case of Mr. João Batista Souza, owner of the Nova Melia farm in Barra do Corda, in the interior of the state of Maranhão. According to the farmer, during the early hours of July 17th, 1977, after being unable to sleep, he decided to take a walk around his property. At one point during the walk, he came across a "ball of fire" flying over the land two hundred meters away. Frightened, he hid behind a bush, from where he watched the luminous sphere land. After the vehicle landed, when its brightness diminished, he could see that the object was shaped like a straw hat. From inside it, through a door, a small creature approximately one meter tall emerged. In its left hand, it held a kind of flashlight emitting a purple light. In its other hand, it held a piece of equipment that the farmer could not identify. It was not possible to see the humanoid's face, only a helmet with antennas, and the rest of its body was completely hairy. Overcome by a strange sensation, Mr. Batista fainted. Hours later, he was found unconscious by his children who took him home where he remained bedridden for several days, without the strength to get up.

42

Fishermen

Local fishermen were the ones who felt the worst impact of the phenomenon. Before May 1977, they would return from work late at night. Many of them had some kind of experience, usually frightening. Many of them were apparently hit by objects that left them with serious burns. Given the intensity and seriousness of the cases, the population avoided going out at night, even to their own gate.

Inácio Rodrigues was a fisherman from Pinheiro and was one of the first witnesses of the phenomenon in the region. He and his friend Genésio Silva were fishing around 1:00 a.m. when they had of an interesting experience:

"I was fishing with my friend Genésio Silva one night in April. Around 1:00 a.m., we saw a small fire in the sky, to the north. It was very small. I was a little worried and asked Genésio to put out the cigar he was smoking. Suddenly, the fire grew bigger and bigger, and we could see that it was spinning. We jumped out of the boat into the water and tried to find a hiding place. The fire grew bigger and closer. We hid under some large bushes so that it would not see us. The object stopped about 100m away from us and stayed there until about five in the morning. We stayed hidden the whole time, because we were afraid to go out. The light was bluish, but when it first appeared, it was a small red ball. It was pretty, but it shone so brightly that I could not look at it for long. Just before dawn, it disappeared, as someone turns off a light. And, where it had been, you could see a kind of shadow, shaped like a refrigerator. When the sun came up, the dark shape disappeared too. I got dysentery and felt sick that whole day."

Light and Heat

A curious detail noticed by the local population, is the apparent interest of the so-called "fire" in sources of light, regardless of their size. Lanterns, flashlights, bonfires, embers, or even a lit cigarette, attracted the attention of these objects.

On one occasion, 26 people were working on a farm building fences. Due to the owner's urgency, the work continued into the night. One of the workers went fishing so that everyone could have dinner. While fishing, a very bright object with blue tones appeared above the boy. Frightened, he dropped everything and ran towards the camp to warn his companions. Everyone could then see the strange object approaching and illuminating everything within a radius of approximately 1 kilometer, scaring the cows and horses that were there.

The next day, the workers moved to the camp because they were afraid to spend the night there. They set up a scarecrow, placed a kerosene lamp on top, and hid. Later that same night, the object reappeared and approached the scarecrow. The object remained there for approximately 45 minutes, illuminating everything around it. The workers, afraid, hid until the object left, and when it did leave, several workers went home.

Another dramatic case occurred in the region of São Bento, southeast of Pinheiro. The protagonist of the case, João Barros, a fisherman, was 41 years old at the time. He was on a river in the region, around 1:00 a.m., with two friends, when an intensely illuminated object appeared over the boat. It was reddish in color in the center and bluish-green on the sides. The object passed close to the boat, behind João Barros. His back felt intensely hot for approximately 3 days after contact.

In another location called 'Mata do Olimpio', Antonio Olimpio had a similar experience when he left his house at night to go to the bathroom, which was outside the house. He crossed the yard and entered the bathroom. It was then that a reddish object appeared over his head, frightening him. He ran back home, shouting to his wife. When he reached the kitchen door, he fell, and his wife pulled him inside. Antonio felt his back, arms and legs were very hot, and he was dizzy. For the rest of the night, his wife had to apply cold compresses to relieve the burning sensation in the affected areas. For the next eight days, his back, arms and legs felt numb.

Due to the constant appearance of these objects, the mayor of Pinheiro at the time, Manoel "Maneco" Paiva", sent a letter to the Air Force reporting the occurrences and asking for action. He did not receive any formal response from the authorities. Apparently the only people interested to the events were some reporters from Maranhão state.

Filming

One of the reporters who was in Pinheiro was Cinaldo Oliveira, who was in the city for approximately two weeks investigating the events. At the time, he worked for a TV station in São Luis.

"About 90% of the people we spoke to had seen UFOs. Many fishermen were even burned. One night, we filmed a strange thing passing through the sky in an undulating motion. It looked like a satellite, but it varied a lot in shape and size. It got bigger and, suddenly, it disappeared.

The thing we filmed flew in a movement that seemed triangular. It came from Crab Island, in São Marcos Bay, and continued to Anajatuba, then to São Bento and Pinheiro. It looked like a star, but as it grew, it changed color to yellow, blue and red.

The next day, about 3 km from where we had been, we spoke to a man with burns on his back. He told us that it was the night before, when the light went out and came back on again, right above him, that he had suffered the burns. I don't know how many fishermen were burned, but we interviewed about 10. They weren't serious burns, but the men were so scared that they didn't want to go out to work anymore. We spoke to some people on a farm that has a building where all the workers live and sleep. This guy in question ran as fast as he could to the building, and the light flew around the building for about 20 minutes."

In Viseu, a city in the interior of Pará, people were talking about nothing else. People who came by boat from Maranhão along the Gurupi River, said that hundreds of residents had seen the lights and that several of them had become ill after being struck by lightning and losing their strength. Over the months, I built up an archive of stories and researched what other journalist colleagues had found and published. The newspaper 'O Liberal', from Belém, in its July 11th, 1977, edition, featured a full-page report. One of the interviewees was João de Brito, a resident of Vila do Piriá, 14 km from the city. He told reporter Álvaro Martins that he had suffered an attack by the "devil's light". Father José Giambelli had coined the phrase and it caught on among frightened Catholics. They thought that the devil manifested himself in the form of a bright light and infinite evil.

João de Brito said that the intense light "hurt a lot, it took away all my strength". He continued: "*I knew it was killing me, and I prayed to God to help me. But it seemed that God didn't listen to me. And death was coming*". João's father, known as Tomaz, and 400 other residents listened to the boy repeat the same story for days. The light in the middle of the forest attacked João when he was alone. The shock of the attack was so great that, according to his neighbor, Anastácio Costa, the victim began to suffer from a strange disease that was sapping his vitality. Adventures like Brito's, the newspaper reported, were experienced by dozens of others — fishermen and hunters — throughout the Bragança region and near the border between Pará and Maranhão, notably in the strip immediately below Viseu, between the Piriá and Gurupi rivers, especially in the village of Piriá and surrounding areas.

"*But while practically everyone in the village of Piriá has seen the strange, luminous flying objects and therefore believes in them, the reporter noted, in the neighboring village of Itaçu — less than 5 km away by air — the story was different: no one has seen anything, but curiously everyone believes, perhaps because there is a lot of communication between the two villages. In the city of Viseu, however, from which Piriá and Itaçu are about 15 km away, opinions are divided and there are sometimes fierce discussions between believers and non-believers about the existence of the mysterious lights. But the mystery is greater on a strange island inhabited by many bats,*" the reporter narrated.

An identical object flew over Colônia Nova on a July night, illuminating the home of teacher Maria Goretti Garia, who was questioned by Sergeant Sabino, a Brazilian Air Force officer who was part of Operation Saucer. She told the officer "*The device was cylindrical and so bright that it lit up our entire house and the surrounding neighborhood*". The nights of July 1977 became restless after the appearance of the "blood-sucking lights." Children who used to play in the streets now stayed indoors at night. Adults avoided nighttime walks, and fishermen took frequent trips to the sea. Some believed that the objects came from the depths of the Atlantic Ocean, as they had been seen on several occasions emerging from the sea and hovering over boats and small coastal communities. Many villages in the interior of Viseu, and later in other municipalities, were on alert and kept a constant sky-watch, using everything from prayers to fireworks, in fact they used everything they could think of to scare away the "Vampire Lights".

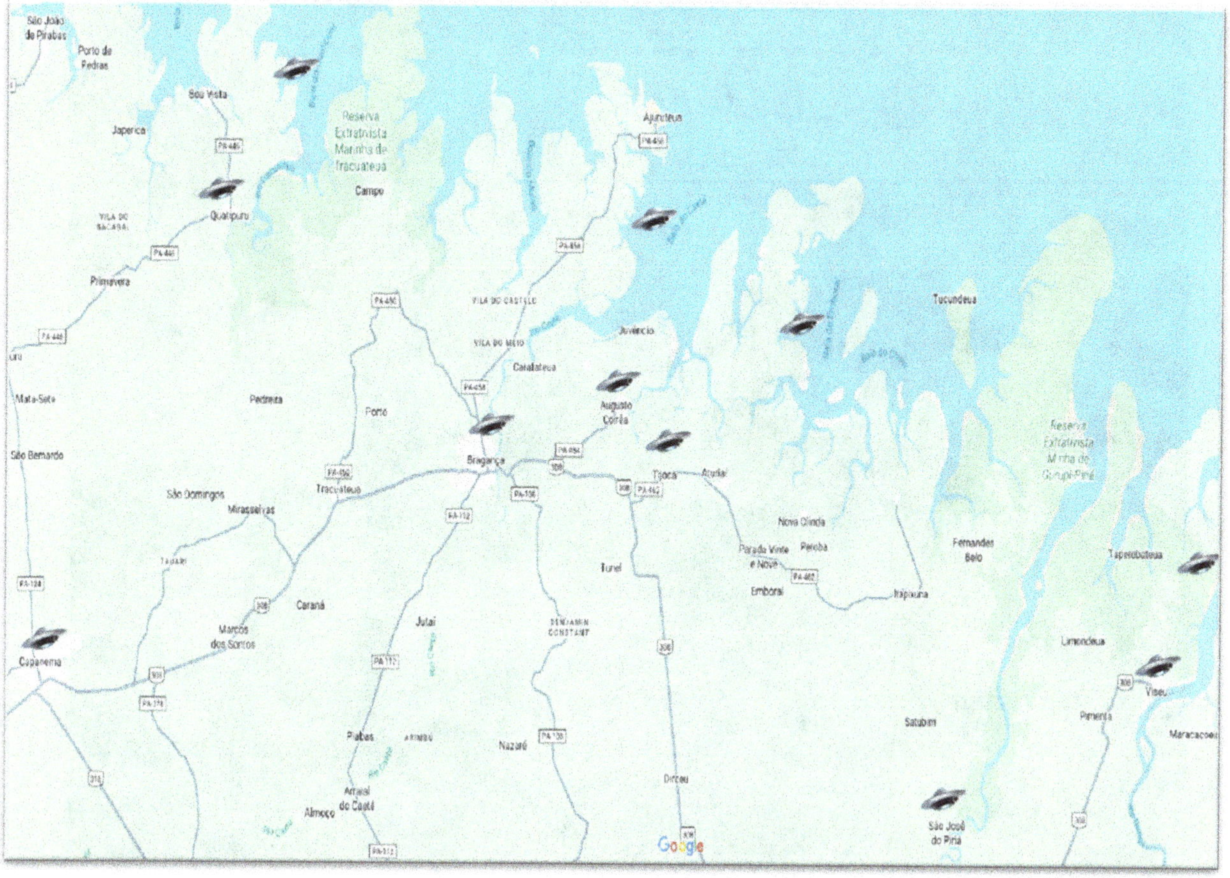

Figure 18. Main areas of UFO sightings in the regions of Bragança and Viseu. (Google Maps).

Figure 19. Cylindrical object described by fishermen Benedito and Simão Siqueira, near the mouth of the Gurupi River, on July 1st, 1977. (Credit: book "Vampiros Extraterrestresna Amazônia". GIESE, Daniel Rebisso).

News from the other side of the Gurupi River reveals that the same objects and table lights were circulating in the Baixada Maranhense region, where they were seen in large numbers, with victims even being reported. During the second week of July, the phenomenon spread beyond the limits of Viseu.

The appearance of the "Vampire Lights" over a period of a few weeks left numerous communities in Bragança, Augusto Corrêa and Viseu, in Pará, and Baixada Maranhense, in the state of Maranhão, in a state of panic. The population was concerned about the attacks of the devices from space, but as usual the authorities remained skeptical.

The police chief of Bragança, Arlindo Dourado, was one of the few authorities who seemed intrigued by the stories of the Light, but he did not have any concrete proof of its existence, despite the growing rumors and reports.

Faced with so many rumors and without an explanation for the attacks and sightings, the population of several cities began to weave their own theories; and one of them was about a monstrous creature known as 'Ataíde'. As for its "existence", only deep "paw" marks were found in the ground and the surrounding vegetation was completely burned and destroyed.

Figure 20. "A jet of light descended from the object so intense that it produced a blinding flash". (Credit: book "Vampiros Extraterrestres na Amazônia". GIESE, Daniel Rebisso).

The strange rumors surrounding the "Ataíde Monster" - a creature with unusual powers, capable of destroying vegetation and burning the soil - allow us to create a possible answer to its origin.

There is evidence that during their landings, extraterrestrial spacecraft produce, among other phenomena, regular depressions in the ground and burns in the vegetation. These phenomena are the same as those attributed to the "Ataide Monster"; which leads us to believe that it is also an extraterrestrial entity. It is understandable that the settlers of the interior of Viseu, upon identifying these signs in their plantations, imagined that they were the result of an extraordinary creature, perhaps a prehistoric being capable of flying. The name "Ataíde" is perhaps linked to the name of the possible owner of the land where the "creature" first landed, hence the term "Ataide Monster".

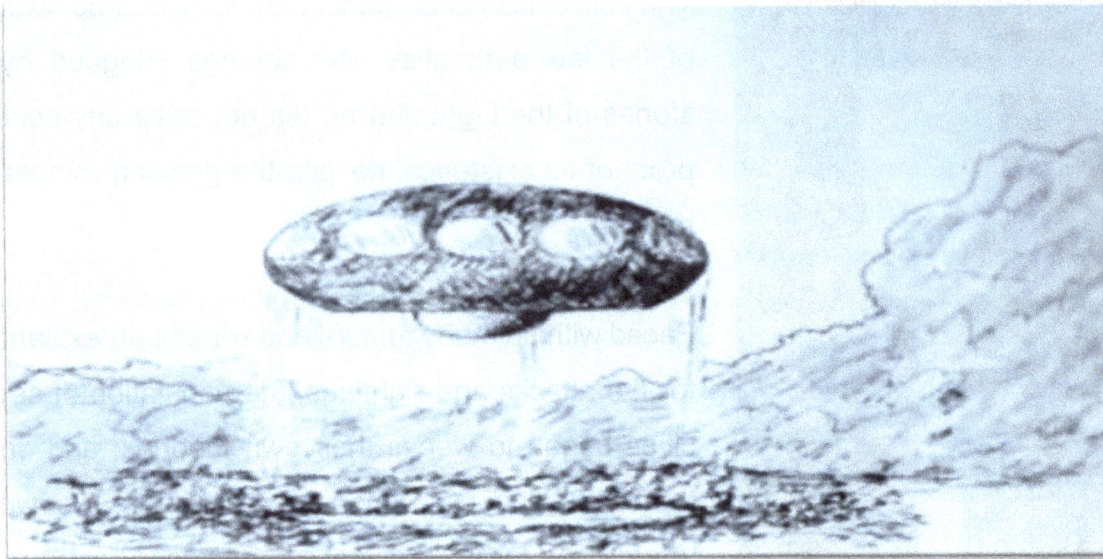

Figure 21. The landings of "extraterrestrial ships" over the region of Bragança and Viseu gave rise to the belief in the Ataíde Monster. (Credit: book "Vampiros Extraterrestresna Amazônia". GIESE, Daniel Rebisso).

Figure 22. Example of the marks found at the Jejo Farm, in São Domingos do Capim, Pará, after numerous UFO sightings over the location. (Credit: book "Vampiros Extraterrestres na Amazônia". GIESE, Daniel Rebisso).

Another, more surprising story is that of the "Woman of the Fishes", an enigmatic entity who temporarily circulated through the region of Bragança and, according to locals, had connections with the beings from the alien ships.

Many people from the ancient city of Bragança, in Pará, spent the month of July 1977 worried about the presence of a solitary young woman with fair hair and white skin, who seemed to be of foreign origin. No one knew for sure who she was; where she lived, what she did, or where she came from. However, it was said that she lived alone on an island near the town of Augusto Corrêa, in Pará.

Popular imagination created countless explanations for the fact that the young "hippie" bought large quantities of fish at the municipal market, usually between 100 and 200 kilos. Many, of course, asked themselves "*why so many fish, if she lived alone?*" "*Well, maybe she's feeding the creatures in the devices*", others replied. One of the fishermen who frequented the market confessed that the solitary "Woman of the Fishes" lived on a remote island, near the coast of Augusto Corrêa, known as Cajueiro Island. Strange things happened there. Some fishermen who ventured to the place had observed her walking on the water, and her house was haunted every night by strange lights.

A woman named Margarida, who lived in the city of Bragança, was walking along one of the lonely roads in the region, when she was surprised by the presence of a beautiful woman dressed all in black, with a long-sleeved blouse that reached her wrists, and gloves that covered her hands. The expression on her face, highlighted by her long blond hair, seemed to "scan" the soul of Mrs. Margarida, who immediately recognized her as the "Woman of the Fishes".

The woman asked Margarida how many children she had, what she did, and why she was not afraid to walk alone in that place; but when she looked at her questioner, she had disappeared as if by magic. Margarida returned home scared and with a headache.

The newspapers produced in Belém, the capital of Pará, printed small news stories about the case of the "Woman of the Fishes" and later, agents from the Intelligence Service of the Brazilian Air Force (A2) and also the Navy visited the region investigating the case. At first, there was the suspicion that the woman was involved in drug or weapons smuggling and even espionage, but the investigations did not reach any conclusion. Years later, an envelope with French markings and addressed to Elisabeth, was found in the lonely cabin where she lived. Many people believed that Elisabeth maintained strong contact with the beings from the bloodsucking ships, because just as suddenly as she had appeared, she mysteriously disappeared, along with the "flying devices".

According to testimony from Doctor Wellaide Cecim Carvalho and farmer Manoel Mattos, a resident of the municipality of Santo Antônio do Tauá, Pará, white women with light hair and singular beauty were observed on board the "spaceships".

However, Mrs. Francisca Costa Silva and her husband, Aurélio, tell a dramatic story. She says that on September 19th, 1977, her two sons, Antônio Élcio, 9, and Raimundo, 11, were in front of the house, at around 7:00 p.m., with several of their friends, all children from the neighborhood. Suddenly, Élcio was enveloped by a strange, yellow light that descended from the sky. He was paralyzed, although the others, who were not under the direct influence of the light, felt nothing. When the light rose into space again at a dizzying speed, Élcio, now without strength, was released. According to the boy's mother, he remained bedridden for three days; he was shaking and had a violent fever. His parents became desperate, fearing that he would die. After that, Antônio Élcio became famous throughout the city of Viseu. However, he no longer leaves the house after 6:00 p.m. Mrs. Francisca also said that she had heard that a woman in Alto Gurupi had also become paralyzed and never walked again after being struck by a ray of light.

Also, another popular story in the city involves two fishermen — a father and son — Benedito Gonçalves dos Anjos Siqueira and Simão Manoel Raimundo Siqueira. Recently they had been on the banks of the Gurupi River, with their nets catching fish near Nova Island — five miles from Viseu — when, around midnight, Simão, who was 17 years old, casually looked up at the sky and noticed that a star was moving. He looked at it again and noticed that it was getting closer. *It's light was so different that it stood out, even in the moonlight,"* he says. He alerted his father. They both noticed that the strange star was moving very fast and that it was flashing intermittently, emitting jets of light. Remembering the stories that a mysterious light had already killed two people by sucking their blood, the two fishermen unhooked the net, leaving it at the bottom of the river and rowed to the shore, where they left the boat. They then ran into the forest, lying down under the low vegetation.

Benedito, the father, was terrified and buried his head in the ground, not wanting to see anything else. Simão, however, remained watching the strange, "very yellow" luminous object move forward until it lost altitude and stopped in mid-air, about four meters above the canoe. It was shaped like a large drum, but a little larger. According to Simão's description it looked like iron and had no doors or windows — and a jet of light came down from it and it remained over the canoe for about 5 to 10 minutes. The brightness was so intense that its reflection produced a blinding glare in the river waters. Then, the object gained altitude again and began to move away. When warned by his son, Benedito, very pale, looked up. Both fishermen claim that the object headed towards Zé da Granja's ranch, at the Ilha Nova (New Island), where the fishermen usually make their base. According to the report, the men were very popular and well-liked figures in the area.

Figure 23. It was common for lights to appear over boats, even the smallest ones, sometimes even chasing them down the river and scaring the fishermen. (Credit: book "Luzes do Medo". MENDES, Carlos).

Romeu da Silva Brito, was 62, a retired sailor, and had seen many things in the world. He was a crewmember on cargo ships that transported goods from the United States to devastated England during the Second World War, always at risk of encountering an Axis submarine during one of these trips.

Born in Viseu, Romeu was the owner of the canoe used by Benedito and Simão. In addition, he says that, because he was not with them at the time of the sighting, he saw nothing. He also said that will not believe it until he has experienced it. He says he wants to find the strange object: *"I will fish it out with my hook."*

For example, João Martinho de Almeida Filho reported having heard from a citizen of Viseu who witnessed a woman emerging from a device that emitted a strong light. She was wearing a silver outfit, and approached the resident and restrained him. The man began to scream and other people appeared to help him. The strange woman let go of him and disappeared along with the light. From that day on, residents began to carry shotguns to defend themselves from possible attacks. And those who did not have a firearm slept with pieces of wood under their hammocks for fear of being attacked.

From Viseu to other cities. More cases emerge.

The terror spread further and further, driving out the population who fled in panic as more and more cities were invaded by the "Vampire Lights". In the cities of Bragança, Augusto Correia, Tracuateua, Quatipuru, Primavera, Nova Timboteua, Capanema, Marapanim, Igarapé-Açu and Curuçá, the terror attacked more than 30 people.

Four cases were amazing. For example, in Augusto Correia, a fisherman said he had been attacked by a luminous object that descended from space when he was anchoring his boat in the Urumajó River. Francisco Carlos Trindade, 56, said that a very bright light left him paralyzed. He was struck by a small ray of light that came from the ship and he fainted. When he regained consciousness, two friends helped him. He said he could not remember anything for several days. "*I even forgot who I was,*" he said. He was treated at the health center in the city of Augusto Correia and underwent tests that confirmed anemia and a burn on his right arm, with small holes in the skin. He also complained of severe headaches. Mrs. Severanda, his wife, said that Trindade "*got confused*" and said things she did not understand. "*I told him to go to church and pray, but you know how it is, he doesn't believe in those things.*"

Three other testimonies, from brothers Eraldo and Erenildo Costa, and from farmer Martinho Pereira Farias, reported that lights hovered over wooden huts and that people, lying in hammocks, felt unwell after seeing a flash of light, and then a small light entered their houses to drain their blood. They had marks on their arms and in an area near their necks. They were treated at the small hospital in the city. The case was registered in a police report, but the local clerk said to throw the police report in the trash, he said: "*the accused was a light and there was no way to arrest her. If I take this further, I will be the one arrested and admitted to a mental hospital in Belém".*

The three testimonies had something in common: the fear of having done something wrong to someone and paying for it. In these cases, the light would have a punitive effect. In other words, in the understanding of those people, they were not chosen by chance, but in a kind of selective judgment, without the right to a defense, by an unknown entity, earthly or extraterrestrial, to pay for something they were not aware of.

Other cases in Ceará state

There have also been some reported cases of aggression in the state of Ceará. Perhaps the best known was that of Alfredo Marques Soares, who at the time was working on a farm in Cardeiros, Ceará. He was attacked by a UFO/UAP in July 1977.

It was already night when Alfredo walked to a friend's house. Suddenly, something hit him in the back of his left leg. When he looked back to see what was happening, he saw a large, luminous, yellowish-white object. Alfredo felt as if the strange object was trying to suck him in. In fear, he held on to a wooden fence. He could not move his right leg, which was in great pain. He felt heat and cold being emitted from the object, which was intensely illuminated and momentarily blinded him. At one point, the object apparently let go of him. Alfredo took the opportunity to run under a nearby cashew tree. The witness saw the object rising and moving away, and with difficulty he returned home, where he asked his daughter to check the back of his leg. It was bluish-black and looked burned. The next day Alfredo appeared traumatized; he was crying a lot and shaking. When he calmed down, he went to seek medical treatment. The affected area on his leg was covered in blisters and infected. In addition, the victim urinated heavily for two days, had stomach pain, diarrhea, and pain both in his back and on the entire left side of his body. For three months, he had great difficulty walking, needing to use crutches during this time.

Abduction

Although little discussed in UFO circles, there are some cases of abduction amid the ""*Chupa-Chupa" Phenomenon*" wave.

In Pinheiro (MA), on July 10th, 1977, José Benedito Bogea left home at around 1:00 am to board a bus. He carried a flashlight to light his way to the bus stop. His fantastic experience began before he got there.

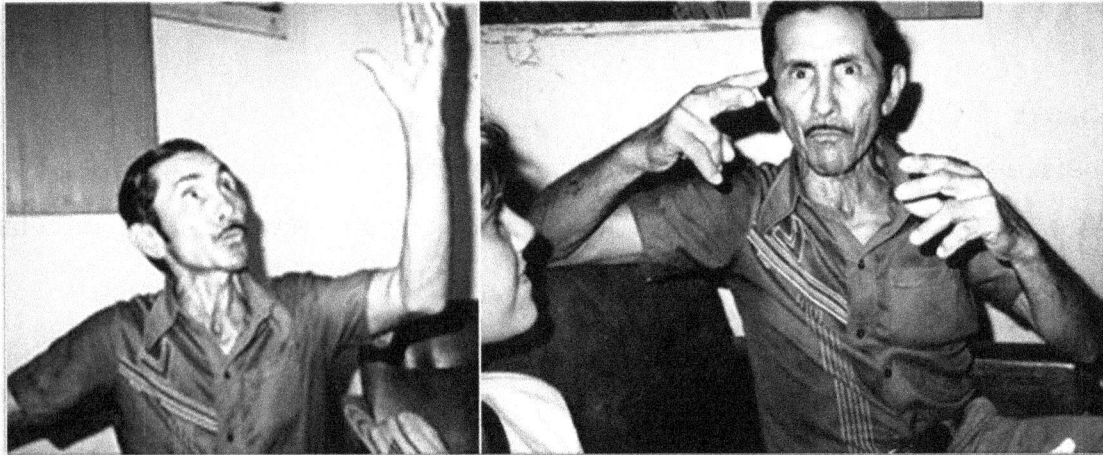

Figure 24. José Benedito Bogea who suffered an abduction during the lollipop wave in the Pinheiro region. (Credit: www.fenomeno.com.br).

Halfway along the path, a luminous, blue-green object suddenly appeared above him and followed him for 200m. After that, the object positioned itself over a bush, from where it emitted a beam of light at José, who fainted. When he woke up, he discovered that he was in a strange environment, where there were small beings with whom he spent a few hours. At a certain point, these beings took him to an object, and he lost consciousness again, only waking up in the morning, around 8:30 am. He was near the Port of Itaqui, in São Luis (MA). A few hours later, he began to feel a terrible pain on the right side of his body, and he returned to Pinheiro with great difficulty.

In the following months he still felt very ill and needed a walking stick to move around. An interesting detail of this experience is that before the contact, Bogea underwent surgery 13 years earlier and as a result of this surgery he had hearing problems and impaired vision. When he fainted after seeing the object for the first time, he lost his glasses and did not realize this when he woke up in São Luis the next day. He only realized it later when he returned to Pinheiro, because now his vision was completely normal and his hearing also improved a little. Before, he was completely deaf, but after contact he could hear the telephone, the sound of the television and also dogs barking.

Chapter 4

Ubintuba, Land of Terror, Reports and Cover-Up

The most notable events attributed to the attacks did not occur in Colares, although even now everyone gives enormous importance to the facts reported in that municipality. This may be due to number of reports from victims and the medical care provided by Doctor Wellaide Cecim at the health center. It is also likely that emotional involvement of the population, who used fireworks and shotgun blasts into the air may have played a part. In addition, given their religious beliefs, they also conducted pilgrimages to ward off what they believed to be aggressive signals coming from space.

The small communities close to the Ubintuba River provided accounts that still impress those who investigate the events that occurred there. More than 80% of the 35 families — about 200 people — who lived in Santo Antônio do Ubintuba at the time fled the area, leaving behind their shacks and small animals. The nighttime attacks led Operation Saucer to set up a tent to provide medical care to the residents who insisted on staying in the area at the insistence of "sheriff" Benjamim Amin Fernandes, who in fact was not a member of the Pará Civil Police, but acted on their behalf.

In those fearful days, Fernandes gained a level of importance he had never imagined within the military circles of Belém. He was "appointed" as an informant by Captain Uyrangê Hollanda, serving as the 'eyes' of Operation Saucer in that location nestled in the middle of the forest, 6 km from the edge of the Vigia Road. He would make a note of everything that occurred and report to Hollanda when the captain arrived at the location. The head of the Brazilian Air Force's secret service and the coordinator of Operation Saucer, Colonel Camilo Ferraz de Barros, even flew by helicopter to Santo Antônio do Ubintuba to oversee the care of the victims. He brought a doctor and psychiatrist with him, and they produced a report. The soldiers went to several of the residents' homes, listened to what they had to say and made notes.

The second half of October was considered by the residents of Santo Antônio do Ubintuba and Vila Nova do Ubintuba to be the worst month of their lives. They had no peace and what caused their fear was not on earth, but in the sky.

In Vila Nova, there is the story of Casimiro Coelho Ferreira, who was 57 years old at the time and the father of six children. He was sitting on a tree stump, at the door of his house, when out of nowhere a flash of light appeared that illuminated practically all the shacks in the neighborhood. It was around 10 p.m. when the light appeared, and his neighbors who were in their houses began to scream, "it's the *chupa*, it's the *chupa*, help me".

Ferreira said he did not have time to run, much less find the strength to leave where he was sitting and help those who were asking for help. "*It was too strong, that green light, it didn't make any noise. You could see that there was something solid behind it*", he said. He was hit by a smaller fire that came out of the "light" He added: "*My vision went black, I couldn't see anything else and then, boom, I fell to the side*". When he woke up, he saw his son Renato, 20, with a wet cloth on his chest, where there was a burn. His son asked, "*What happened, dad*?", but Ferreira didn't know what to say. Renato said: "*I know, it was the 'Chupa-Chupa', yes, that was it*".

At this point, other relatives and neighbors also helped victims who had been attacked. Conceição de Oliveira was one of them. According to her, the light attacked her in the kitchen, where she was making coffee. The kitchen window was open and when she saw the green light enter, it illuminated the entire room. She was unable to make a sound and the scream was blocked in her throat. The light hypnotized her, leaving her motionless, but Conceição did not faint. "*It all happened so quickly*," she said. It appeared suddenly and left just as quickly, but it left a mark on the right side of her chest, high up, near her breast. Her brother, upon arriving in the kitchen and seeing her sitting there trembling, ran to help her. A relative of one of the neighbors who was delivering fruit to Belém arrived in a van and took the victims to receive medical attention.

From the official report of Operation Saucer on the events in Santo Antônio do Ubintuba and Vila Nova do Ubintuba, I suspect many things are still being kept secret, including photographs and films taken by the military, because, according to the journalist Carlos Mendes, the military was in both locations. They knew that some really interesting phenomena were occurring there, more so than in other locations. Carlos heard from a retired Air Force officer, recently in Belém, that the events that occurred in the villages close to the Bituba River, which is, a tributary of the Guajará-Mirim River — left the investigators impressed. The victims' testimonies fit like a glove and provided valuable clues for Captain Hollanda's people. It was on the Guajará-Mirim River that the captain reported in his sensational interview for Revista UFO Magazine, that he had seen, in the company of his man who were also on the boat, a huge craft shaped like an American football, leaving everyone perplexed. By the way, the super 8 mm films they took have 'vanished'.

Figure 25. Unidentified flying objects often appeared in broad daylight, increasing fear. (Credit: book "Luzes do Medo", MENDES, Carlos).

It is striking that in an area rich in sightings and attacks Sergeant João Flávio Costa obtained the testimonies of only 10 residents, as reported in the less than 200 pages released so far, (thanks to pressure from the campaign "UFOs: Freedom of Information Now",) launched by the Brazilian Commission of Ufologists (CBU) through Revista UFO Magazine.

In the National Archives in Brasília, there are hundreds of files about UFOs, including some from the Operation Saucer. In one of them, Adelaide Ferreira da Silva, 37 years old gave her testimony on October 16th, 1977, at 9:00 p.m. She said she noticed that her bedroom was lit up by a reddish light that spread throughout the house. With the doors and windows closed, Adelaide opened the window a little and saw a bright blue body moving slowly, about 30m above the nearby trees —. "*I felt intense pain in my eyes and numbness throughout my body, and this sensation lasted for several days,*" she said. A few days later, she had another contact with the light, but this time she was directly hit by it and was hospitalized in the city of Vigia. Carlos Mendes tried to find out more details about this second appearance but was unsuccessful. He learned that the Brazilian Air Force had obtained privileged information. What information was this? And why was it not disclosed in the partial report?

What follows leaves either involuntary or deliberate confusion. The witness is Maria Celeste Pereira da Silva, 20 years old and literate. The date on which the incident occurred, as stated in the military report — October 18th, 1977, at 9:00 p.m., 48 hours after her mother, Adelaide, saw the light, on the 16th. Was this just a simple error by the writer, Sergeant João Flávio Costa, or does what is written below refer to the second time Adelaide was attacked, but this is not in the official document? According to the report, "*at the same moment that her mother Adelaide noticed the light that invaded the interior of the house, she felt intense pain throughout her body, as if she were being strongly compressed. There was a numbness starting from her feet, intense heat diffused from her right shoulder to her head. She believes she was hit directly on the right side of her body by a reddish light source.*"

Well, it is worth asking if Maria Celeste believes she was hit by the red light, how is it that her mother, Adelaide, who opened the window and was the first to notice and come into contact with the light, was spared? Was the light selective, choosing Celeste and not Adelaide? Again, I cannot rule out the possibility of a typing error in the date, but what occurred during the attack still draws attention. And what about the time when Adelaide, days later — or on the 18th, who knows? — was hit and had to be admitted to the hospital? There is something wrong with this report involving mother and daughter. Emanuel de Souza Farias, 21 years old, and illiterate. Statement also taken on October 18th, 1977. "*She was in the same room as Maria Celeste and saw the light, but says he did not feel anything physically*".

Maria Francisca Furtado, 20 years old, elementary school education, testified at 9:30 pm on the same day, the 18th. She lives some distance from the place known as Vila Nova do Ubintuba. Due to the high incidence appearances by the light, she and her husband travel on a daily basis to the residence of Mr. Miguel Arcângelo Soares, where several families live. She was hit by the light on the day and time mentioned above. There is nothing to add to this testimony.

With the lack of testimonies and the fact that there are so many cases of people being attacked, it is really worth asking whether it was in fact a mistake on the part of the A2 team by not including this in the reports, or whether it was done on purpose to leave these reports out, or whether they were recorded, but remain in the secret documents of the Brazilian Air Force.

Chapter 5

A Priest and the "Chupa_Chupa‹ Phenomenon

No one in Colares dared to make fun of or disrespect Father Alfredo de La Ó, an American from Texas who had come to Pará thanks to the then Archbishop of Belém, Dom Alberto Gaudêncio Ramos. On the contrary, most of the island's 1,700 residents liked and admired the priest. Others feared him, especially because of the priest's dishonorable record in the Vigia region. In 1976, he was arrested by the Civil Police for covering up the criminal activities of the German Erich Smith, who was involved in arms smuggling and accused of murdering a man who had reported him to the police.

In view of this, the Federal Police also began to investigate not only Smith's conduct, but also the priest himself. When the German was arrested, he was hiding in Alfredo's house. The Catholic Church became aware of this and, to reduce the damage to its image, decided to remove him from priestly duties in 1976. In his statement to the police and the Public Prosecutor's Office, Smith took full responsibility for the crime of smuggling, thus clearing the priest.

Dom Alberto Ramos knew more about the American's shady life, although he rarely mentioned it to his subordinates in the Metropolitan Curia. Closely linked to the military and a supporter of the 1964 movement that deposed the then state governor, Aurélio do Carmo, in Pará, the archbishop managed to find Father Alfredo a job in a public agency responsible for the construction and maintenance of roads. The position of chaplain was created for him. His work consisted of celebrating masses for the employees but this was after Father Alfredo left the church in Colares.

In the case of the mysterious lights, the priest - who told the Operation Saucer team that he had seen luminous objects on two separate occasions - was found to be friends with several foreigners who had visited the region to investigate the sightings. Journalist Carlos Mendes reported hearing on multiple occasions, even from agents of the former National Intelligence Service (SNI), that the priest had connections with the United States Central Intelligence

Agency (CIA). This espionage agency, representing the world's most powerful nation, had a particular interest in the events unfolding in the northeastern region of Pará.

The most unusual episode involving Father Alfredo occurred at the end of November 1977, in the early evening, near Igarapé do Tubinho. The priest appeared and pointed a gun at Carlos Mendes, threatening to shoot him. The incident happened when he was interviewing council member Manoel Ferreira and his wife, about a bright light that descended on some houses. At that moment, Father Alfredo appeared holding a shotgun. He began shouting, visibly out of control, and ordered Carlos to leave immediately. The council member tried to intervene, saying that he was doing his job as a reporter, but the priest did not stop. He pointed the shotgun and, stating that Carlos was not welcome in Colares, announced that he was going to pull the trigger. *"Get off my land, get out, this is mine, you're invading what's mine,"* he said, irritated. Manoel Ferreira argued that the land belonged to the municipality and asked the priest to calm down and let Carlos stay there to finish the report.

The priest, still with his gun in hand, said that Captain Hollanda had already told him who Carlos was. *"The captain was right, you came here to stir up the people who are scared by what is happening. Get out of here, that's the best thing for you."* Encouraged by the support of council member Manoel Ferreira, Carlos told the priest that if he shot him, without any reason, he would be arrested and convicted of the crime. The council member approached Father Alfredo and mumbled something that Carlos could not hear, but what he said made the priest lower his hands, turn and leave.

The council member asked Carlos to be careful with the priest, advising him not to contradict him. The politician from Colares did not want to get into trouble with the influential local priest. A few days later, the priest saw Carlos at the city hall, talking to the Mayor Alfredo Ribeiro Bastos. He approached Carlos and, in somewhat slurred Portuguese, apologized for what had happened in front of the house near the Tubinho creek. He said that the events in Colares, the nights of skywatches and the fear of the residents had gotten on his nerves. Therefore, Carlos had no choice but to accept the apology.

Figure 26. Artistic representation of Father Alfredo de Lá Ó's sighting of a huge mysterious light that passed over his head (Credit: book "Luzes do Medo", MENDES, Carlos).

Chapter 6

Hunting the Vampire Lights

Two reporters, Biamir Siqueira and José de Ribamar dos Prazeres, were working for the newspaper '*O Estado do Pará*' in the area of Mosqueiro Island, 85 km from Belém, and the nights were full of surprises. From the bright, starry sky emerged a spectacle of strange lights, which alarmed the residents who locked themselves locked in their homes. They were not satellites, nor the planet Venus, much less weather balloons or atmospheric anomalies. The lights, of different sizes, shapes and colors, descended with astonishing speed, zigzagged across the sky, stopped suddenly and then rose to disappear into the sky again. On some occasions, they hovered over the shacks, very close to the ground, emitting jets of light that penetrated the bodies of people who were there, and sucked their blood. The phenomenon, which from September to December had made life hell for residents on the other side of the river, in Colares, Vigia, Santo Antônio do Tauá, and dozens of villages and riverside communities, was now manifesting itself more intensely in Benevides, Santa Bárbara, Mosqueiro, and Baía do Sol.

Operation Saucer, at least officially, ended in late 1977. However, a larger group of military personnel - including agents from the National Information Service (SNI) and the Brazilian Air Force, such as Captain Hollanda and Sergeant João Flávio Freitas Costa - along with civilian pilots and businessmen, all armed with Super 8mm video cameras and still cameras, continued conducting investigations on their own. Brigadier Protásio de Oliveira, head of the 1st Regional Air Command (COMAR), following instructions from the Air Force Ministry, ordered that no official vehicles be used in UFO investigations. Anyone wishing to investigate outside of working hours could travel to Mosqueiro, Baía do Sol, or Benevides in their own vehicle. Public funds would no longer be used for any expenses. Hollanda and the SNI were aware, from reports in Pará, that others were on their trail, working diligently to investigate new cases of sightings and attacks.

For Siqueira and Riba, what was happening in Baía do Sol was abnormal and had no explanation. "*It's not normal, I know it's not normal,*" Siqueira commented. For Riba, it looked as if what was happening was deliberately designed for them to report on.

At the beginning of May, they were witnesses of a terrifying luminous apparition during the night, when they were both taking a short nap inside an old VW Beetle they were using. This was near the Carananduba road, in Baía do Sol. "*I got the biggest fright of my life, woken up by an unusual flash, much bluer, with a gray tone. I wasn't dreaming, It woke me up, it was very strong, and exerted a force that terrified me,*" Biamir said. Almost paralyzed, he shook Riba awake and he was also perplexed by the intensity of the flash. Riba, typical of a photojournalist, opened the car door and, with the camera in hand, tried to photograph what he was seeing.

Siqueira told me that a spaceship, estimated to be about 50 meters in diameter and no more than 20 meters high, suddenly turned off the intense glow, which illuminated the entire area, and sped away, disappearing into the sky. Siqueira, who had also gotten out of the vehicle, said he shouted for Riba to take a picture of the object, but Riba seemed immobilized and did not have the strength to raise his arm and take the picture. After the spaceship left, Riba, according to his story, said he felt nauseous, like he wanted to vomit. He and Siqueira drank an entire bottle of water that was in the back seat of the car. They got back in the car and quickly left the area, Riba said it was around 1:30 a.m.

Twenty minutes later they were in Vila de Mosqueiro and could no longer sleep. "*I felt a very strong heat in my head, it felt like it was going to explode*", even after that strong light had gone, said Siqueira. Riba added that at the time, all he could think about was photographing that ship, but his hand would not obey and remained lowered, holding the camera. Only after the object had departed did he regain the stability in his arms. "*It was as if I was all tied up, I just couldn't move*", added Riba.

According to Riba and Siqueira, several hypnotists and parapsychologists from Belém tried to perform memory regressions. Some were certain that the two journalists had experienced close encounters of third kind with the occupants of the "devices" they saw and photographed in Baía do Sol and Mosqueiro. Others believed that a regression would help them recall their possible contact with the beings. Two psychiatrists, Durvalino Braga and Orlando Zoghbi, with whom the two reporters had maintained contact, have since passed away, but they told them that they did not believe the extraterrestrial theory of the *"Chupa-Chupa"*. They attributed everything to the power of the mind, which is capable of anything, even creating events that never occurred and believing in them. They also said that it was all nothing more than collective hysteria.

Another man, Biamir Siqueira, had an incredible sighting that was published by *'O Estado do Pará'* newspaper. He said that the event happened in the early hours of June 7th, 1978. "*In the clear sky of Baía do Sol (Sun Bay), inexplicably, a cloud began to move in the opposite direction to the wind, expelling small multicolored rays. It was the first time that something unusual had been seen in the sky during the month of June.*" It was precisely 3.25 a.m. Then, minutes later, a fantastic spectacle was witnessed by people who were in the village square.

While everyone was still commenting on what they had just seen, two unidentified flying objects crossed the sky at incredible speed. They had a rotating lamp on their top surface and were below the clouds. They darted around in the sky in complete silence. After 10 minutes, they unexpectedly stopped and just hung there motionless. Then, without any warning, they moved at breakneck speed and disappeared into the sky. Nothing more happened for several days, then on the 13th, they reappeared. The unidentified flying objects, flew below the clouds for 30 minutes. Then something else occurred and six smaller ships emerged from the mother ship through an opening in its side.

Then the mother ship headed straight up at high speed and stopped, it looked just like a star. However, the smaller craft that came from it began to dart around the sky performing inconceivable maneuvers. Moments later, they simply disappeared. Three days later, another event was recorded using a 16 mm telephoto lens. An incandescent body appeared in the sky and approached the square at high speed then abruptly stopped. Completely visible to the naked eye, it unexpectedly released six small, unidentified flying objects. While the six UFOs begin to make a series of movements, the mother ship, still glowing, passed over the square of Baía do Sol at moderate speed. Then it suddenly stopped and shot straight up into the sky.

Figure 27. Military personnel participating in Operation Saucer were constant witnesses to UFO sightings on roads, beaches, fields, everywhere. (Credit: book "Luzes do Medo", MENDES, Carlos).

It was the first time the spacecraft had been so close. All its movements were carefully observed. It was stationary on the horizon, in terms of diameter and brightness as looked like the planet Venus. After a few minutes, the craft glowed even more brightly. The impression one had was that it was surrounded by a halo of light, then it expelled — this was carefully observed — rays. The craft remained in like that for exactly two hours and thirty-seven minutes, then it disappeared into the sky.

On the beaches of Colares, Baía do Sol and near the ravine in front of the Arapiranga neighborhood, in Vigia, were the locations that recorded the highest incidences of UFO phenomena similar to those that had already been observed. What really caught people's attention was how the craft suddenly stopped in space. This morning, in addition to reporters Biamir Siqueira and José Ribamar dos Prazeres, the unusual spectacle above the Baía do

Sol square was witnessed by the notary of Salinópolis, Raimundo Emir d' Oliveira, brother of the journalist and Walmir Botelho the editor-in-chief of the newspaper *O Estado do Pará*.

Taxi driver José Rodrigues Lopes and another person, student Raimundo Campos Pinto, were two more victims of the Vampire Lights. The lights attacked around 2:00 a.m., according to neighbors who ran to help them when they heard their screams for help. Lopes said he was asleep and woke up to a penetrating beam of light coming through a hole in the roof. The beam immediately hit his chest, and paralyzed him, he was unable to speak. His body went numb, and it was several minutes before he regained his senses and saw that the mysterious light had gone. Screaming in despair and not knowing what had attacked him, (there was a thin trickle of blood coming from the spot on his chest where the light had hit him,) Lopes woke up his entire family and then his neighbors, telling everyone what had happened. Quite nervous and in pain from where the light had struck him, he decided to go to the hospital, where he was given medication. The next day, now calmer, he went to the police and reported the incident. Lopes were traumatized and said he would not return to his house and that he would sell it.

The second victim, a resident of the Campina neighborhood, was Raimundo Campos Pinto, 19 years old. He was sleeping on the porch of his house. According to his father, André Pinto, his son received medical attention immediately after the light attack and was taken to the IML, where he underwent a forensic examination. André and the boy's mother, Teodora, said that on the Tuesday night, many people saw a light appear from behind a large mango tree in the backyard of the house. The light disappeared after a few minutes. This happened around 10:00 p.m. After that, the residents of the area stayed awake until after midnight. After two in the morning, a scream from came from where in Pinto lay in his hammock. His parents ran to him and saw he had a small wound on his left arm.

Figure 28. The unexpected encounter with a ball of light, which was a shape taken by the "Chupa-Chupa", was a source of despair for most people. (Credit: book "Luzes do Medo", MENDES, Carlos).

The following statement is from José Maria Albernaz, a former policeman, who was also attacked in the municipality of Benevides, 30 km from Belém. He reports that it was almost 30 minutes into the morning on November 17th, 1977, when, as was his custom, he began to patrol the city, riding his bicycle. *"As I was calmly riding down the street in front of the police station, I saw a very bright light coming towards me. I started to feel cold and immediately thought of the mysterious light. So, I turned the corner more quickly and pedaled faster. The flash of light passed over me, and that was when I felt a severe headache."* He told reporters that many people in that city have already been hit by the "fire", but they don't say anything to anyone — only to their family members — for fear of being attacked again, since they heard from a radio station that if anyone continues to say they were hit, they will be sought out again by the "thing." The police officer, luckily, was spared any loss blood.

On November 19, 1977, Oberlando de Almeida Teixeira, a student at what was then the Federal Technical School of Pará - now the Federal Institute of Education, Science, and Technology of Pará (IFPA) - In Portuguese: Escola Técnica Federal do Pará, now the Instituto

Federal de Educação, Ciência e Tecnologia do Pará (IFPA) arrived at the offices of the newspaper O Estado do Pará and left everyone stunned. Visibly shaken, on the verge of tears, and unsure of what to do, he recounted how the 'focus,' 'disc,' or whatever it was had attacked him.

Terrified that the attack might happen again, Teixeira even considered suicide. The mysterious light enveloped him while he was studying during the early hours of the morning, it seemed to give him great strength and he broke three windows, a sheet of plywood, and violently threw his sister and brother-in-law out of bed.

The incident took place at house number 13 on Travessa 9 de Janeiro Street, between Oliveira Belo and Antônio Baena Street. "*To this day, people remain terrified and stay at their front doors until late at night, with a mixture of fear and curiosity in case there is another outbreak,*" said Teixeira. The boy saw a doctor from the Technical School, but the doctor did not really believe him. Anyway, just in case, the doctor recommended that he take a radiation test at the Air Force base.

"I don't know. I asked God to leave a mark on me, so they could say it is true. Many people think it is a lie, but it is not. I am used to studying at my sister's house, constantly, with a classmate, but that day my classmates did not come, and I was left studying alone. My sister went to her room and the house was all messed up. I felt something penetrate; I remember it very well. You know when you take a glass and put it on top of any animal? As soon as it feels like it is running out of air, it runs, right?

It penetrated like that, you know, inside me, in my head. It did not manage to penetrate all the way, because I ran. I really ran and screamed. It was the 'darkening heat', a strange force, really, I do not know how I managed to drag my brother-in-law and my sister out of bed and throw them against the wall. She says I hit her all over, but I do not remember having gained that much strength. All the windows broke. I broke two windows and there was also some plywood. I punched the plywood so hard that my fist went right through it.

Afterwards I wondered how I did that. I still did not believe it, but someone came up to me and said: 'I saw the light'. At that moment, I said: I want to leave. I want get away from here,

I was afraid, everyone is terrified there. My sister is in a terrible state of nerves because she experienced it, in fact she fainted.

Since that day I haven't slept. I have insomnia, I stay at home but I am restless, I try to walk, I don't feel hungry. I feel a 'pressure' inside me, a tightness. I went to the doctor, but he did not believe me and sent me to take a radiation test at the Air Force base. My dad says he will try to help. I am terrified, trying to figure it out, to find out. I think it was the light, like a hot spot that makes people go crazy. And that heat gives a person strength. I screamed, but I can't understand why people 10 meters away from my house heard it not anyone inside."

Chapter 7

Before the FAB, the Armed Forces Had Already Been Alerted

Few are aware that, before the Brazilian Air Force (FAB) dispatched its mission to Colares, the Brazilian Navy had already logged reports of mysterious lights over the region's rivers - objects seen plunging into and emerging from the water and even incidents involving attacks on local fishermen. Report 45 dated November 23rd, 1977, from the command of the 4th Naval District of the Ministry of the Navy, in Belém, deals with the appearance of unidentified flying objects in Pará. It says the following: 'since April 1966, newspapers in Belém have been publishing reports about the appearance of UFOs in various regions of the State of Pará and northern Maranhão'.

In the regions where these UFOs have been sighted, residents live in fear, speaking of mysterious lights said to cause both deaths and hallucinations. A Navy report circulated among military intelligence agencies such as the Navy Information Center (CENIMAR), the 8th Military Region, the National Intelligence Service (SNI), the Public Security Secretariat, the Military Police, and the Federal Police compiled numerous newspaper clippings documenting these accounts. Among them was the chilling case of a man named Firmino, found dead with severe burns on his chest inside a vessel anchored near Crabs Island, close to the Port of Itaqui, in Maranhão.

The official document contains sections that indicate the locations where UFOs are said to have appeared most frequently. One of these sections shows a drawing of the shape described by *"several natives of different regions"*. Also included is a request from the City Council of the municipality of Maracanã, in the northeastern region of Pará, addressed to the Navy General Staff. It is an appeal for the military authorities to take action the regarding the appearance of the lights that were terrorizing the residents. One of the attached clippings is of two news items from the newspaper *'A Província do Pará'* with statements allegedly made by Colonel Camilo Ferraz de Barros, head of A2, the Air Force's secret service in Belém, which deny that of UFOs appeared in the city of Vigia. In an interview with the author of this work, the colonel denies having made such statements. "What they published is a lie, I never said that," said Colonel de Barros.

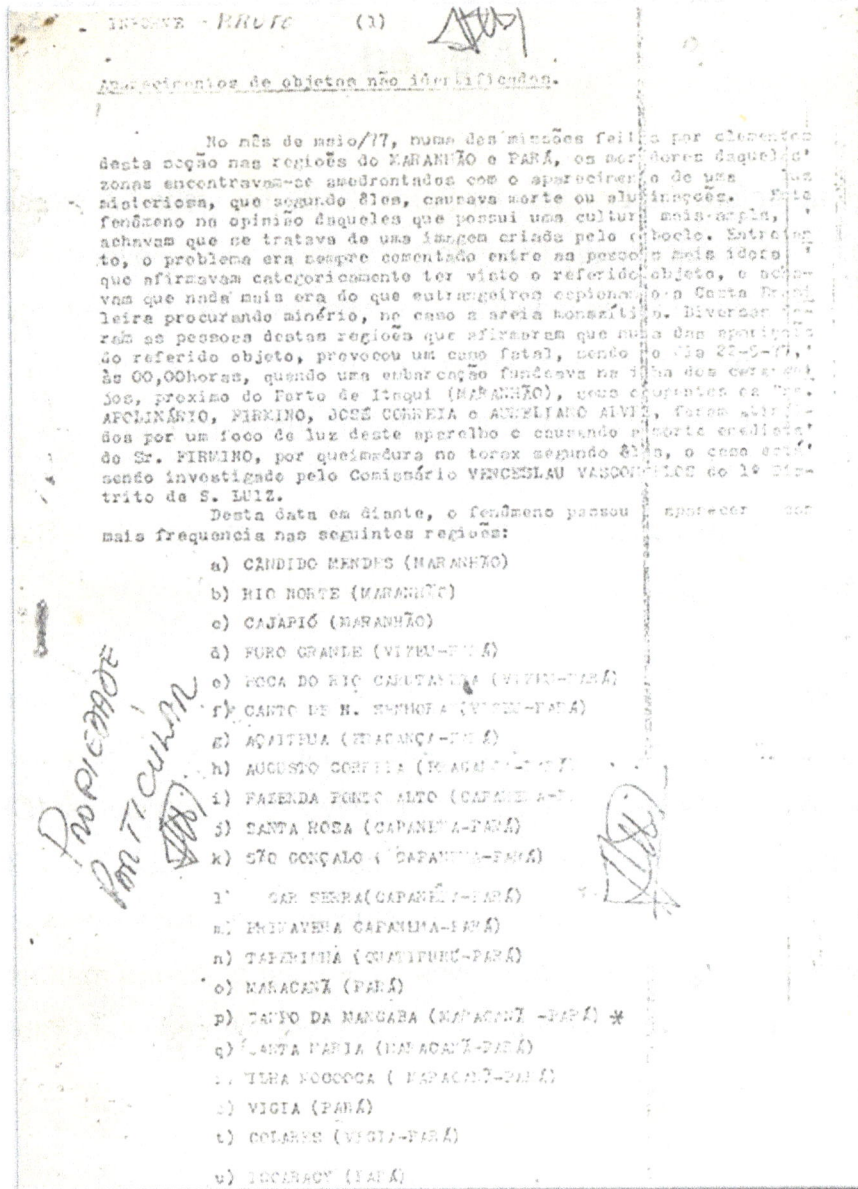

Figure 29. A 5-page "summary" report, where the Navy lists several cities where the "Lights" were seen. (Credit: Author's files).

Em 12/8/77

Tapéirinha - Quatipurú - Milton Borges da Costa e esposa Joana Dart da Silva - na terça-feira (dia 19/8/77), ás 03,00 horas da madrugada, um aparelho do estilo de um balão andando nas proximidades de sua casa, numa altura de uns 10 metros, todo iluminado. No dia 12 as 07,00horas da manhã houve uma explosão em frente da barra de Jepirica, dai em diante aparecido muito peixe morto. E comentario que está aparecendo um submarino entre Quatipurú e Japirica (bem fora de barra) e tem sido visto por pescadores daquela área.

RAIMUNDO LISBOA DOS SANTOS 48 anos pai de 8 filhos irmão de CARLOTA LISBOA DOS SANTOS, residente em Quatipurú, no início de agôsto, foi apanhado pelo aparelho ás 02,00hs de madrugada no lugar Peruquara, só conseguiu sir 06,00horas da manhã. Foi trazido pa Quatipurú em estado de choque sendo-lhe aplicado sôro pelo Vice - Prefeito, Sr. ANTONIO MORAES DO NASCIMENTO. Este senhor ainda se encontra com alucinações quando dorme.

PAULICINHO : - Lugar Santa Rosa fazenda Sr. Jaime.
Apareceu diversas vêzes o aparelho não identificado.

7ª Travessa - Capanema: (São Gonçalo).
Sr. Jacó - Viu um clarão mas não viu o objeto.

Lugar Serra - Sr. Severiano - foi perseguido pelo objeto.

d) APARECIMENTO DE OBJETOS ESTRANHOS NA REGIÃO:

Alguns moradores de Carutapera disseram que no Município de Cajari8 (Maranhão), tem aparecido constantemente uma luz misteriosa, visto inclusive por quase toda a população daquele município. Frisaram ainda que a mesma luz teria dias atrás, seguido um motorista na Estrada Pará-Maranhão. Disseram ainda, que em maio deste ano, a referida luz causou a morte instantanea de um tripulante de uma embarcação ancorada ao largo da ilha dos caranguejos, próximo ao Porto de Itaquí.

No dia 22-5-77, após as 00:00 horas, os Srs. Apolinário, Firmino, José Correia e Aureliano Alves, todos do município de Alcântara (Maranhão), se encontravam numa embarcação carregada de madeira para construção de um "Rancho", quando foram atingidos pela referida luz brilhante, causando queimaduras no tórax e a morte instantanea do Sr. Firmino. Este caso consta que está sendo investigado pelo Comissário Venceslau Vasconcelos do 1º Distrito de São Luis. Do mes de junho em diante o mesmo fenômeno vem se repetindo, deixando em pânico os moradores dos seguintes pontos:

a) Furo Grande - Viseu (foi visto pelos filhos do velho pescador - Cachorrinho).

b) Colégio Abel Alves - Viseu (foi visto por José Pucheiro e esposa).

c) Boca do Rio Carutapera - Viseu (visto pelo Sr. Cornélio).

d) Campo de Nossa Senhora - Viseu (visto seis vezes pelo Sr. Manoel Pereira de Araujo).

e) Candido Mendes - Maranhão (João Galego viu às 21:30 hs do dia 19-7-77, no Rio Morto).

f) Apaiteua - Bragança (visto pelos pescadores daquela região).

g) Augusto Correa (Maranhão)- Bragança (visto por pessoas consideradas idôneas no local).

h) Fazenda Ponto Alto - Capanema (visto por diversas pessoas, inclusive moradores em Belém).

80

Além destes outros ainda estão sendo visitados pelo objeto conforme recortes de jornais anexos,

Aparentemente o problema parece ser simples manchete, mais na realidade o caso merece ser estudado com mais carinho, uma vez que o problema existe e está causando prejuizos aos pescadores e agricultores que alegam temer represália ou serem atingidos pelo fenômeno.

A maioria dos caboclos, acham que êste aparelho é manobrado por um submarino e que por traz disso está uma francesa que reside na ilha do MEIO (AUGUSTO CORRÊA), pois acredita ser a dita cuja uma simples "espiã".

O detalhe mais importante é que no dia 12/9/77, as 07,30 horas houve uma explosão na costa de QUATIPURÚ e dois dias depois uma grande quantidade de peixe começara a boiar e morrer. Dias depois, de madrugada, um objeto estilo submarino teria sido visto entre as pontas de JAPIRICA e QUATIPURÚ (fora da barra), andando todo no escuro.

Como se ve o caso merece atenção e esta seção esta acompanhando atentamente para poder chegar a uma conclusão. Em anexos croquis das áreas com pontos assinalados onde houveram os aparecimentos e recortes de jornais sôbre o fenômeno.

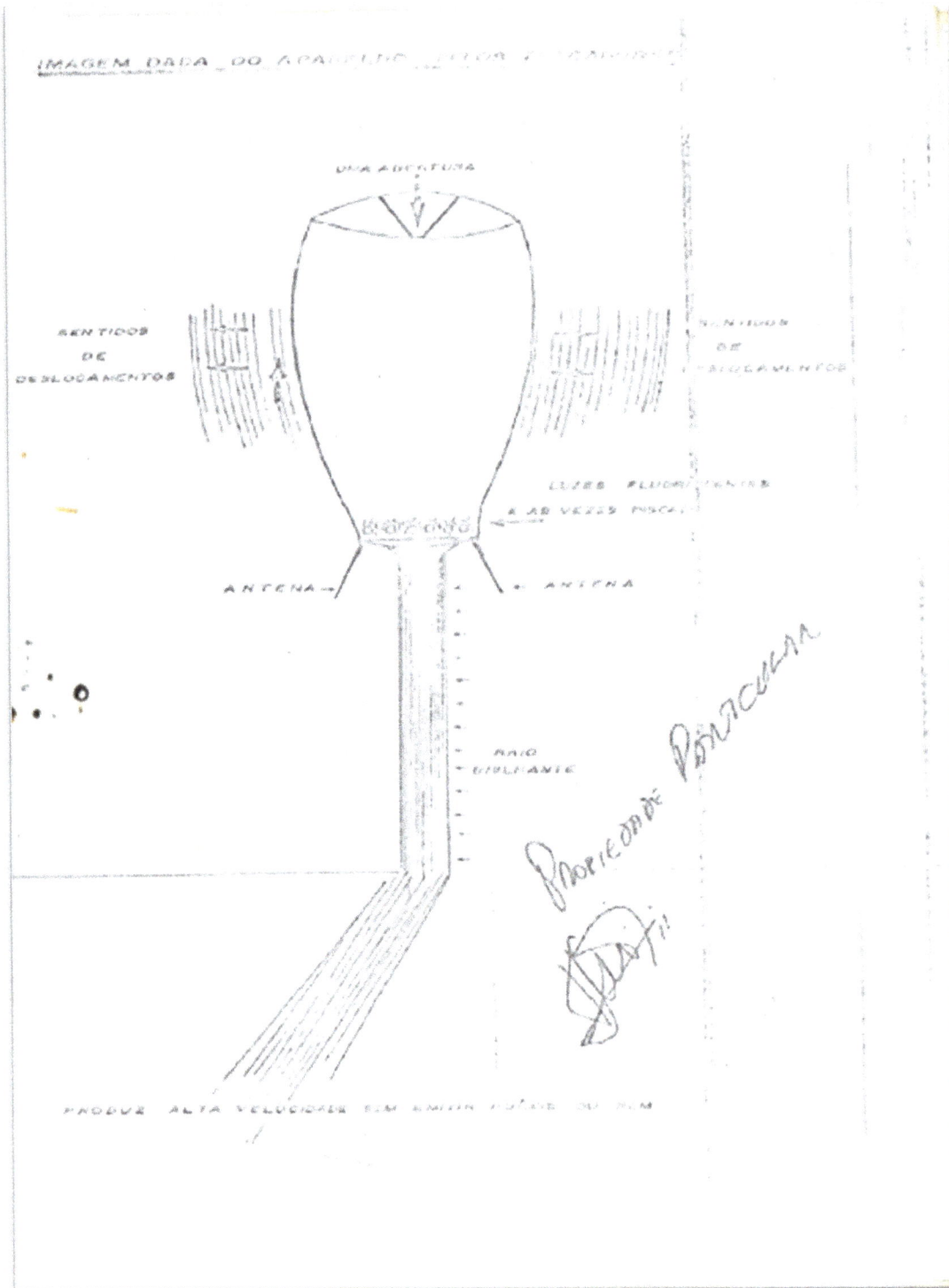

In the short Navy report, which is 15 pages long (the summary report has five pages, as seen above), some of which are illegible, Captain Paulo Fernando da Silva Souza also attached press clippings about the appearance of UFOs in the city of Belém and its surroundings. "*Fishermen in the region speak of a large explosion off the coast of Quatipuru, causing the death of many fish. According to the fishermen, days after this explosion, an object*

resembling a submarine was sighted between the points of Japirica and Quatipuru. The fishermen link the presence of this object with the appearance of UFOs in the region," concludes the military report.

Paralyzing green light in Santa Maria do Caripi

A desperate call for help reached the Armed Forces General Staff (EMFA) in Brasília and the military commanders of the Army, Navy and Air Force in Belém, in the form of a request signed by councilwoman Dalgisa de Alcântara Garcia, of the Maracanã City Council, dated October 18th, 1977. The request, fully, is as follows: *"To the Command of the General Staff of the Armed Forces, Brasília, command of the 8th Military Region, command of the 1st Air Zone and commander of the IV Naval District. Strange unidentified flying objects have been attacking the town of Santa Maria do Caripi, in this municipality, for several consecutive days, using a mysterious beam of paralyzing green light, according to witnesses, whose effect has caused some victims to visit the Maracanã Mixed Unit. The residents of this town are in a panic and are afraid to go out to do their daily activities in the field, because even during the day they are attacked, which is why armed men are on standby to confront the mysterious enemy".*

"Not only in Santa Maria do Caripi, but in other locations in the municipality, some people have seen this thing and have also been attacked in Vila do 19, Porto Alegre and towns along the river. The statements are borne out by many witnesses, and we are in danger because we have no defense against this thing. In accordance with established procedures, I wish to formally and urgently request that the authorities send help through the security agencies and also send a team of doctors to evaluate the effects of the mysterious green ray in the village of Cajueiro. I also ask that that the State Legislative Assembly be informed of the content of this letter." Session room of the Municipal Chamber of Maracanã.

What the victims of Colares have to say and other notes

The military team went to Colares at 7:00 p.m. on a day in October 1977, arriving at 8:15 p.m. After contacting the mayor, Mr. Alfredo Bastos, who because of his health, asked council member Manoel Costa to accompany the team. Several people who had been hit by the "light" were interviewed, as listed below:

José Jorge dos Santos, 48 years old, elementary school education. Date and time of the incident, October 23ʳᵈ, 1977, at 1:00 a.m. *"He said that on the day and time mentioned above, while inside his home, he noticed a beam of light coming through the roof and it hit him on the neck. As he said that at first he thought it was animal, like a bat, and he hit it with his hands as if trying to scare it; then he felt intense heat and numbness in the right part of his neck. Screaming for help, his wife, who was sleeping nearby, came to his aid and helped him get out of the hammock. Because of the noise they made, they believe they drove the "animal" away."*

"In the morning, they were surprised to find that their daughter Domingas' nightgown had blood splatters on it, there was also a drop on Jorge's shirt. There was also blood on the hammock where Domingas and her sister Selma were sleeping. After this, he and his family began to sleep in another house. Jorge mentioned a strange numbness and drowsiness he felt when he noticed the "light" in his house. After this, he had a headache that lasted for a week."

Domingas Maria dos Santos, 22 years old, primary school education. Date and time of the incident: October 23ʳᵈ, 1977, at 1:00 a.m. *She said she had not felt anything when her father José Jorge dos Santos suffered what they call the "animal attack". She was sleeping with her sister, Selma, in the same hammock, when they were both woken up by their mother. At the time, her father was very nervous, but he did not mention anything about his injuries. Her sister Selma was also present, but she did not say anything.*

José Zilton Aranha, 26 years old, elementary school education. Date and time of occurrence: October 21ˢᵗ, 1977, at 1:00 a.m. *"While sailing in the Quiririm river, he found a gray, conical, tubular plastic object adrift, it had a rubber visor. He said he had not seen any boats near the place where he found it during the six days he spent at sea."*

Emídio Campos de Oliveira, 50 years old, elementary school education. Date and time of occurrence: October 23rd, 1977, at 2:00 a.m. *"He said he woke up in the middle of the night feeling weak; he tried to get up, but was unable to, and fell asleep again. In the morning, he noticed that he had a purple mark, with a sensitive spot in the center, located on the inside of his right thigh. He did not pay much attention to the fact, but connected it with recent events, assuming that he had been "sucked" by the "light". However, he did get alarmed and continued to work normally. He says he told other people in the city about what happened. Emídio said he had been hospitalized at the "Juliano Moreira" hospital when he was 16 years old, and had not suffered any mental problems after that age.*

Neusa Pereira Aragão, 25 years old, 5th grade elementary school. Date and time of the incident: October 26th, 1977, at 10:15 p.m. *Neusa was with her younger sister in the kitchen of her home reading a prayer book, when she noticed the 'bright light' already mentioned by the other people. She was immediately attended to by a member of the local hospital emergency team who provided medical care for her.*

***Maria Beatriz Leal Ferreira, 42 years old, literate. Date and time of the incident: October 27th, 1977, at 12:35 a.m.**, when she was getting ready to sleep, Dona Beatriz sensed the brightness already described, and suffered a nervous breakdown. She was taken to hospital and promptly treated by the medical team.*

Chapter 8

Military see a "meteor" - More Cases Emerge

The cases were piling up, with more and more people telling the military about their sightings and saying they had been attacked by the "Vampire Lights". The situation was becoming increasingly worrying. The disbelief among the members of the Operation Saucer team - if any still doubted that what was happening was a real and unknown phenomenon - vanished when two teams, in different locations, saw the same object.

On October 26th, 1977, at 8:05 p.m., a meteor of unusual size was observed, with a nucleus diameter of 20 cm calculated from a distance and a trajectory perpendicular to the horizontal plane with an inclination of approximately 30 degrees. It was observed as it ran up to the tree line. It is interesting to note that the object's luminosity was only visible when it was at a low altitude, and it left no trail. Its tail, however, extended to an estimated 40 cm and emitted intermittent sparks. This occurrence took place while the team was nearing the junction of the main road from Vigia and the road from Santo Antônio do Ubintuba. The same phenomenon was also witnessed by two members of the operations team who remained in Santo Antônio do Ubintuba.

The military then went in search of other reports and witnesses who could provide more information and there was no shortage of witnesses. In every house they knocked on, every person they talked to on the street and every person who came to them looking for help, the story was the same. "*We are being attacked by the Chupa-Chupa.*"

The military received a statement from Father Alfredo de Lá Ó. At 3:25 a.m., on October 21st, after being woken up by the insistent barking of dogs in the houses near the chapel, he saw an object that caught his attention, because it emitted a strong light and was moving from the sea to the land in a north-south direction. The object was moving at a great speed, greater than that usually observed in airplanes. It was flying in absolute silence at an altitude of approximately 20m, it was very quiet at that time of the morning, and he did not hear any noise coming from the device. At the top, the device emitted a powerful red light and at the bottom there was an intense light that lit up the entire area as it passed over. Note: the electricity in the city is turned off at 10:00 p.m. The approximate size of the object in relation to where Father Alfredo de La Ó was, according to his statement, was about 50cm, the size of size of an oil drum lid. The approximate distance between the objects and the priest was 75m.

Another witness was Antônio Acácio de Oliveira, 53 years old, who also had a sighting on October 19th, 1977, at 7:30 a.m. He observed the light passing by, which was at the height of the trees just in front of his residence. The object's light decreased in intensity, becoming less bright and comparable to a lit cigarette. He said that after stopping briefly, the "device" quickly flashed three times, like a car headlight, and headed towards the city; moving from southwest to northwest; he said that at that moment he fired two gunshots at the object.

On November 1st, 1977, at 7:00 p.m., the event happened a second time and he observed the light moving, flashing intermittently, and then momentarily disappearing. He said that after the FAB helicopter took off, he noticed that the aforementioned light was moving overhead, following the helicopter; after that it began to turn to the right, and disappeared. He then sought out the helicopter support team and was referred to the head of the 2nd Section of the 1st Regional Air Command (1st COMAR).

At 7:00 p.m. on November 1st, 1977, for the first time members of the 2nd Section observed a light moving in a southwest-northwest direction at an estimated altitude of less than 6,000m, according to observations by Sergeant Flávio it was at precisely 3,000m. It was emanating an intense, pulsating blue-green light, with a reddish semicircle at the top. It was moving at high speed with a straight trajectory, and it began to curve to the right after traveling a few hundred meters, at which point it became a small, reddish luminous spot. The event was observed by Lieutenant Colonel Camilo, Sergeant Flávio and some of the locals who were positioned near the city cemetery. The team members, Sergeants Almeida and Pinto, who were providing information for the helicopter's pilot during landing and takeoff, observed the light from their position, as did the crew members, Lieutenant Colonel Gonçalves, Lieutenant Kuster, and Sergeants Roberto and Dourado.

Here we return to Wellaide Cecim, then 24 years old and a doctor at the Colares Island health unit. When interviewed by members of the military team, among other statements, she said that to preserve her ethical and professional reputation, she did not provide a more complete report regarding four cases she had seen at her workplace. She stated that, in addition to nervous breakdowns, her patients presented with other symptoms, such as paresis [partial numbness of the body]. There were also headaches, asthenia (physical weakness), dizziness, generalized tremors and, what she considers most important, all had first-degree burns, as well as small punctures marks. The men had the marks in the neck region, more precisely over the jugular, and the women, in only one case, on the breast. She read a communication that she was going to forward to the Secretary of Health of the State of Pará to the team, but in the end did not do so to avoid looking ridiculous.

Figure 30. Doctor Wellaide Cecim was one of the witnesses of an airship (Credits: Author's file).

Doctor Wellaide stated that on both October 16th and 22nd, 1977, at 6:30 p.m. and 7:30 p.m., respectively, she saw a luminous, metallic object, approximately 3 x 3m in size, which was moving at high speed over city of Colares. She describes this object as having a cylindrical, almost conical shape, with a narrower part. She said she saw the artifact clearly, in the company of other people who were present at the hospital unit. She described the developments in an almost "comical" manner: the object moved with pronounced lateral swaying and did not rotate. However, occasionally it seemed to stop and turn around a little. On condition that that she was not quoted, she expressed her opinion that she believed the reports of what had been occurring in the area. In addition, she said that she had not observed anything to suggest that the victims had been had blood forcibly extracted. However as for individuals having been hit by a 'ray' or 'focus of light of unknown characteristics'; based on the examinations she carried out on the patients she treated, she believes this is possible.

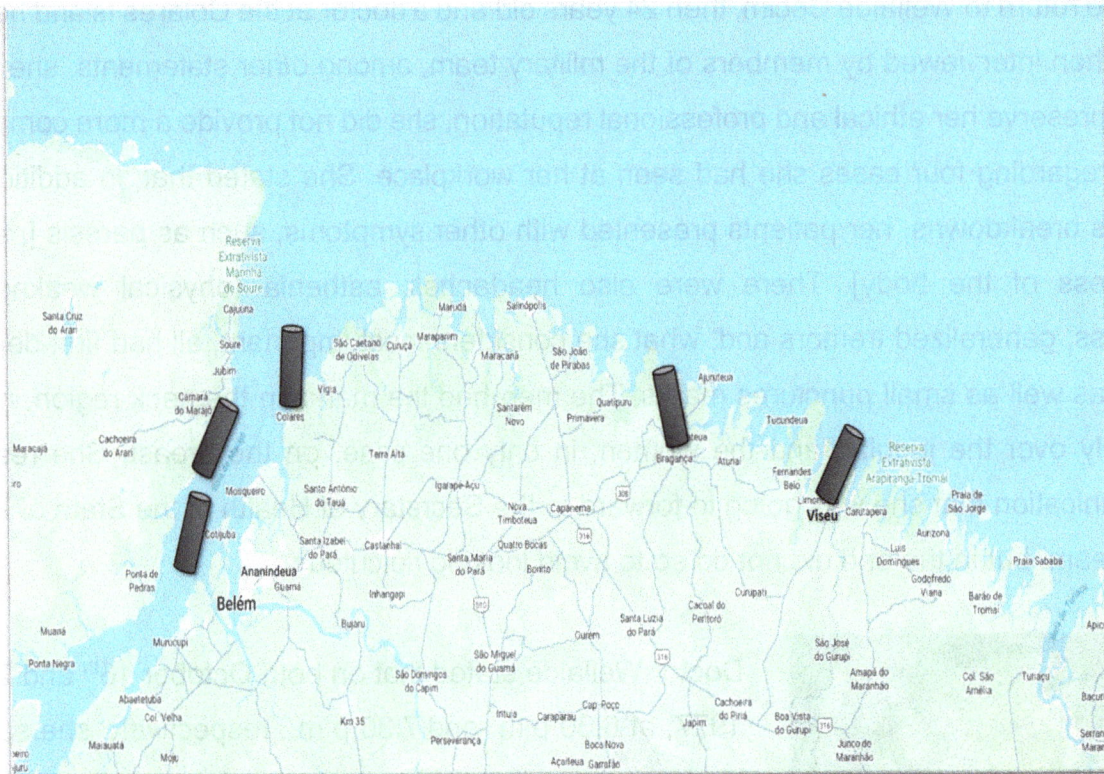

Figure 31. Cylindrical object observed by Doctor Wellaide Carvalho, on October 16th, 1977, at 6:33 pm, in Vila de Colares. Objects of the same shape were seen also in other cities: Bragança, Viseu, Contijuba, Colares e Baía de Marajó. (Google Maps).

Humanoid fired a light gun and hit Claudomira

Claudomira Rodrigues da Paixão, 35 years old, literate, gave a statement to the military on October 18th, 1977. After being interviewed by the head of the 2nd Section of the 1st Regional Air Command (COMAR), the already mentioned Colonel Camilo, she began to talk about the experience she had gone through.

She said that she was awake, lying in a hammock with another lady and her children. She felt a light that ran through her entire body, settling on her left breast, sucking it, then moving down to her right hand, when she felt a pain, as if it she had been pricked by a needle. It was then that she managed to scream for help. Previously, she had tried to scream several times, but she could not make a sound, and she became completely paralyzed. She said that a green light illuminated the entire room and that she felt a strange numbness, she was awakened by the voice of her companion, who called her attention to a child who had urinated in the hammock, saying: "*I am the one who is already ruined*".

Claudomira said that the "animal" had already sucked her blood, and she felt a great heat in her right breast and a sharp pain in the back of her right hand. She also had a headache and numbness on the left side of her chest. She was examined by Dr. Wellaide Cecim, who did not give her any medication, and referred her to the Forensic Medical Institute (IML), in Belém, where she was examined, and told that she should return for more tests. Above the incision made during the examination at the IML, a slightly burned area can be seen on her left breast, as well as an almost imperceptible mark on her right hand.

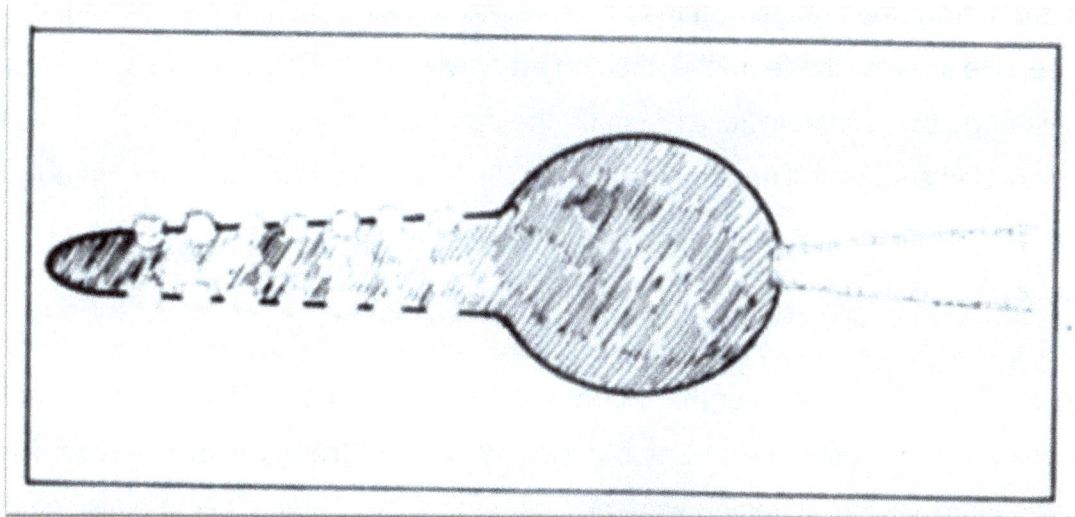

Figure 32. Ships, similar to tadpoles or rays, were frequently observed over the region of Vigia and Colares Island. Above, the 'aircraft' described by farmer Oswaldo P. deJesus, over the village of Coração de Jesus, in Vigia, in the state of Pará. (Credit: book "Vampiros Extraterrestresna Amazônia". GIESE, Daniel Rebisso).

In 1986, Claudomira revealed that she saw a light-skinned being with narrow eyes and large ears emerging from an umbrella-shaped object that emitted light. The being wore green clothes, he looked as if he was Chinese or Japanese, and he held a pistol in his hand from which emitted a beam of light that hit her left breast. She died in the 1990s. She always complained that after this episode she was never in good health again.

Raimundo made fun of people and got hit in the thigh

Raimundo Galvão Trindade had his sighting on October 15th, 1977 at 4:00 a.m. Interviewed by the 2[nd]Section team, he gave the following account: he was sleeping in Mr. Eduardo's house, in one of the upstairs bedrooms. At approximately 4:00 a.m., he woke up to find the room brightly lit by a greenish light. He felt something like a sting on the inside of his right thigh and said he looked outside the house and noticed nothing unusual. He was certain that it was not the light from the moon that had given him this impression. In the morning, when he woke up, he felt weak, his right thigh was numb. He also had a headache and dizziness, and there was a hot, purple spot on his thigh.

He did not pay much attention to what he felt; because he disagreed with the people who said they had been "sucked" by the device. He did not seek medical attention or publicize the incident, although a friend of his who works at the newspaper 'O Liberal' insisted on reporting it. He said that from then on he began to believe, having already had the opportunity to see the "device". He showed the team the affected area, which was described as follows: an oval area measuring approximately 20 x 12 cm on the darkest part of the thigh, which was brown or cinnamon, and a surrounding area measuring 2 x 5 cm with a border of a markedly lighter color. On the outside, the epithelial tissue was noticeably peeling and had a burnt appearance.

Antonino de Souza, a retired sergeant from the Military Police of Pará, and a resident of Colares, also had an experience on October 23rd, 1977, at 3:00 am. He said he did not believe what people were saying. When he went out to the backyard, he noticed a bluish light in his room. Checking outside, he noticed a very bright star. The light was in another position; he commented to his wife: "It can't be the Moon". He went back inside the house and felt that he was not himself, as if he had been hit by a "breath of air". When he got to bed, he lay down and pulled the quilt to cover himself. It was then that his wife, frightened, said: "Look at the light!" He heard her speak, but he could no longer see. It seemed as if lightning had struck him. He felt completely paralyzed. At that moment, he shouted for his wife, but she did not hear him, it was as if his voice had become stuck in his throat. After this, he was taken care of by his wife, who told him: "The Sucker has already got me, it almost sucked me".

On October 26th, 1977, at 1:30 am, after the experience described above, Antonino began to pay more attention to the people who constantly commented on the appearance of lights in the city. On the day and time mentioned, he observed a luminous object moving from southeast to northeast, it was hovering at the corner of the hospital at a height of 40m, well above a mango tree. When the locals nearby raised the alarm, at first the device moved slowly, then it accelerated, rapidly gained altitude, and disappeared. On November 6th, he was repairing a television set. At first he thought he had fixed the problem although there was still a lot of interference, so he went to check the condition of the antenna, but he had to climb onto the roof to do so.

When he was on the roof of the house, he turned and was facing northeast, when he saw a slightly illuminated object, it was snow-gray and almost white in color, it was circular in shape with a dome on top. It had two small diameter tubes on the front which were emitting a green and red light. He estimated its size was around 1.50m, at an estimated distance of 500m. The following night, he saw it again, but this time there were other people there. At 1:30 a.m. on October 25th, 1977, he noticed a light directed toward a house on the corner on the main street. An alarm sounded and the light moved and focused on a nearby store. He noticed that the device was undulating and displaying a bright yellow light with other smaller lights. He said it was shaped like a stingray, round, with a raised spot on top, which Antonino thought might be a gas tank. He sketched the shape of the device, and he mentioned the remarkable speed at which it travelled. Antonino is certain he was saved from harm by his son's scream, because the device shot away when he heard the noise.

On November 3rd, 1977, from 10:20 p.m. onwards, this was when the tide on the Guajará-Mirim River was best suited for the ferry crossing. The rest of the night passed uneventfully. On November 4, along with others, they informed the members of the 2nd Section team that the previous day, around 10:00 p.m., the light was seen several times above the city of Colares, more precisely over the cemetery.

At 11:00 p.m., a light was observed at medium altitude moving from Ponta do Bacuri towards Soure. On November 5th, at 1:00 a.m., another light was observed hovering close to the water over the bay near Joanes. Soon after, it moved over the sea at the same altitude, crossing to Ubintuba and Baía do Sol. It gave the impression that it had emerged from the water approximately 15 kilometers from the observation point. At this distance its size was estimated to be around 10cm.

A trawler that was close to the site used a searchlight and the light, after moving over the water, headed north. At 1:15 am, the light was again observed resting on the beach in front of Colares, at Ponta do Cajueiro. The light gradually increased and decreased in intensity, ranging from reddish-orange to pale yellow. The estimated distance was 5km and its convex shape was estimated to be around 3m in size, this was estimated by comparing the artifact with a tree.

Humanoid leaves ship and chases fisherman

On November 2nd, 1977, from 7:00 pm to 9:00 p.m. There was another occurrence when a luminous body appeared and moved below the level of the trees, once again it emitted a bright light. Due to the intense light the witness was unable to observe the structural details of the object in any detail. It seemed to be transparent with a circular shape and a reddish dome on top. At an estimated distance of 70m, the witness estimated its dimensions at 3m in diameter by 2m in height. At the bottom, a hatch opened through which a humanoid shape emerged, it was short in stature, but had a strong, stocky build, and it seemed to float. It was wearing a tight-fitting dark-colored uniform and a red light came from its hands and shone on the reporter, who quickly left the area. As he left, he observed that the area where he had been standing was being carefully examined.

Then, the humanoid returned to the object which moved towards him, as if searching for him. The witness, after running through the mangroves to escape, arrived at the boat where his companions were. As he told them what he had seen, the object reappeared, but it was a different color: it was reddish on top and blue-green on the bottom. Everyone fled when the humanoid reappeared and examined the boat, then it returned to the object, which quickly disappeared behind the trees. The person who reported the events was examined out by a psychiatrist, but he appeared to be normal. .

The procession in Colares, fireworks and an eye in the sky

The National Intelligence Service (SNI) was also involved in the investigations into the sightings that were occurring in Pará, as well as investigating the witnesses and journalists who were covering the events. The National Information Service (SNI) report stated that for some years the country's press had been reporting the appearance of strange objects in the Brazilian skies, these became known as unidentified flying objects (UFOs).

Approximately six months ago, the press in Belém reported that the population in Viseu was living in fear due to a strange object that came silently from the sky, sometimes it hovered in the air and, sometimes it landed, but it always emitted a very intense light. The population of this area was not very well educated, and, above all, they had an intensely mystical outlook. The stories that were told about their beliefs would be worthy of being included in folklore. For this reason, no one payed much attention to their claims. Later, it was reported that the unidentified flying object had also been seen in Bragança, Vigia, Colares and Mosqueiro. As the press kept reporting these sightings, the Air Force, through the 1st Regional Air Command, organized a team of intelligence officers to investigate what was happening.

The team went to the city of Colares, where the problem among the population was becoming more serious. In this city, at night, the population would hold a procession, light bonfires and set off fireworks in order to scare away what they called the "beast". According to the residents, the beast would appear sporadically from 7:00 p.m. onwards. It was seen flying above the city at high and low altitudes. In the latter situation, it would emit a very intense beam of light, sometimes aimed at people, those that the light touched became inert and trembling, and unable to speak. The 1st COMAR team, after hearing the reports of people who said they had been hit by the UFO, set up a device to photograph it when it appeared. Two teams were distributed to locations where sightings of the UFO were most frequent. After a few days of skywatches, the team managed to photograph what could have been the vehicle. The person who photographed it explained that the object was at an altitude of approximately 3,000m and seemed to be travelling at approximately 30,000 km/h.

The developed film showed some spots of light, as if it were only light, not allowing any conjecture about its shape, but tending towards a circle. When this object was photographed, the other team also stated that they had seen it. On another occasion, when it was photographed, the developed film showed some black spots, as if the film had burned. The equipment used in the photographs was a Minolta SRT-181 camera. One of the photos, precisely the one that was most impressive, was identified by the team leader as the photo he had taken of the Morning Star or planet Venus. The similarity to the others led him to doubt that the photos actually showed the UFO. Having concluded this first mission, the 1st COMAR team returned to Belém, keeping quiet about what they had seen. There is no consensus among the team members about what was seen, but it seems that attitude is closely related to the fear of being ridiculed in front of their colleagues.

Figure 33. It was common for unidentified flying objects to cross the dusty roads of Colares Island and other locations in Pará. (Author's files).

It is worth noting the data taken from the report of one of the members about the statements given by the doctor from the Colares Hospital Unit, Dr. Wellaide Cecim de Carvalho, and the parish priest of Colares, Alfredo de La Ó, the then mayor of Vigia, José Ildone Favacho Soeiro, a high school and college teacher, told to the newspaper "*A Província do Pará*", that at 6:45 p.m. on October 18th, 1977, he heard rumors on the street about the sighting of the UFO. Having reached the window, he saw a strange object crossing the sky at an astonishing speed, emitting a yellow light without making any noise.

Two minutes after that sighting, another object appeared in the opposite direction and then disappeared again. When the spectacle seemed to be over, two objects identical to the first one appeared from the Colares Island and Tapará Island, which crossed the city and then disappeared. "*The excessive coverage of the subject by the press, including the transcription of fantastic accounts of people who had been struck by the ray of light — which sucked people's blood — was already leading the population of Belém to disbelieve and make fun of it. The 1st COMAR was organizing a new mission to continue the investigations,*" the report said.

Orlando Fontenelle Trindade, 40 years old, with primary education, said that one of his fishing companions had called his attention on October 26th, 1977, at 4:00 a.m., when he had observed a very intense blue light with a reflective glow — he compared it to the sun's rays — shining on a windowpane; the object had suddenly appeared near his canoe. His companions suggested that they "run sideways," approaching the boat from the side to get a closer look, and that they should flee if necessary. Rather than agree to that plan, Orlando recommended safety measures. He ordered one of the headlights to be turned off and the other to be lowered into the hold, as this would help to attract less attention from the likely crew. After that, they immediately left the area. Orlando said that the light was exactly the same as the one previously observed but added that it was moving at a steady speed in an outward direction, to the north. Another statement by Orlando Fontenelle Trindade was given to *Operação Prato* about a sighting on November 1st, 1977, around 00:30.

He was fishing in the Canal do Navio when he noticed, at the height of the Mossoroca bridge, on the coast of Marajó Island, a light moving at medium speed in an outward direction, to the north. At first, he thought it was a boat with only its top light on. When the light reached Areia Vermelha beach, during the shallows, it stopped. At that moment, a canoe passed nearby, beginning to set the net, or "netting". The light then moved towards the canoe. Then, they observed, coming from the direction of Soure, to the northwest, a luminous "device" of a reddish-yellow color, moving at an average speed similar to that of an airplane, at a low altitude above the water, about 20 m. As it approached the first light, the second one stopped. At that moment, he calculated that they were in the shallowest part of Coroa Vermelha beach. The light remained visible towards the sea until approximately 5:30 a.m.

Elias Oliveira, "Fi", 42 years old, with primary education, also had an experience on November 1st, 1977, between 1:00 and 1:30 am. He was throwing his net in the river, when he saw a luminous body moving at a low altitude over the water at medium speed, without making any noise. A bluish light approached which looked like a headlight on the mast, which was in Coroa Vermelha, making several evolutions and landing nearby, when it finally disappeared. The blue lightheaded towards the sea and slowly moved away, being visible until 5:30 a.m., when Elias told his companion to climb the mast to observe; he reported that he did not see any boat nearby. He calculated that his position must have been 300 m away from the light.

Roberto dos Anjos Silva, 32 years old, literate, also had an incident on November 1st, 1977, at 5:30 a.m. He had been keeping watch since the previous night in front of his house when he and the people who accompanied him saw a luminous object moving at a low altitude, about 2,000 m, at low speed, without making any noise, coming from the west towards Belém. Its color was reddish-yellow. The object made a turn to the right, increasing its speed, and disappeared behind the trees. On November 6th, around 5:08 a.m., in the same place, he saw a luminous body moving at a low altitude, 3 km away, apparently 0.2 inches in size, with a strong yellowish color, emitting very strong bluish flashes — like electric welding — in a south-northern direction. It rose and disappeared high up.

"Captain, I didn't see the creature, I was sleeping when I was attacked"
Colares dweller, fisherman Newton de Oliveira Cardoso, nicknamed "*Tenente*" (Lieutenant), who was 21 at the time, still remembers perfectly and fearfully the attack he suffered from the light. The most interesting thing is that on the day of the incident, he, who lived in the city of Colares, fearful that the lollipop would appear, since his acquaintances and neighbors were terrified and never ceased to talk about it, decided to spend the night at his girlfriend and current wife's house, in Mocajatuba, 3 km away. "*I was lying in the hammock and my mother-in-law was in another hammock, next to me. I didn't see it, but the neighbors saw it and said that a luminous object had already flown over the house, and they started screaming*", he says. Shortly before that, Cardoso says he felt a strong "heat" in his body and then a sting in his neck, as if he had been bitten by an animal.

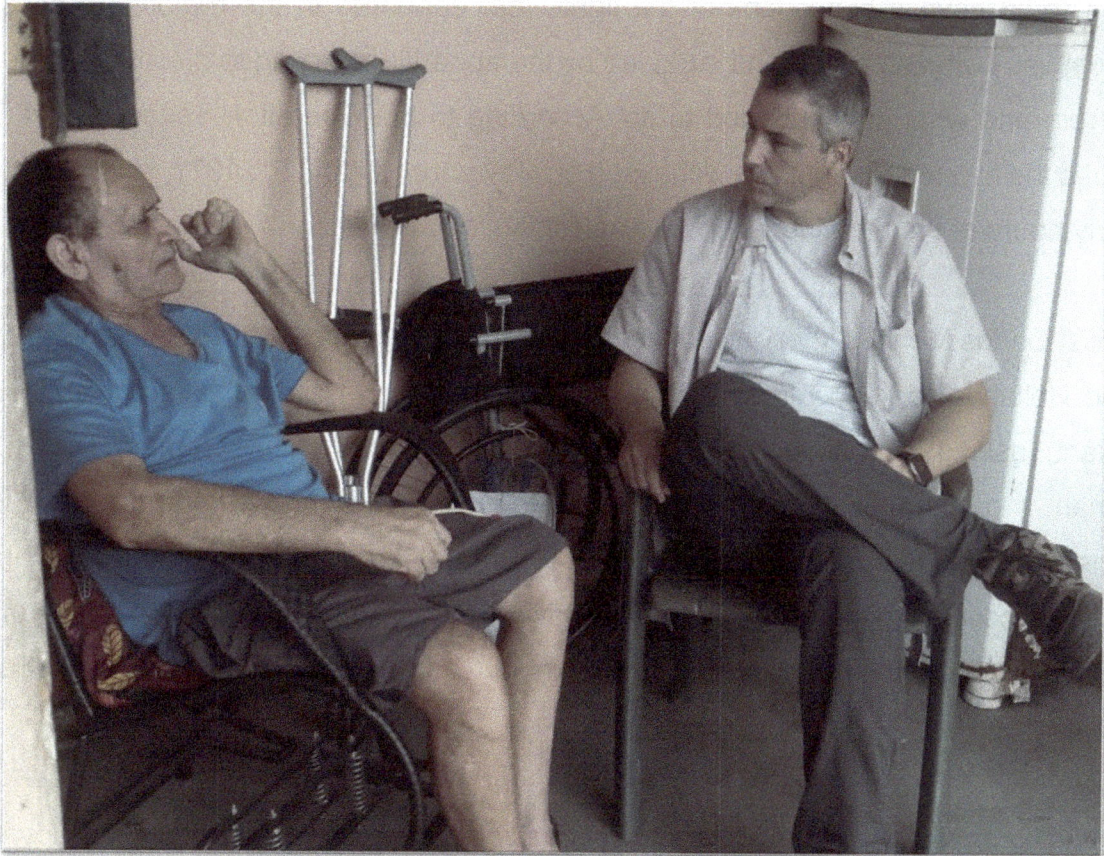

Figure 34. Newton Cardoso and Thiago Ticchetti, in Belém do Pará, in 2024. (Credit: Author's file).

After calling for help and hearing no one, he fainted. When he woke up, his neighbors were already inside the house, claiming that Cardoso had been attacked by the lollipop because they had seen the light. Traumatized for several days, he says that he even slept under a counter, thinking that he would be safer from the attack. The military from Operation Saucer went to question him about the episode and Captain Hollanda, with some drawings of aliens in his hands, asked him: "*Lieutenant, which of these creatures here came to attack you?*" Cardoso told me that he turned to Hollanda and said: "*Hey, Captain, I won't and can't tell you which of these creatures sucked my neck. Damn, I was sleeping, I didn't see anything, I just felt it and woke up. I'm not going to say I saw it because I'm not a liar.*"

Hollanda's boatman knows a lot and keeps it a secret

Fisherman Rosil de Oliveira, 36 years old at the time, with a primary education, was the man who, in the city of Colares, became friends with Captain Hollanda. For more than a month, he served Hollanda, taking him in his boat, along with other soldiers, to islands near Colares so they could have a better look at the lights that appeared in the region. He says that Hollanda was both fascinated and intrigued by the apparitions, trying to discover where they came from and what they intended to do. According to the Operation Saucer report, Rosil described the mysterious lights he witnessed on November 4th and 5th, 1977, six times, during the early hours of the morning when he was fishing.

Journalist Carlos Mendes interviewed the fisherman, who gave detailed descriptions of facts that led Hollanda to keep many of Rosil de Oliveira's stories confidential, in a classified document, asking him never to tell anyone anything that should remain secret. The fisherman, it seems, has kept his commitment to Hollanda to this day. What follows in the report is what was released as part of Rosil's testimony that could be authorized by the Air Force. His account, recorded by Sergeant Flávio Freitas Costa, gives an example of how these lights were present every day in the region between Marajó Bay, Mosqueiro and Colares.

"I was fishing on the rocks, in front of the Colares lighthouse, around 4:30 a.m. on November 4th, when I saw a light moving at low altitude over the water, from north to south, light yellow in color, at an average speed comparable to that of an airplane. It made no noise." Earlier that day, at 11:00 p.m., Rosil was fishing in the Canal do Navio when he noticed a luminous object moving at medium altitude, crossing the bay in a north-southern direction. The light increased in speed and disappeared in the glare of the lights near the city of Belém. An hour later, at midnight, he was still fishing, this time in front of *Praia do Machadinho* beach, when he saw another luminous object over Marajó Island, heading north at great speed. It rose and   quickly   disappeared.

On the 5th, around 1:00 a.m., still in the *Canal do Navio* area, Rosil saw an object flying over the bay — the Colares shore — at low altitude and flying low, in a south-to-north direction. It had a very intense blue color and stopped in front of the Colares lighthouse but then accelerated and disappeared up in the sky. Another luminous object, coming from the south at low altitude, landed in the bay in front of the lighthouse, according to the boatman, still during the early hours of the 5th, at 3:00 a.m.

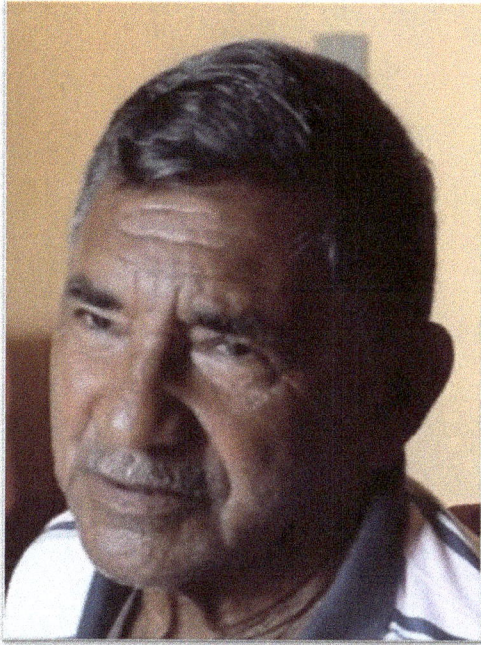

Figure 35. Rosil de Oliveira, a key person in Operation Prato for being the owner and pilot of the vessel that was requested by Captain Hollanda for all military incursions carried out in the waters of Marajó Bay and Amazon rivers in search of the "Vampire Lights". (Credit: Author's files).

"*A boat approached and used its headlight, trying to locate the object. After about 20 minutes, the object reappeared in the same position, but moved towards Soure, at a low altitude, and disappeared.*" Rosil reported another sighting, still that same morning, at 4:00 a.m. The fisherman had traveled from *Canal do Navio* to *Pedras*, another fishing area, when he observed a luminous body moving from south to north. Upon reaching *Ponta do Cajueiro*, it stopped. Its coloration was constantly changing, and its reflections were felt at an estimated distance of 1,500 m, causing a change in color — an optical effect — on the boatman's shirt, which changed from pumpkin color to a purple hue. A few minutes later, in the same position, the luminous body disappeared.

What few people seem to know is that, in addition to filming and photographing UFOs, the military also recorded the sounds of these aircraft. Rosil himself claims this. "*The military set up a camp on the islet in Colares and from there, whenever a light passed near them, heading towards land or coming from the mainland, they would point a type of radar at it and record the sounds emitted by the objects. I saw this several times.*"

Figure 36. The Humaitá Beach lighthouse, on a small island, became the military monitoring base for Operation Saucer. From there, the military recorded the sightings and movements of UFOs using video, photos, radar and sound equipment.

UFO or submarine? That was the question posed to Ivan de Oliveira, 36 years old, known in Colares by his nickname "Perema," when he observed a very intense blue light, visible from a great distance, which remained stationary over the water at an estimated height of 2 m. "*I didn't notice any movement in the water caused by the light moving. At first I thought it was a boat. I got closer to the spot and the light simply disappeared,*" Ivan said. At first, he thought the object was a submarine, but later he learned after talking to other fishermen that they had also seen a light with those characteristics. One of the fishermen mentioned was Orlando. Although it was visible from a great distance, Ivan described the light as almost dull and without any variation in intensity.

Chapter 9

Military Report points out increasingly incredible cases

The report produced by the Operation Saucer team contained comments, drawings and also the witnesses' accounts. The population coined words to describe the objects and lights that were being seen, such as "Light, Object, 'Device', 'Animal' and 'Chupa-Chupa'" in several cities in Pará. The witnesses came from different social and cultural backgrounds — they were fishermen, farmers, doctors, aviators, priests and physicists.

In one of the most important passages, the document states "that it feels it has not reached a fully satisfactory conclusion. There was still doubt and a lack of explanations for some details in the occurrences, among so many of them".

But one record draws attention and shows that even they, the Brazilian Air Force military, were not prepared to deal with such unknown technology. *"At 7:00 p.m. on November 1st, 1977, a luminous and shapeless object was observed, moving in a southeast-northeastern direction, reddish-yellow in color, rotating in a semicircle, initially at 4,000 m away and 2,000 m high. Its apparent size was estimated at 2 cm. Its flight was "dipping" in a gentle curve to the right. Upon reaching half the distance, it began to recover, ascending in a curve to the left, crossing vertically at 1,500 m high, when its apparent size was estimated at 8 cm. The object quickly lost its luminosity, morphing into a tiny red dot, at an altitude of over 6,000 m — in spite of any estimation errors that may have occurred. The whole sighting lasted only 45 seconds"*. Considering that the shortest time within the maneuvers described was spent between the exact vertical point in its trajectory and the point at which it disappeared, it is assumed that such a body was propelled by a uniformly accelerated movement, while going from a subsonic speed, around 800 km/h, to supersonic; within the known concepts the so-called sonic wave, or boom, would necessarily be heard, which did not happen during this occurrence.

On November 5th, 1977, at 6:26 p.m., a luminous object or body was observed with electrostatic and magnetic effects. No voices or static were heard within a 30-degree range for 20 minutes, according to the vibration of the theodolite's magnetic needle. On the 6th, between 5:20 a.m. and 5:25 a.m., three luminous bodies were sighted, all of which at low altitude and with very similar characteristics. The occurrences on the 5th and 6th differ only in the direction of movement and angles obtained by the theodolite. *"And other cases left us in doubt and without an explanation, based on our knowledge standards. As for the people who were affected, although there are some discrepancies regarding the color of the light source, the symptoms described are too coincidental for us to deny their evidence."*

The report goes on to call the press "irresponsible" and even use "derogatory terms" when referring to the lollipop, in addition to treating the nervous breakdowns of people who reported seeing the lights as natural. *"Nervous breakdowns, yes, in the presence of the unusual. There is no cause without effect, although we must recognize the negative influence of the press, to the point of irresponsibility in causing panic among entire populations — Colares, Ubintuba and other locations — and the disrespectful use of derogatory terms."* However, the symptoms of the first people affected were not made public, otherwise the stories might have spread among the town dwellers, often without any sensible connection.

Now, it was not the press in Pará that created the expression lollipop, but the very people who lived in the places where the sightings and attacks were most intense. At country parties, during the early hours of the morning, with heavy drinking and sensual dancing, the country folk usually say that it is time for the *"Chupa-Chupa"*, that is, the moment of the most intense caresses and kisses that precede the climax of sexual intercourse. The analogy with the lights that sucked people in, or pierced their necks*, although strong, did not fail to make sense to the countrymen and women. Since people in Brazil were living in a time of press censorship, there was a military man who defended that the Federal Police of Pará should prohibit the use of the expression *"Chupa-Chupa"* in Belém's newsflash contents.

The military reported what they saw but never forgot to attack the press. *"Psychosocial and economic aspects. In its entirety, the region where the appearance of optical phenomena or luminous bodies of unknown origin was observed is inhabited by people with the lowest cultural, socioeconomic and health background, coupled with superstitions and simple education, easily influenced by the media, not always used by scrupulous people and up to* the task of public information. To enhance that scenario, certain authorities allow abuses, such as fireworks — pistols and rockets —, unregulated consumption of alcoholic beverages and riots. And most importantly, they neglect their community duties, denying even the smallest support, which would be at least a word of guidance to the less enlightened." The report continues:

"The city of Colares lives in a state of mass hysteria. Its residents, shocked by the appearance of mysterious lights of unknown origin, do not sleep, do not fish — the population's main activity — and above all, they weaken themselves with drink, spending their meager resources on fireworks and cachaça. From nightfall to dawn, bonfires are lit, and there is a daily procession. Fireworks and shots are fired constantly, as if to scare away an enemy who nobody knows when or where will attack. Groups of 20 to 30 people — mostly men — roam the city in all directions. The population lives in terror; every now and then there are screams of terror and the news that the device has attacked a certain person. The people affected suffer what we can classify as a severe nervous breakdown — unless I am mistaken —, reporting almost unanimously: partial or total immobilization, loss of speech, chills, dizziness, intense heat, hoarseness, tachycardia, tremors, headache and progressive numbness of the affected parts — the vast majority complains about it.

If we think that the current situation will continue or worsen, we foresee problems of various kinds, including the possibility of self-destruction by the weakest of spirits because of fear of the unknown. As a suggestion, the following preventive measures could be taken: prohibition of the sale of fireworks and alcoholic beverages; instructing the population on how to maintain skywatches, that is, in a more objective and rational way, with rotation; divide and distribute groups of no more than 10 men to specific zones, following a rotation. The rest of the population would carry out their normal activities. In Ubintuba, Campo Cerrado, Vila Nova and other smaller towns, the situation is almost identical, with one advantage: several families are gathered in a single residence — we counted 36 people in one small house. This is solidarity in the face of need. By often offering words of comfort and solidarity, the team made these simple people see that they were not completely abandoned to their fate. This was beneficial to them."

Another important excerpt from the military report: *"The city of Colares — where we spent most of our time — has a new 'atmosphere'. Its residents, for the most part, have learned to live with the problem. Perhaps our lectures, contacts, and slide presentations have contributed to comfort them, not as the most important factor, but we believe they have been significant. The lights continue to appear and, surprisingly, they do so on a schedule. The locals no longer seem so frightened. However, the question remains: the monster created by the press — the sucker, in its blood-sucking action, possibly not real — has left a mark on those minds with fear and a distorted and adverse image of reality."*

The text continues: *"The existence and presence in the region of unidentified flying objects or lights is evident. We have not confirmed anything. We have seen, yes, luminous bodies moving at different altitudes and directions, performing complex maneuvers, indicating that they are intelligently controlled. Our certainty is based on our personal observations and on the reliable reports of people in whose actions and behavior, if properly analyzed, we can trust. Our film and photographic records do not reflect our certainty, since we are very lacking in technical resources, materials and personnel — only at the end of the period did we use a type of high-sensitivity film —, and we left much to be desired. On other occasions we missed the opportunity, photographing with inadequate material; we believe that with better resources we can reach a reasonably satisfactory level."* The report is signed by João Flávio de Freitas Costa, 1st sergeant, head of the Air Force A2 team.

Human-looking crew members and light beam

The second part of the report contains the testimony of Manoel Espírito Santo, who was 20 years old at the time and had had barely finished primary school. He said that at 11:30 p.m. on October 12th, 1977, he was in front of his house with some friends — Júlio, Paulo, Deca and Carlito — when he noticed a yellowish light moving from east to west, slowing down and almost stopping about 20 m from the group. He said that he then noticed that the light was being manned by two human-looking individuals, with the "man" on the left side and the "woman" on the right side of the craft. Both were wearing different-shaped glasses and intercom equipment.

The person on the left put his hands on his glasses as if observing the group of people more closely. At the same time, the other, through a tube on the side, directed a red light beam towards the group. When hit directly, Manoel felt a strong shock — like an electric shock — starting from his feet to his head. He then became paralyzed in his upper and lower limbs and semi-unconsciousness. The device gradually moved away, increasing its speed. Manoel began to move again but felt numb for a few minutes. The description of the object is that, from a distance, it resembled a reddish-yellow star and changed colors from light yellow to red.

When he got closer, Manoel noticed a bluish light on the upper front part. It was shaped as a barrel, with a smaller, reddish tube in front of it, and another thinner one on the side, at about 45 degrees, which was intended to emit a beam of bluish light. Its apparent size was 1.20 by 1.40 m. It appeared to be transparent in the bluish luminous part, with a division between its crewmembers. Its movement was a straight ascent with an undulating movement, like a leaf in the wind, until it reached a certain height. Its speed was variable, with sudden impulses and changes. The descent was straight and without undulations, with the larger diameter tube functioning as support. It moved after gaining height. It had the smaller diameter tube in front that emitted a bluish light, and it made the same movement during the ascent. It did not produce a loud noise. The interesting details are that the goggles were protruding and the intercoms were a pair of headphones, with a headband around their heads.

Another report by the military team is also quite interesting and describes the experience of Raimundo Francisco das Chagas, 36 years old, semi-literate, on September 22nd, 1977, at 9:30 pm. He was returning from the village of Santo Antônio do Tauá to his home, at Km 12 of Highway PA 16, about 6 km from the highway. He was walking and absentmindedly smoking, when he noticed a very strong yellowish light — the size of a car headlight — coming down from above in his direction. He ran, leaving the path and heading into the woods, being chased by the light. At a certain point, when he came across a clump of tucumã (palm tree with thorns), he jumped and remained there, motionless. The device, as if searching for him, emitted bluish flashes in different directions around him.

Raimundo compared the flashes of the object to a very powerful flashlight. He was unable to observe the shape of the device because he was too frightened. He did not hear any loud noises produced by the object as it moved. He heard a slight hissing sound that resembled an irrigation pump on a farm near his house. The device moved at a low altitude towards the village. Initially it was moving from east to west. Raimundo lost all the shopping articles he had purchased and was carrying with him when the object attacked. He says he felt a slight numbness in his body that lasted for a few days.

The following narrative reveals another form of attack from space, that of the "translucent light", which pierces the roofs of houses to suck the blood or energy of the victim.

On October 12th, at 11:30 p.m., the same time that Manoel do Espírito Santo saw two beings in a barrel-shaped device attack him, fisherman Manoel dos Santos, said that he was inside his house, along with his youngest children, sleeping with the lights off when he woke up to notice "*a strange light inside the house, of unknown origin, filtering through the roof.*" Santos tried to get up but could not. Feeling paralyzed, he tried to scream for help, but was unable to do so immediately. After a few minutes, he managed to get up and make himself heard by his neighbors, who came to his aid. He spent eight days with the left side of his body numb, in addition to being hoarse, after his experience with the light. Santos also said that he had seen lights passing by at low altitude in front of his house without making any noise, unlike airplanes. The lights almost always come from where the Sun rises. The objects move at a variable speed, sometimes stopping in space, and other times accelerating abruptly, disappearing quickly.

Ms. Amélia also saw it, but she thought it was beautiful

Heard by the military, Amélia Martins da Silva, 77 years old, literate, reported that at 10:00 pm on September 2nd, 1977, she was walking from Santo Antônio do Tauá towards her home, at Km 19, on the same highway PA-16, accompanied by Odete, one of her daughters. When she reached about two kilometers after passing the village, she noticed a reddish-yellow light coming from her right and ahead, with a very bright glow, similar to a car's headlight. Initially, the light moved diagonally across the highway, but suddenly changed direction by 45 degrees, coming exactly towards Ms. Amélia and her daughter, who quickly crossed to the left shoulder, seeking shelter under a rose apple tree.

After stopping for a few moments, the device moved at low speed, returning to its original direction, and then made a gentle turn to the left, following the road towards Vigia-Belém. Ms. Amélia did not feel any physical abnormality, nor was she frightened. She described the device as "very beautiful" and said that she would like to see what she had seen again. The altitude of the luminous object was 30 m. It was 1.30 m long, compared to a table in her house, and one meter high. It was rounded in shape, like an upturned plate with a pronounced protuberance or beak, from which a reddish light was emitted, while around it was bluish, from which a strong luminescence emanated, giving the impression of being transparent.

Ms. Amélia did not see any human forms or anything similar inside the device. It moved at low speed, but sometimes it accelerated to the speed of an airplane. The report on Operation Saucer concludes that Amélia is "*quite lucid, despite her age,*" and states that she "*was very firm in her account.*" Odete Martins da Silva, 35 years old, literate, Ms. Amélia's daughter who witnessed the UFO sighting alongside her mother, is described by Air Force personnel as "*a very shy woman.*" She confirmed her mother's account in its entirety and added nothing.

Alzira Farias de Campos, 55, told members of Operation Saucer that she had seen a "great light coming from above over a mango tree, about 20 meters away" when she went out into the backyard of her house, around 11:30 p.m. on October 13th, 1977. She was frightened and ran back inside the house. As she ran, she was hit in the left leg by a reddish light, and she fell over a bench in the kitchen. Alzira reports that after being hit, she felt a progressive numbness throughout her body, starting with a chill that ran from head to toe. She was helped by her daughter. She began to feel headaches, tremors and numbness for about eight days. Ms. Alzira told the press what she had omitted from the Operation Saucer team. She said that after seeing the light, its brightness grew in size until it completely enveloped her. Terrified, she still had the strength to run, trying to reach back into the house, but felt herself fainting when she fell to her knees on the door frame, with her back to the light. It was the noise of the fall that made her 25-year-old daughter wake up and run to help her. After she regained consciousness, she noticed the burn mark from the light on her left leg.

There seems to be a general consensus that one should not shoot at UFOs or their crews. There are dozens of cases where the reaction of visitors, even though in most cases it was not lethal, was to retaliate. However, what happens when a psychiatrist defends the "crazy guy" who shot at a UFO?

In his book "Lights of Fear", journalist Carlos Mendes presents dozens of cases that he investigated during the period of the Chupa-Chupa attacks. Carlos often arrived at the scene to interview people even before the military team from Operation Saucer came up. One of the most interesting and shocking cases is that of Manoel Matos de Souza, who fired at a UFO!

The story told by farmer Manoel Matos de Souza, "Manoel Coronha," 38 years old and illiterate, involves a mix of belief from those who knew him and mockery from the men from Operation Saucer who took his statement. Carlos Mendes met Manoel Coronha and interviewed him at his home in the municipality of Santo Antônio do Tauá.

The incident occurred on October 18th, 1977, at 2:00 a.m., in the place known as Monte Serrado. In the October 20th, 1977 edition of *A Província do Pará*, he said the following about what the reporter considered an unusual apparition:

"I was sleeping, it was about two to three in the morning on Tuesday, when through the gaps in the wall of the shack, above the door, I saw rays of bright light penetrating with great intensity. I got up and, opening the door, saw a strange object standing about 5 meters away from me and, inside, a couple laughing out loud as if they were joking. Even though I was blinded by the red rays that were reaching me, I quickly went into the house and grabbed my shotgun. I pointed it towards the 'ship', pulled the trigger, but to my surprise the gun didn't fire. At that point, the right side of my body started to go numb. That's when I screamed for my wife and children, causing panic inside the house. Meanwhile, the device disappeared without making a sound, disappearing into the sky."

Figure 37. Manoel "Coronha" tried to shoot a UFO, but the gun did not work. Soon after, the right side of his body went numb. There were two beings inside the object. (Credit: book "Luzes do Medo", MENDES, Carlos).

The farmer added that the bulk of the device was painted with red and black stripes, and its shape was described as a "*flying police van.*" As soon as it arrived at his house, the UFO emitted a yellow light, then a red one. Its front part, where the couple was, had a huge transparent windshield and right in the center, an opening, with a black circle around it, from which a very strong light came out. Regarding the occupants, he said that their features were the same as those of an ordinary person. The woman had blond hair and a strange smile, and the man had a hat similar to a cowboy's. The entire sighting lasted about 5 minutes.

Brazilian Air Force Sergeant Flávio Costa, on the other hand, noted the following statement from Manoel Coronha: "*When he woke up inside his house, he noticed a strong light. He grabbed his shotgun and left the house, then came across a bluish light hovering at a low altitude, about 20 m, over nearby trees. He aimed his gun to shoot; a beam of reddish light hit him and paralyzed him. Feeling as if he had been struck by an electric shock, which spread as heat starting from his feet, he also felt a tremor in his flesh (sic). The 'light' moved away in a slow, undulating movement up to a certain height, then suddenly increased its speed, disappearing towards the west"*.

He described the shape of the object as a flour oven, with a blue "spout" on top and black stripes on the bottom. Its estimated size is 1.40 m in diameter. It did not make a loud noise, but rather a slight hissing sound, "zim, zim, zim, tim". *It rotated*, as if it were a wheel. Manoel Coronha felt such a great tremor that he was bedridden, unable to work. He later returned to work, but still referred to tremors in his flesh, with a shaken nervous system. He was unable to draw the shape he had observed, but his tremor was visible. He reported that his wife also felt numb. The military personnel of Operation Saucer state in the report that Manoel *"is shy and has a mongoloid type (cretinism)"*. In an excerpt from another report, information appears that the farmer had, when he was still young, been admitted to the Juliano Moreira Hospital, a former psychiatric center located in Belém that was deactivated in the late 1990s.

Carlos Mendes, in his book, wrote that he disagreed with the military's assessment when they labeled Manoel Coronha as a *"mongoloid and a cretin"*. Carlos did not notice any symptoms of mental illness in him. On the contrary, his narrative was reliable and rich in details. The fact that he was illiterate does not disqualify him morally. He told Carlos exactly the same things he told the reporter from the newspaper *A Província do Pará* and the Brazilian Air Force military. What's more, Carlos Mendes looked for information at the time on Juliano Moreira and found nothing. He did not even have a medical record or file.

Psychiatrist Durvalino Braga, a doctor at that hospital, when informed about the light that attacked Manoel Coronha and his attempt to shoot the luminous object, even joked: *"There is nothing crazy about this guy, he just wanted to defend himself from that sinister thing."* Braga did not want to go into further detail about whether or not he believed in aliens and flying saucers, stating that the subject was none of his business. *"If he is crazy, so are a lot of people in Belém and in the small towns who claim they are seeing the same lights and even losing blood to them,"* Braga said.

Chapter 10

UFO lands on farm and another object destroys crop

In an interview given on March 5th, 1979, at 1689 Curuzu Lane in Belém, to Brazilian Air Force agent Ernesto, neurosurgeon Pedro Rosado, a professor at the Federal University of Pará (UFPA), reported that he owns a farm in the Quatipuru region, on a side road to Tracuateua. While visiting his property, he observed on several occasions unidentified flying objects — not mistaken for airplanes — moving in different situations in terms of altitude, speed and directions. Rosado states that he has witnessed "*impressive maneuvers*", and even mentioned that he saw "*a single object split into three, taking different directions at different speeds*".

One night in October 1978, while they were talking in front of the farmhouse, Rosado and other people were intensely bathed in a quick bluish flash, which allowed him to observe details of the terrain that could only be seen in daylight. The doctor said that the dazzling vision had been very brief. He stated that he had heard of a case involving a teenager living in the area, the son of one of the cowboys, Mr. Antônio Galdino, who after being struck by a beam of light emitted by one of these objects, was paralyzed on the left side of his body for almost 10 days, with a perfectly limited division of the side of his body where he had been struck. There were no symptoms of burns, but rather a pronounced pinkish rash that greatly affected him. He also mentioned Mr. Simeão de Aviz, a town dweller the area, who observed one of these objects very closely, which he described as having the shape of a rectangular box, with a very bright headlight on top, pointing vertically upward.

According to military officer Ernesto, Dr. Rosado promised to send all the information regarding the case of Mr. Antônio Galdino's son, and to that end he would contact the A2 chief by telephone. The doctor avoided discussing the matter because he knew that people who see or have seen such objects are mostly treated as visionaries or unbalanced. The report concludes by saying that Dr. Rosado is considered a man of few words, "*almost a difficult person to deal with*," but "*serious and of impeccable professional reputation, very frank and firm in his statements*."

Wherever it appeared, the unknown light left a story to tell. One story, in fact, is fantastic. At Km 29 of Santa Isabel-Vigia highway, Ms. Amélia Sarmento was awakened by a bright light coming from the roof. Startled, she left the house and saw, with absolute clarity, a strange device emitting a strong light that could easily pass through the roof of the house.

Another testimony is that of Antônio Santino, a market inspector in Benevides. He, who was even the target of an investigation by official agencies, on a night with a clear moon saw an unidentified flying object in the backyard of his land removing the vegetable plants from his vegetable garden, one by one — the craft hovered about six meters above the ground, opened a door and emitted a light that destroyed his crops.

In Colares, one of the UFOs' favorite places, its population is constantly visited by the strange objects, and always in a panic. In an uproar, the villages of Juçarateua, Jenipapo, Tupinambás and Fazenda organized litanies. In these towns, as in the others visited by UFOs, the unbelievable has become commonplace, a fantastic reality in the clear early mornings.

Civilian pilot reports what he saw in Belém

There are newspaper reports from several people who saw a strange light crossing the sky at high speed in Belém. One of them was a doctor and a civilian pilot, Direito Álvares. He was in front of his home, located on *Governador José Malcher Avenue*, when he saw a strange object in the sky. "*I can't say for sure that it was a flying saucer. What I saw was a red light, tending towards lilac, that remained stationary in space for about five minutes. It was so bright that I could not look straight at it. To tell the truth, I was afraid to look directly at it, because the light was so bright,*" the pilot reported.

The unidentified flying object was seen by the doctor around 7:15 p.m. "*I was with my wife. When I least expected it, the light shot at high speed towards the Manoel Pinto da Silva Building, leaving four smoke trails, just like those expelled by our jets*", he states.

Direito Álvares believed in the possibility of the existence of flying saucers. "*Is Earth, which is small, that only inhabited planet? I have my doubts. However, I can't say for sure, because*

I'm like Saint Thomas: I'll only believe it if I see it. Many friends of mine saw an unknown light across the sky in Mosqueiro. A colleague also saw the light on the same day as I saw it in Belém".

Although he did not say it directly, the doctor pilot made it clear in his words that people who don't believe in the existence of UFOs are being hasty. *"I can guarantee one thing: the light was too strange to have been emitted by a man-made device. As far as I'm concerned, I am unaware of any device capable of producing such a high speed as the strange light did."* As a pilot, Álvares had more than 2,700 flight hours.

In the 1977 testimonies of those who had seen the strange objects in the sky, there are numerous coincidences. The most common one is that the flying object crosses the sky at dizzying speed, stops suddenly and, unexpectedly, shoots towards the horizon.

During the early morning hours when he had the opportunity to watch the aerial shows of the unusual devices in Sun Bay (*Baía do Sol*), a reporter from *O Estado do Pará* newspaper not only confirmed the veracity of this detail, but also observed that the objects resembled bells sailing in space.

During an interview to the newspaper *O Estado do Pará*, Colonel Rocha Santos, of the Brazilian Army, claimed he believed in the existence of UFOs and mentioned another coincidence.

"The devices are built for various climates and pressures, and there are also gigantic types, similar to large, inverted bells and with an enormous circular platform. As already reported in previous articles, bodies of relative size emerge from the mother ship and begin to cut through the sky at incredible speed".

Locals of Benevides and Genipaúba also witnessed the same thing. The coincidence is that no one saw the smaller objects being collected by the larger spacecraft. Another curious detail: the mother ship, after releasing the small devices, glows until it disappears.

Young Edilson Ribeiro de Souza, 11 years old back at that time, lived in Marituba with his parents, José Gomes and Maria Ribeiro. The fisherman Benedito da Conceição Silva was born and lived with his family in Genipaúba, in Benevides. The distance between the boy and the fisherman, however, does not prevent them from reporting the same event. The details match in several points — the incredible adventure they lived when they saw a strange greenish light that suddenly became motionless in the sky.

Maria Ribeiro da Silva, the boy's mother, has not forgotten it to this day. It all happened in early 1977. The time, she remembers perfectly: 7:30 p.m. "*I swear it was true,*" she emphasized. According to her, "*we decided to go for a walk. Edilson was a little behind us. When we noticed, a light from the sky illuminated him and he started screaming. It seemed like he was being pulled. José and I tried to get closer and felt a kind of heat in our bodies*".

"*The first thing I felt was my body getting hot. The light coming from the sky scared me. I tried to run and felt myself being pulled by the ray of light. When I screamed, the light let go of me*", said Edison.

Benedito, the fisherman from Genipaúba, has a faulty left arm. And he doesn't forget one of the events that contributed to this.

"*I was in my canoe with Álvaro, my brother-in-law, when I felt my body heat up suddenly. I think it was 5:00 a.m. When I looked up, I saw a light in the sky.*" Insisting that his brother-in-law witnessed everything, Benedito says that he tried to run but couldn't. "*I felt like I was being pulled by the light. Álvaro then started screaming and at one point pulled my shirt*." The fisherman admits that after everything he had been through, he was completely disoriented. "*I ran like crazy. I crossed the forest and before I knew it I was at the house of an acquaintance. It was already 11:00 am. With my arm bleeding profusely, terrified as I was, I told him what I had seen. Because of the strange light, I spent three days in the hospital in Benevides.*" Another fisherman from Genipaúba, Domingos Rocha, saw the strange light. While hunting, he was surprised by a huge flash of light in the middle of the forest. He tried to run away but could not. A day later he was found very weak. According to Benedito, the hunter was unconscious for three days.

One of the most curious and at the same time inexplicable aspects is that the rays from the UFOs passed through the roofs of the houses as if they did not exist.

In one attack that occurred in the Reduto neighborhood, in the center of the capital of Pará, the taxi driver Valnor Ferreira Ramos, a Catholic and a married man, reported that he saw and felt the light that terrifies the population.

"Last Monday it was very hot. So, I set up my hammock and lay down sideways, trying to sleep, though I was not at all sleepy. Suddenly, I began to feel a kind of heat, a strong heat growing in my back, towards my shoulder. I turned around and saw a focus of light coming down from the ceiling in my direction, exactly towards the 'heat', a color between red and yellow. With great effort, I managed to turn around — because I lost my strength for a moment — and then get up. However, the light was short-lived. A fraction of a second. It disappeared and I was immobilized for a few moments. I returned to normal, but I was not afraid. I did not say anything to anyone that night. I got up to see if everything was normal with the children and the whole family. It was only in the morning, when I told my wife about it, that she noticed marks on my back. A kind of scratch or bite, I do not know. She treated it with andiroba oil and it disappeared."

Valnor showed the reporters the place where the light had struck. There was no gap in the roof. The lining prevented any ray of light from outside the house from penetrating. He added that in the morning, when he woke up, he had a headache and difficulty moving his arm.

Chapter 11

Beings attack pregnant woman and dog stops barking

In Tapiapanema, a town on the banks of the Pratiquara River, on Mosqueiro Island, Benedito Campos Trindade, 24, and Silvia Mara, 17, husband and wife, were resting from their daily chores in a hammock. The year was 1977. They were alone. The other people who lived in the house had traveled to the village of Mosqueiro, 16 kilometers away, with access by sea. Shortly after 6:00 p.m., through a crack in the window, covered with a piece of plastic, the two noticed that an oval, silvery object was emitting a greenish, spotlight-shaped light in the direction of the room where they were lying down. Curious, the couple turned completely towards the place, and that was when the spotlight passed through the crack and hit Sílvia, causing her to go into a kind of trance, and leaving her numb. Concerned about his pregnant wife, Benedito tried to protect her quickly, carrying and dragging her from the place before she fainted and fell to the ground.

But the terror would not end there. According to Benedito, two characters came out of the house, with a golden object — like a flashlight —, shining it through the many cracks in the house, once again hitting Sílvia's body, this time on her left arm, specifically on her wrist. Her veins seemed to come out of her body, so swollen were they when they were hit by the light. Distressed and already screaming loudly for help, Benedito took his wife to the living room, hiding her behind a wall. At that moment, his neighbor José do Nascimento Sobral, who had heard the screams, ran with a shotgun, shooting at the place where the characters were and managing to scare them away. He did not see when the escape occurred, because his concern was to know if everything was okay with the couple. Benedito and Sílvia were taken to Sobral's house, about 50 meters away from their own. When the two men were trying to calm the young woman down, afraid that she would miscarry, the strange object appeared again, this time flying very low. Benedito ran to the front door, courageously trying to check what the device looked like. The beam, however, hit him, leaving his body momentarily paralyzed. Sílvia was not harmed. As he said later, Benedito believes that the attackers returned to take revenge for his interference. At this point, Sílvia's family arrived. She was taken care of by her mother, Osmarina Souza Nascimento, while her brother Raimundo took care of Benedito. Soon after, the whole family headed for Vila do Mosqueiro in a motorboat. On the way, the mysterious object appeared in the sky once again, blatantly following the boat until it docked at the Escadinha da Vila port. This time, however, the greenish focus did not threaten anyone, disappearing as quickly and miraculously as it had appeared. Only on one occasion, Benedito says, did they become more fearful: when the focus hit the water, causing the noise of a large stone being thrown to the bottom.

The couple Benedito and Sílvia Trindade were hospitalized at the Vila do Mosqueiro Health Unit. Benedito was very depressed, cried a lot every time he talked about the episode.

In the episode involving Benedito and Sílvia, the testimony of young Maria Raimunda de Souza, 18 years old, their neighbor, stood out, as well as what happened to the dog Vitória, also hit by the beam of light. Maria Raimunda was a few meters from her house, around 8:00 p.m., when she heard a noise *"like wood being processed in a sawmill"*, she said. She ran to the place where the object that made that sound was and saw the dog Vitória, who belongs to her, running after a figure, being hit by a beam of light on the head and fainting. The dog never barked again. Its skull has become a little protruding, although it is gradually returning to normal. Maria Raimunda was extremely frightened by what happened and didn't like to talk about it. She spends hours and hours leaning out of the window, looking at the sky, thinking that this will happen again. The strong light left her in a state of shock.

Benedito's and Sílvia's relatives, alarmed by what had happened to them, acted against the incident by first contacting the Mosqueiro District Police Station. The local police chief, Ordrado Pantoja, told them that there was very little he could do, because he did not know how to fight such a mysterious light and could not even locate it. However, he registered the incident in a "private report", which was sent to the Police Station in Belém, with a request for urgent guidance on how to act. After contacting the police, the couple's relatives went to the village local agent and received a similar response.

They were very alarmed and intended to leave the area where they lived, since the object, late Sunday afternoon, passed quickly behind village dweller Manoel Silva's house; both he and his daughter, Maria das Graças, saw it. The locals, together, began to pray at Benedito and Sílvia's house, asking that the object and its light leave them in peace. Those prayers were redoubled when they learned, unofficially, that Benedito might have the left side of his body paralyzed because of the focus that hit him.

Figure 38. An unidentified craft flies over the community in Tapiapanema, inland from Mosqueiro Island, hitting Benedito and Sílvia Trindade on October 29th, 1977. (Credit: Author's file).

Other people, in localities far from the village of Tapiapanema, were also hit by the mysterious fire. On the São Francisco beach, a certain doctor Medrado was shot and sought medical attention. In Baía do Sol, also in Mosqueiro, another case occurred with an elderly man. In Porto da Escadinha, on the Tamanduá creek, in the early hours of the day, also in Vila, sailor Antônio Maria da Silva, when he was preparing to ride on the boat "Amazolina", owned by merchant Florentino Sales Corrêa, known as "Pelé", had his attention drawn to a greenish fire that came from the sky in the shape of a snail, towards him. He immediately threw himself into the water, reaching the riverbank, fleeing at full speed. When it missed its target, however, the fire hit the water and it seemed like a hot iron had penetrated it, causing a characteristic noise. Antônio said he would go to the police to report the incident.

On November 2nd, 1977, with the lights also reaching the area of Belém, the incident in Tapiapanema again was the headline of *O Estado do Pará*. The information said that the atmosphere was tense among the city dwellers, especially between 5:00 p.m. and 8:00 p.m. They were waiting for the object with the greenish light to appear again, as it had done the previous weekend. In the late afternoon, they gathered at the home of farmer Benedito, one of the victims of the strange object, to pray. To date, no official action had been taken regarding the matter in the area. The population remained terrified, even threatening to leave their homes if the danger increased. The locals believed that the *"Chupa-Chupa"* was from another world, or even a product of the supernatural. When asked about the possibility that they were men from another planet, they showed complete ignorance of the matter and clarified that the expression *another world*, for them, means the afterlife. It was learned that the couple Benedito and Sílvia — she, pregnant — would be discharged from the Mosqueiro Mixed Unit hospital, where they were admitted shortly after being hit by the greenish fire. Little was known about their clinical condition, but it was believed that they had already recovered, ruling out the possibility that Sílvia might lose the child.

The newspaper *O Estado do Pará* concluded its series of reports showing the presence of unidentified flying objects in the area of Baía do Sol, Colares and Benevides, which cost the newspaper 41 days of waiting, taking photographs with sensitive film, a lot of work and amazement. The reports sparked controversy and debate about a subject discussed and recognized by international experts — and witnessed even by the president of the world's largest power. "*No one, not even the official bodies, denied the reports we published. Our films were analyzed by experts and nothing faked was found*," wrote journalist Carlos Mendes in one of the articles.

Figure 39. Elderly people, children, pregnant women: in short, no one was spared from the attacks of the terrible "vampire light", one of the names that the local population called the lollipop. (credit: book "Luzes do Medo", MENDES, Carlos).

The reporters in charge of this coverage were astonished by what they saw. The people who accompanied them to witness the events still discuss the challenges UFOs pose to physics and the speed at which they fly, which could disintegrate our most resistant metal. Flying saucers, lollipops, dead satellites traveling in space, reflected spectral forms? Many names have been put forward, but no one has managed to come up with a single explanation, at least reasonable, for the phenomenon.

Chapter 12

The "Chupa-Chupa" Phenomenon and the Sun Bay phase

As the Phenomenon evolved, things became calmer in Maranhão and the focus of occurrences turned to northern Pará, in the so-called Baía do Sol Phase.

As the Phenomenon progressed, cases in the state of Maranhão began to become rarer and less aggressive, and the focus of occurrences turned to northern Pará. The so-called Gurupi Phase ended towards the end of July 1977 with a decrease in cases in the Baixada Maranhense region, where incidents were occurring almost on a daily basis. On the other side of the river, in the state of Pará, the number of cases which had already been occurring for some time, increased, bringing back the situation of fear and astonishment previously seen in Maranhão. The difference now is that the cases have become even more impressive, forcing the Brazilian Air Force to intervene.

The cases were concentrated in an area 300km wide, involving 30 small municipalities that were directly affected by the Phenomenon. The cities with the greatest prominence during this phase were: Colares, Vigia de Nazaré, Santo Antônio do Tauá, Vizeu, São José do Pintá, Augusto Correa, Bragança, Santo Antonio do Ubintuba, Capanema and the state capital, Belém, with some sporadic cases. The total number of "Chupa-Chupa" phenomenon victims in the neighborhood is estimated to be over two thousand people, but as several communities lived in isolation and were never visited by ufologists or military personnel investigating the ""Chupa-Chupa", this number could be much higher, including deaths.

The village of Santo Antonio de Ubintuba, located inner Vigia, was made up of a few dozen families who lived off fishing and farming. Their only contact with other villages was via rivers or the narrow stretch of land that reaches km-32 of the PA-140 state highway.

The first news of the second phase, after months of complete silence, came from Ubintuba. On October 8th, 1977, *O Liberal*, a local paper, showed the headline "*A Sucking Bug Attacks Women and Men in the Village of Vigia*", describing in detail the surprising experiences of

Ubintuba. Its residents openly declared the existence of flying objects with light beams, capable of passing through obstacles, such as the roofs of houses, and targeting people who remained momentarily paralyzed due to the "Light".

Later, on October 15th, the same source published new reports from Ubintuba, and for the first time the term "*Chupa-Chupa*" was used. This name conveyed the idea that the objects somehow sucked the blood and energy of their victims. For this reason, the residents of Santo Antonio de Ubintuba sought to sleep at relatives' houses, while the adults guarded the streets armed with weapons and fireworks. They also commented that the ships, during their nighttime attacks, preferred to attack lone victims. In the small settlement known as Coração de Jesus, in inner Vigia, there were also UFO sightings, mainly towards the end of October 1977. One of the witnesses, Oswaldo Pinto de Jesus, reported what he saw:

"*During the 'Chupa' era, we heard a lot of talk about the device that was circling the village of Santo Antônio de Ubintuba, until it appeared in Vila Coração de Jesus. It was a weekend party, when in the early hours of the morning, my mother, Maria Assunção, saw the device. She called our attention. It was flying slowly and without making any noise. It was not very high and from below it looked like a helicopter. On the sides of the tail there were many colored lights and, on the tip, a very bright spotlight. The object seemed to have sensed our presence and suddenly turned off all the lights, disappearing into the darkness of the early morning*".

The description of strange ships like the one from Vila Coração de Jesus was persistent from Baixada Maranhense to Baía de Marajó, in Pará. According to witnesses, in addition to the luminous spheres, ships that looked very much like helicopters, kites or even stingrays were observed.

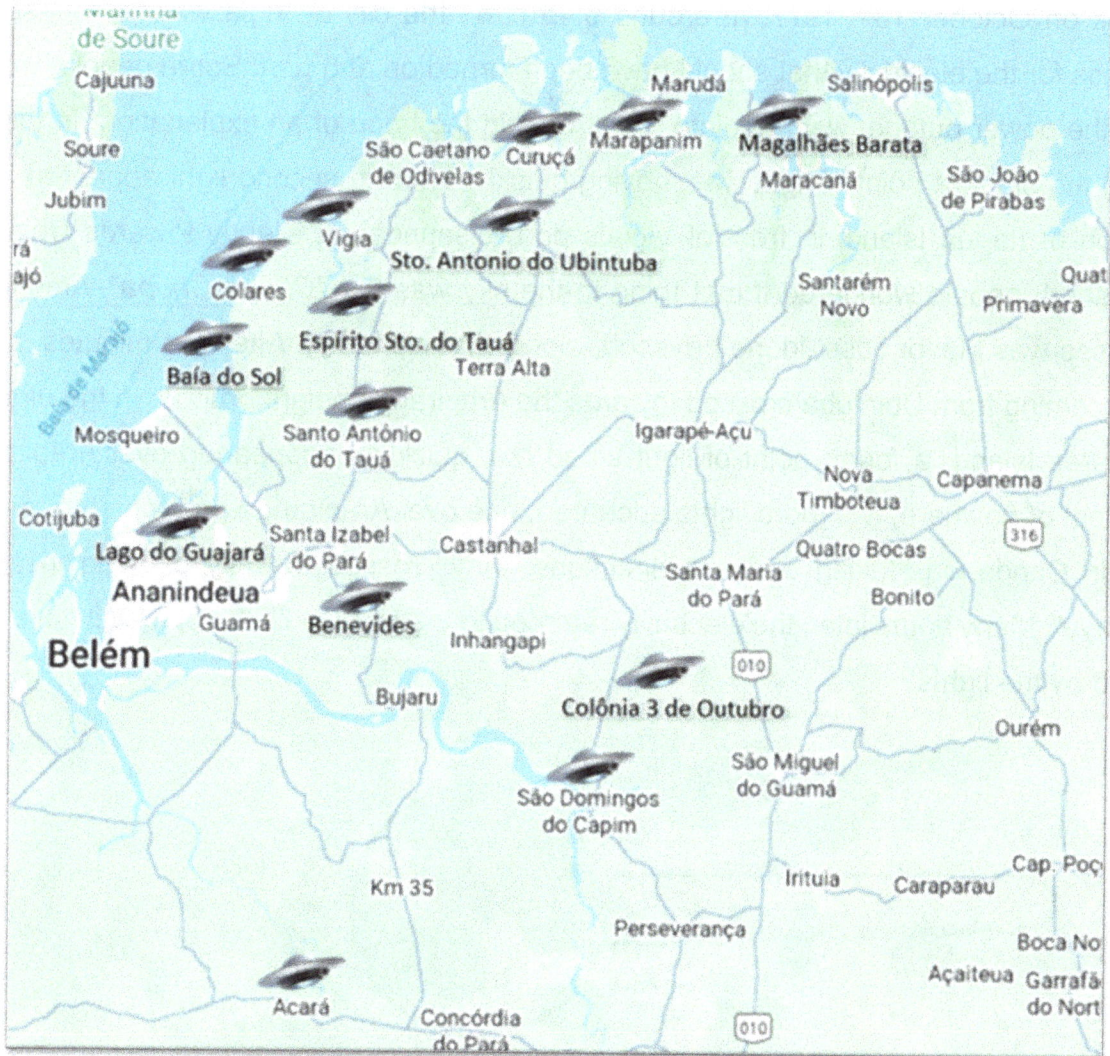

Figure 40. Areas of intense UFO activity from October to November 1977.

Vigia de Nazaré

One of the most affected cities was Vigia de Nazaré, located in the Salgado region, northeast of Belém. In the early cases of the ""Chupa-Chupa" phenomenon wave, several people were attacked, and their reports were identical to those verified in the Gurupi region. The witnesses saw a strange light in the distance that seconds later was very close. Then they felt paralyzed by a beam of light emitted by the so-called "devices". At that moment, they felt weak and in severe pain, while a thin beam of light was emitted by the object towards the victims. This beam, which came from inside the larger beam, produced a wound through which blood was supposedly extracted.

At dusk on October 18th, 1977, at around 6:45 p.m., the city of Vigia waited in unsettling darkness for the electricity that should have been turned on at 6 p.m. Some people, worried about the power outage, went out into the streets in the hope of an explanation. In the sky, among the stars, a point of light was moving quickly. Then, a second light appeared in the direction of Itapuá Island, in front of Vigia's port, disappearing silently towards Ubintuba. Astonished, people wondered if that thing in the sky was the "*Chupa-Chupa*". Among the witnesses was Mayor José Ildone Favacho, along with his family. After two minutes, a third object coming from Ubintuba crossed towards the Arapiranga neighborhood. In the direction of Colares Island, a fourth point of light stood out, quickly disappearing over Arapiranga. Then, out of nowhere, two more lights appeared; one over Arapiranga and the other moving towards Candeuba. Taking opposite directions, they crossed the city of Vigia in a final "goodbye". Many hours later, the electricity returned to the city. Could the blackout have been caused by the lights?

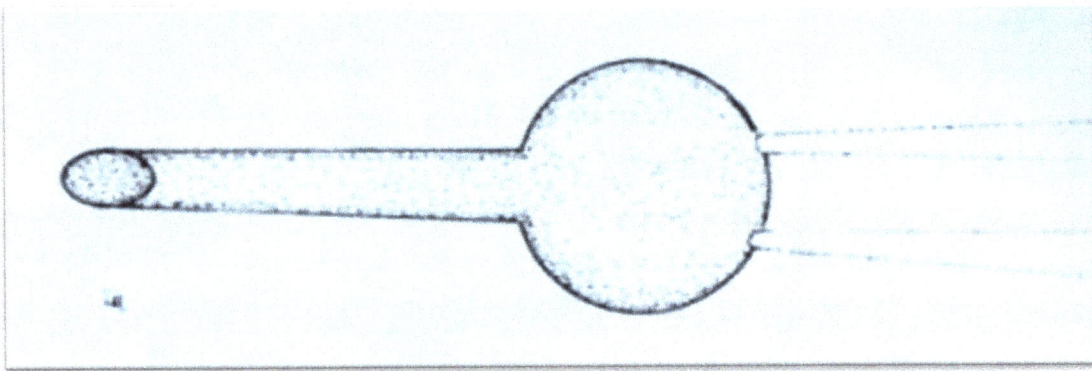

Figure 41. The UFO described by fisherman João Francisco da Silva, near the coast of Vila de Colares on November 11th, 1977. (Credit: Author's file).

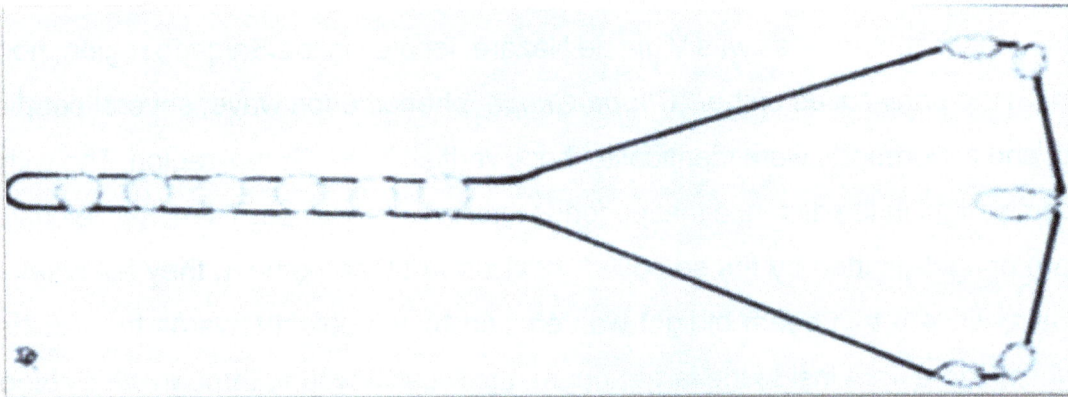

Figure 42. The UFO seen by farmer Vicente Gomes, in São Bento, on July 14th, 1977. (Credit: Author's file).

With the higher number of incidents in the neighborhood, the population began to become very afraid, avoiding going out at night. As the cases continued to pile up, signs of panic and collective hysteria arose throughout the region. Some religious people believed that the end of the world was imminent, and processions were held to obtain some spiritual comfort. Since none of this worked, people began to gather in large groups in small, protected places, where they would spend the night praying. The situation became so critical that all the residents of the municipality would gather in just three houses in the community to pray. It was not long before some people abandoned everything they had and moved out of the area. This situation led the mayor of Vigia at the time, José IldoneFavacho Soeiro, to send a letter to the Air Force reporting the facts and asking for help.

"The people were terrified because this beam of light at night had already attacked several people. The entire community would huddle together in just three houses. They would pray, sometimes they would sing religious songs. People were in a panic... The Colares health unit became almost a playground of miracles" – José Ildone Favacho Soeiro, Mayor of Vigia de Nazaré, in 1977.

Chapter 13

Colares Island, the new hotspot for UFOs

The wide coastal strip from São Luís, the capital of Maranhão, to the city of Belém, the capital of Pará, comprises numerous islands, some of which are habitable, one of which stands out for its great UFO importance: Colares!

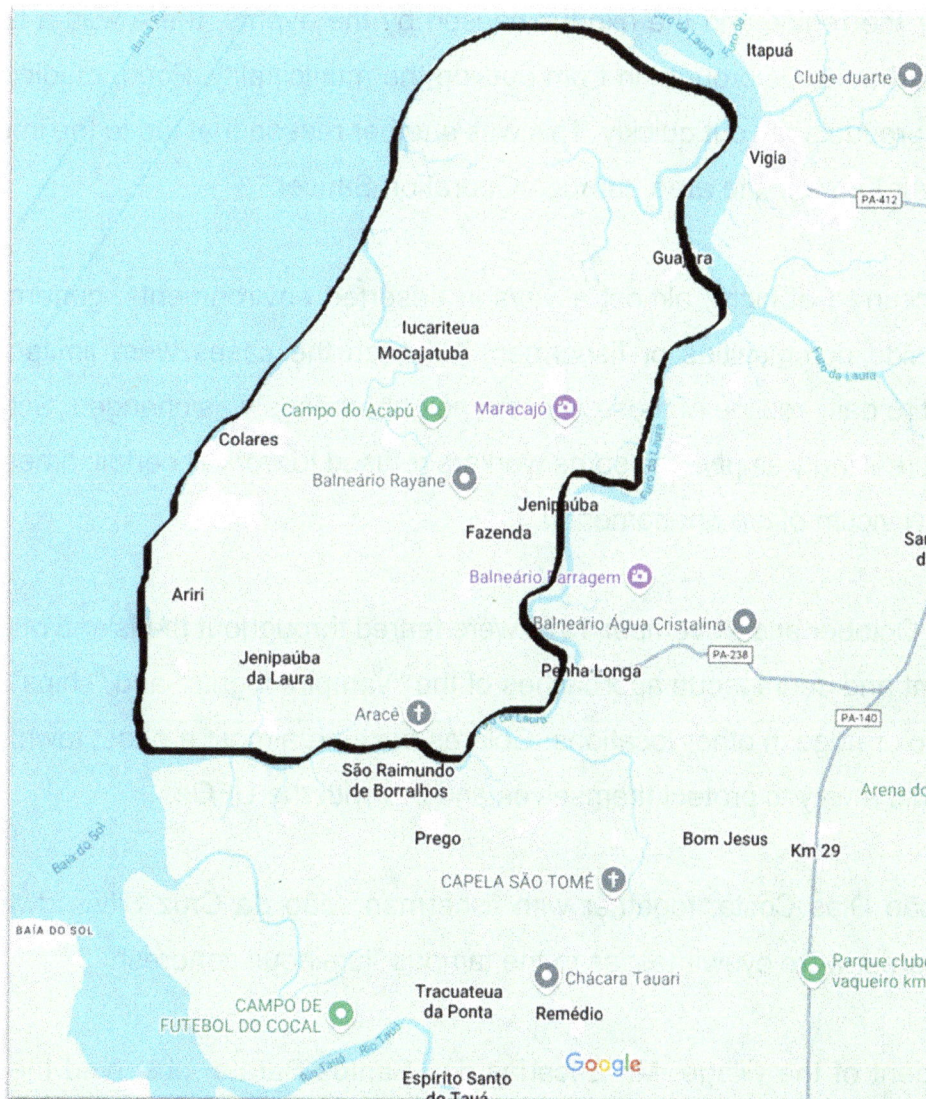

Figure 43. Colares Island.

Isolated from the mainland by the waters of the Guarajá-Mirim River, Colares is relatively close to the municipalities of Vigia and Santo Antônio do Tauá. Its communication with other areas is by river or sea transport, which allows the sales of fish to the main ports in the region.

The island is home to several communities in its 290 km², such as Mocajatuba, Fazenda, Jaçarateua, Arari, Guarajá and many other villages which are not easy to reach.

Colare, too, was greatly affected by the strange phenomenon. The municipality had a health unit and ended up receiving several victims from neighboring areas. Due to these two factors, the situation in the municipality was even more critical. In addition to the widespread panic that took over the region and the exodus caused by the events, there was a lack of basic provisions that had to be brought in from outside the municipality. Food, medicine, hygiene and cleaning products ran out quickly. This was another reason that led to the intervention of the Brazilian Air Force in the area, through Operation Saucer.

The cases occurred at night, almost always in deserted environments, generally affecting isolated riverside communities or fishermen. Although the cases were limited to specific times, the entire daily routine of those affected neighborhoods was changed. Schools began to become increasingly empty, and some workers refused to work at certain times due to fear of being a new victim of the phenomenon.

The nights of October and November 1977 were feared throughout the Island of Colares due to the insistent and courageous approaches of the "Vampire Lights" and "ships". Part of the population took refuge in other locations. Colares became almost a ghost town. Those who remained found a way to protect themselves and live with the UFOs.

Carpenter João Dias Costa, together with fisherman João da Cruz Silva, dwellers in the village of Colares, were eyewitnesses to the famous "luminous spheres".

Another resident of the village, Mr. Zacarias dos Santos Barata, observed the mysterious balls of light for two nights. On the first night, the object came from Marajó Bay and quickly disappeared inland; on the second night, a new luminous sphere, blue in color, flew over the soccer field. "*That light lit up all the trees around the field and disappeared into the village*", Mr. Zacarias reported.

Figure 44. Zacarias points to the place where he saw the sphere of light, in Colares. (Credit: book "Vampiros Extraterrestres na Amazônia**". GIESE, Daniel Rebisso).**

Near the beach, Mr. Sebastião Vernek Miranda, known as "Zizi," described his experience: "*I was with my wife Palmira Miranda, in front of the church, right there on the seafront, when we saw, around 8 p.m., an intense orange light coming from the sea towards the village. As it approached the island, it gained altitude and in a rapid movement, disappeared into inner Colares*".

Barber Carlos Cardoso de Paula was another witness, and one who had closer contact with the Lights:

"*Everyone was sleeping and I was smoking my last cigarette, when suddenly, near the roof of the house, a small ball of fire entered. It started to circle the room until it came close to my hammock. It went up my right leg to my knee, without touching my skin. I watched with great curiosity, when it passed to the other leg. I began to feel weak and sleepy. The cigarette fell from my hand and now I was scared and screamed. The ball quickly disappeared, and everyone woke up. I think it was looking for a vein in my body but had no luck. When it grew brighter, I felt a kind of heat".*

Fisherman Manoel João Oliveira was heading to the beach at dawn with other companions for another day of fishing. Before reaching their boats, they saw an umbrella-shaped object, approximately three meters in diameter, standing about four meters above the ground over the beach of Rio Novo. An intense white light came out of its lower part. From where they were, about 50 meters away, they did not hear any noise and silently the object moved towards Machadinho, suddenly turning off its light.

The lights emitted by the objects were well defined and perfectly directed at any target such as houses, people, boats, trees and even Brazilian Air Force helicopters present on the island during Operation Saucer.

Craftsman Raimundo Costa Leite was another witness to the "Lights". He said: "*Around 4 a.m., along with my friend "Baixinho" (Orivaldo do Malaquias Pinheiro), we went fishing on Cajueiro beach. I remember that at that moment Baixinho said 'here comes the animal' and ran away leaving me alone on the beach. The device was the size and shape of a helicopter, it didn't make any noise, and it didn't fly very high. I could have shot it if I had my shotgun. I was scared when the device turned on a kind of spotlight over the beach. That light scanned the ground, illuminating everything. It was a bluish light, like a cold light. I could see that the device had several small, reddish lights under the front part. The device seemed to be looking for something. I was afraid of being hit by the light and, despite my physical handicap, I managed to run a little, but Baixinho soon came to my rescue. The object came towards the sea and moved away from the island".*

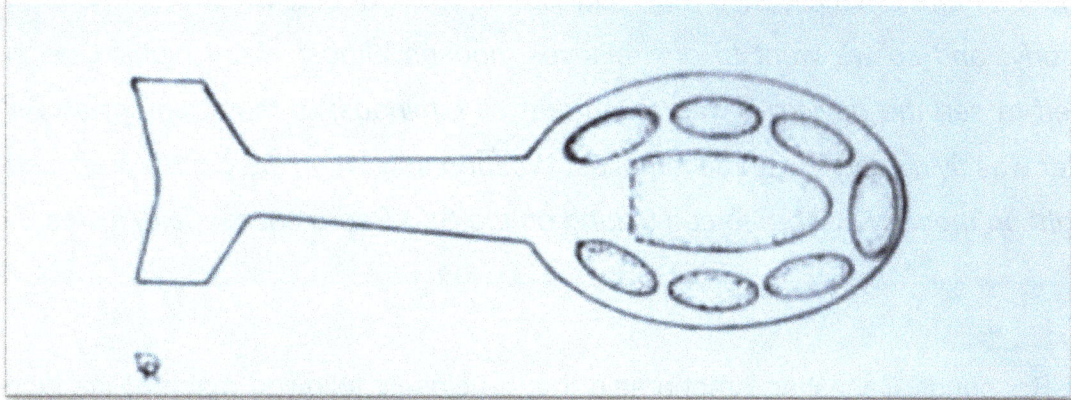

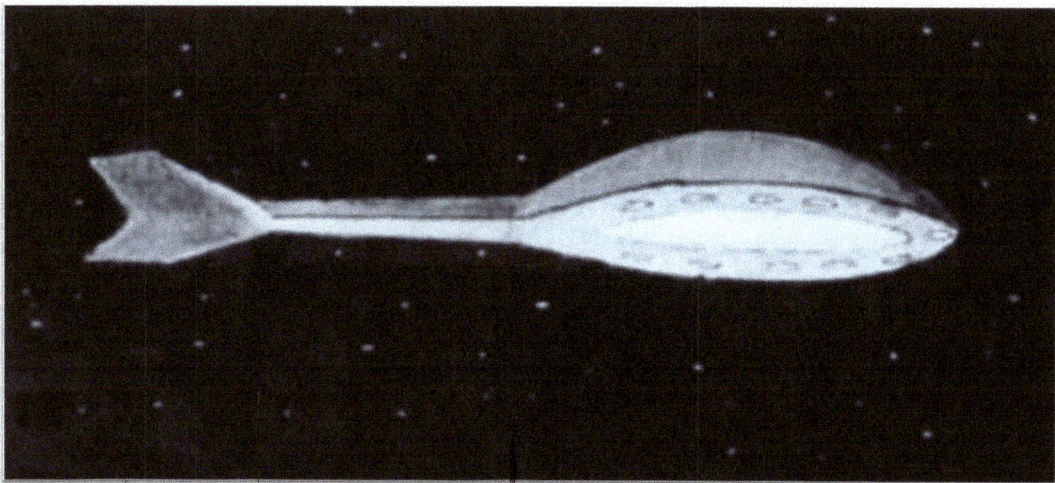

Figure 45. The UFO observed by craftsman Raimundo Leite, in Colares. (Credit: Author's file).

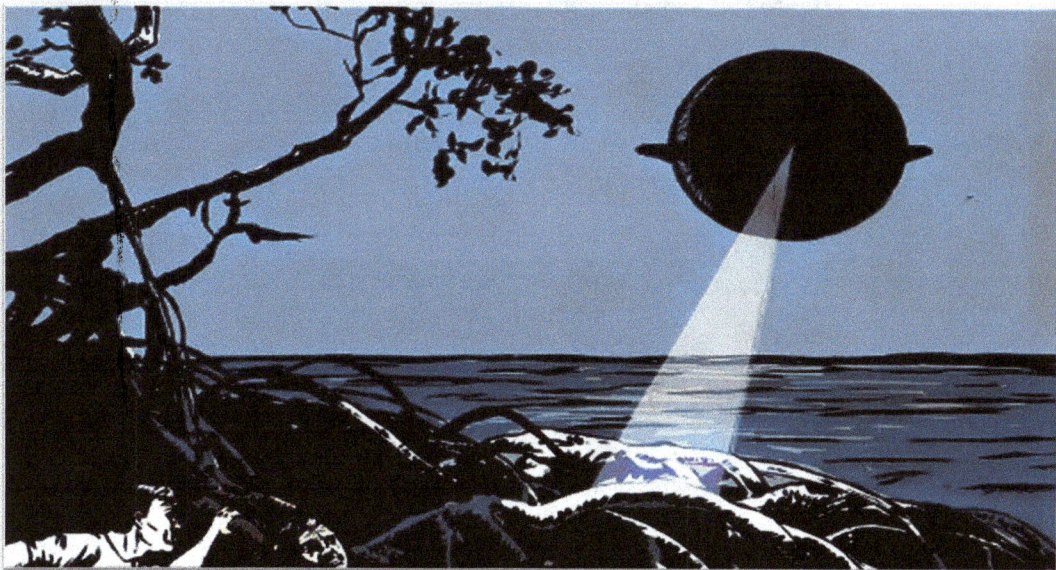

Figure 46. Raimundo Costa Leite and his friend "Baixinho" witnessed a spherical artifact with a flap perform a kind of scanning of the exposed roots of a tree. (Credit: book "Luzes do Medo". MENDES, Carlos).

Mrs. Alba Câmara Vilhena stated that: *"During the 'Chupa' era, we were very scared and slept poorly; and so we went to our relatives' houses almost every night. One time we happened to see the device. It was round and all luminous; at that moment an Air Force helicopter was flying very close to the house. We then saw when the device projected a very bright light on them, who were forced to land on the São Pedro soccer field. It was 8 o'clock at night"*.

Teacher Raimundo Sebastião Aranha said that he closely followed much of the work done by the military mission of Operation Saucer. There was no shortage of equipment, including cars, helicopters, radio transmitters, video and photo cameras, binoculars, etc. Aranha recalls that in addition to the Air Force personnel, there was also a foreigner in the group. *"The helicopters that, at times, seemed to be transporting people and material, tried to chase the UFOs, but without success. What actually happened is that they were the ones being chased by the objects"*.

Figure 47. Ships that looked like umbrellas or soup plates were seen near the beaches of Colares in 1977. (Credit: Author's file).

Figure 48. The UFO saw by Antônio Acácio de Oliveira in Colares, on October 19th, 1977.
(Credit: book "Guia da Tipologia dos UFOs". TICCHETTI, Thiago Luiz)

On October 19th, 1977, in Colares, Antônio Acácio de Oliveira, after being awakened around 3:25 a.m. by the insistent barking of dogs from nearby houses, saw an object that caught his attention due to the brightness it emitted. The strange aircraft was coming from the river Pará towards land at great speed, much faster than the usual planes that crossed that region. The UFO was flying at an altitude of approximately 20m above the ground, in absolute silence.

Since the city's electricity was turned off at 10:00 p.m., its movement was clearly visible in the darkness of the night. As it approached Oliveira, the object stopped and began to spin clockwise, revealing itself in detail to the fisherman. "It was as if it had a dome on top, a thin plate in the middle and a shallow bowl," reported the witness.

The UFO's dome was phosphorescent white. The center, like a ring, was light gray. The lower part was dark gray and had a row of yellowish lights. Its size was estimated at 4m in length.

As Antonio watched, the UFO began to become brighter. The brightness was so intense that it illuminated an area 50m in diameter as if it were the midday sun.

"Suddenly, a luminosity came out from underneath it, like a reddish flame, from an exhaust fan. The light moved down towards the ground, but without touching it." After a few seconds, the UFO began to lose its luminosity and the "flame" suddenly went out. The object began to rise and accelerated towards the city center. This was just one of the cases recorded by the Operation Saucer team.

On November 6th, at around 7:00 p.m., José da Luz was repairing a TV set and, after making all the adjustments, he noticed that it was still experiencing strong interference. So, he left the house to check the external antenna. When he climbed onto the roof, he saw a UFO flying 40m above the ground near his residence and 500 m away from his location. "*It was gray, almost white, with light reflections all over its structure, apparently measuring about 1,5m,*" da Luz reported. The slightly oval and circular shape caught the man's attention, as he was not familiar with any type of aircraft with that shape. The object's movement was wave-like, regardless of which way the wind was blowing.

Figure 49. The UFO seen by José da Luz appeared to have an intelligently controlled beam of light.
(Credit: book "Guia da Ṭipologia dos UFOs". TICCHETTI, Thiago Luiz)

On top of the UFO there was a small transparent dome, also ovoid, and an "extension," which José da Luz said was a "gas deposit." But what mostly caught the witness's attention were two thin tubes placed on the sides of the UFO, which emitted green and red lights. As the object moved, José da Luz observed more details. When the aircraft made a turning movement, the witness saw a reddish circle "as if it were a brazier" on the lower part of the craft. After a few seconds of performing several maneuvers, the UFO slowly moved upward and disappeared through the trees.

One of the protagonists of the events occurring in Pará in 1977 was Doctor Wellaide Cecim Carvalho. She was the one who treated all the people who were attacked by the mysterious lights also known as ""Chupa-Chupa" phenomenon". In addition, she herself witnessed the appearance of strange objects on two occasions.

According to what she reported to the Air Force military agents who were part of Operation Saucer, on October 16th and October 22nd, at 6:30 pm and 7:00 pm, respectively, a luminous object with a metallic glow appeared maneuvering over Cajueiro beach, which is in the front part of the city, at an altitude of about 100 m and at a distance of 1.500m. It was completely silent.

Wellaide described the UFO as having a conical-cylindrical shape, with a narrower upper part, measuring 3 m long by 2 m in diameter. Its movements were irregular, in a vertical position, with pronounced lateral sways. Every now and then, it would make brief stops and turn around. In both sightings, the object's characteristics were very similar, which led the military to believe that it was the same object.

Impressive Cases

During that phase, there were some impressive cases, some of which ended tragically, with the death of the witness. The cases began in mid-July, sporadically, increasing in the following months. In September, the incidents became even more aggressive and several victims of the ""Chupa-Chupa" phenomenon were treated at the Colares Health Unit by Doctor Wellaide Cecim de Carvalho. Two people treated by the doctor died shortly afterwards. One of them was a 45-year-old woman who was attacked in September and immediately taken to the health unit. Due to the severity of the victim's condition, she was sent to Belém, where she died eight hours later. Another fatal victim was a fisherman, attacked in October, who had the same strange burns on his chest. She attended to and spoke to the fisherman, who returned home after being treated, where he died hours later.

Doctor Wellaide closely followed dozens of other similar cases, in which the victims fortunately recovered. There are many other cases that she did not know about or that were treated in other places. Any researcher willing to go to the affected areas will certainly find countless other cases…but perhaps not the medical records. Why? The journalist who followed and investigated dozens of cases of the ""Chupa-Chupa" phenomenon" in Pará, Carlos Mendes, claims that all the medical records of the patients who were victims of the "Vampire Lights" attacks simply disappeared from the hospitals and health centers. According to Carlos, a US intelligence agency would have received these documents.

Another tragic episode occurred in a small village in the center of Mosqueiro Island, in a place called Tapiapanema. The village is made up of a few houses and its residents lived off fishing and farming. In this place, Silvia Maria Trindade, 17 years old at the time, was five months pregnant. Around 6:00 p.m. on October 29th, 1977, she and her husband were lying down, resting in their house. Night was falling when Silvia woke up and saw a luminous object in the sky. A beam of light came out of this object and hit her arm. Frightened, she screamed, waking up her husband and all the neighbors who came out in time to see the object in the sky. One of them shot at the object, which disappeared shortly after. Silvia fainted. When she regained consciousness, she was very nervous and agitated. Her husband Benedito decided to take her to a hospital in Mosqueiro. The journey is made by a rowing boat and it takes about one hour across a river. During this journey, the UAP/UFO appeared again.

"We were in the boat and the UFO flew over the river several times. It followed us and shot a light into the river for about 10 or 15m. It didn't make any noise and didn't throw the beam at us, only into the river. It was about 80m away. Then the UFO flew over the forest and disappeared."

As a result of the contact, Silvia was left with a bruise on her left elbow. She remained hospitalized for two months and ended up losing the baby she was expecting. Silvia was very shaken by the event. Later her marriage ended.

On the day the object appeared in Tapiapanema, the dog Vitória, owned by one of Silvia's neighbors, barked intensely at the object. In response, a beam of light appeared from the object and struck the dog, which stopped barking. In the following days, she began to waste away and died four weeks later.

Manned Flying Objects

Most cases involved only the observation of luminous objects, which paralyzed witnesses and emitted beams of light that caused burns. However, there were incidents in which crew members were seen inside these objects. One of the best-known cases involved Mrs. Claudomira Paixão, who on the night of October 18th, 1977, woke up to an intense light over her house in Baía do Sol (Sun Bay).

Claudomira Paixão, one of the victms of the "Chupa-Chupa". (Credit: Author's file).

"The light was initially green, it touched my head and passed through my face. I woke up at onde and the light turned red. I could see a creature, like a man, wearing a diving suit. He had an instrument like a pistol. He pointed it at me and the object flashed three times, hitting my chest on all three occasions, almost in the same place. It was hot, it hurt me, and it felt like needles were being stuck in all three places. I think they drew blood from me. I was terrified, I could not move my legs. It was terrifying."

After the contact, Claudomira had a headache and pains in her body and weakness that lasted for a few days. She went to the Colares Health Unit, where she was treated by doctor Wellaide and later sent to Belém, where she underwent additional tests at the Renato Chaves Forensic Medical Institute.

Claudomira also had burns on her chest, where she was hit by the beam of light. There were three small circular marks, in the shape of a triangular perforation above her breast.

Figure 50. Claudomira Paixão was struck by a light from the "Chupa-Chupa", while sleeping in a hammock inside her house. (Credit: book "Luzes do Medo". MENDES, Carlos).

"*It was hot and painful. It was like being pricked with a needle. The three points were bleeding. The moment this happened, I became very thirsty. I was terrified, but I could not move my legs. I was paralyzed. I screamed and screamed out of fear. My cousin, Maria Isaete, was sleeping in the same room. She woke up and saw the light and started screaming too.*"

Another case, this time with an uncertain date, involved three fishermen who were on the Guajará River, near Belém. Two were on a boat and one went into the forest where he set a trap to capture small animals. At dusk, he saw a luminous object that positioned itself over the tree. The fisherman, named Luis, hid and watched to see what would happen. A small door opened in the object and a small being emerged through a ray of light. After a few seconds, the stranger returned to the object through the ray of light. The fisherman, frightened, decided to run to his friends on the boat. The object followed him.

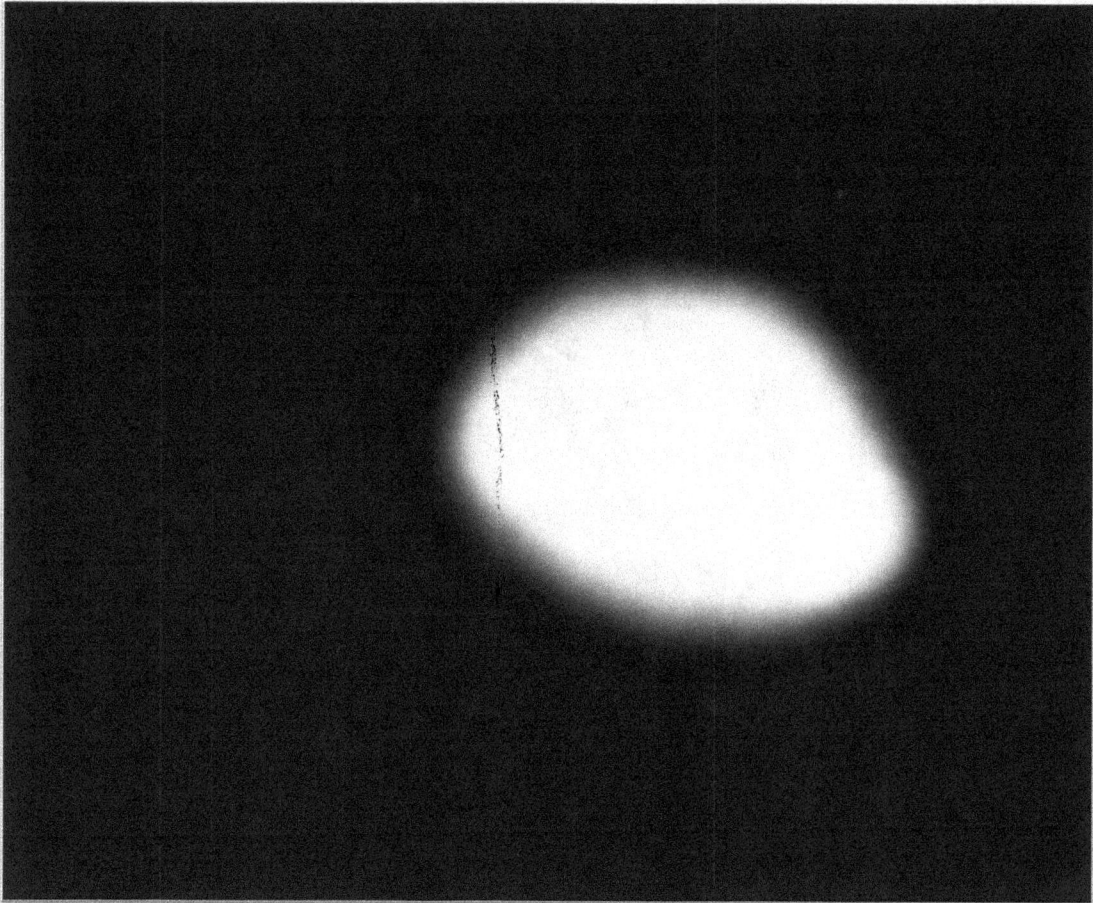

Figure 51. Photograph of a UFO taken during the Bahia do Sol phase. (Credit: Author's files).

When he got to where his friends should be, he could not find them. He shouted after them and discovered that they were nearby. Because of his shouting, his friends returned and found Luis, frightened, who described the strange encounter. Soon, everyone saw the strange object approaching, illuminating the entire region. Desperate, they jumped into the water and hid among some aquatic plants in the area. The object positioned itself over the boat and the hatch opened again, through which the small creature emerged. It was about 1.50m tall and wore a type of dark clothing. Through a dome, they were able to observe the presence of another crew member who remained inside the object.

On October 12th, 1977, at around 11:30 p.m., Manoel Espírito Santo was in front of his house with his friends when he noticed a yellow light moving across the sky, slowing down as it approached the group of men, until it stopped 20m away. The witness said that the light was manned by two human-looking individuals, with the "man on the left side and the woman on the right side of the 'device'. Both were wearing glasses, of a different shape, and an intercom device," he told the officers of Operation Saucer.

Figure 52. The terrible experience with lollipops that resident Manoel Espírito Santo, then 20 years old and with primary education, had with some friends. (Credit: book "Luzes do Medo". MENDES, Carlos).

According to his testimony, the "man" put his hands on his glasses as if observing the group of people more closely. At the same time, the other individual, through a tube on the side, directed a red beam of light towards the witness and his friends. When struck by the light, Espírito Santo felt a strong shock, like an electric shock, from his feet to his head, which left him paralyzed and semi-conscious. The object had a bluish glow on its upper front and was shaped like a barrel, with a smaller, reddish tube in front of it and another thinner one on the side, which emitted the ray of light that hit Manoel. The center of the object was transparent, through which they could see the two beings. The UFO began to move away slowly, with undulating movements until it reached a certain height, when it accelerated and disappeared. Only after the object had gone did the witness start moving again, but for a few minutes, he remained in a "drunken" state.

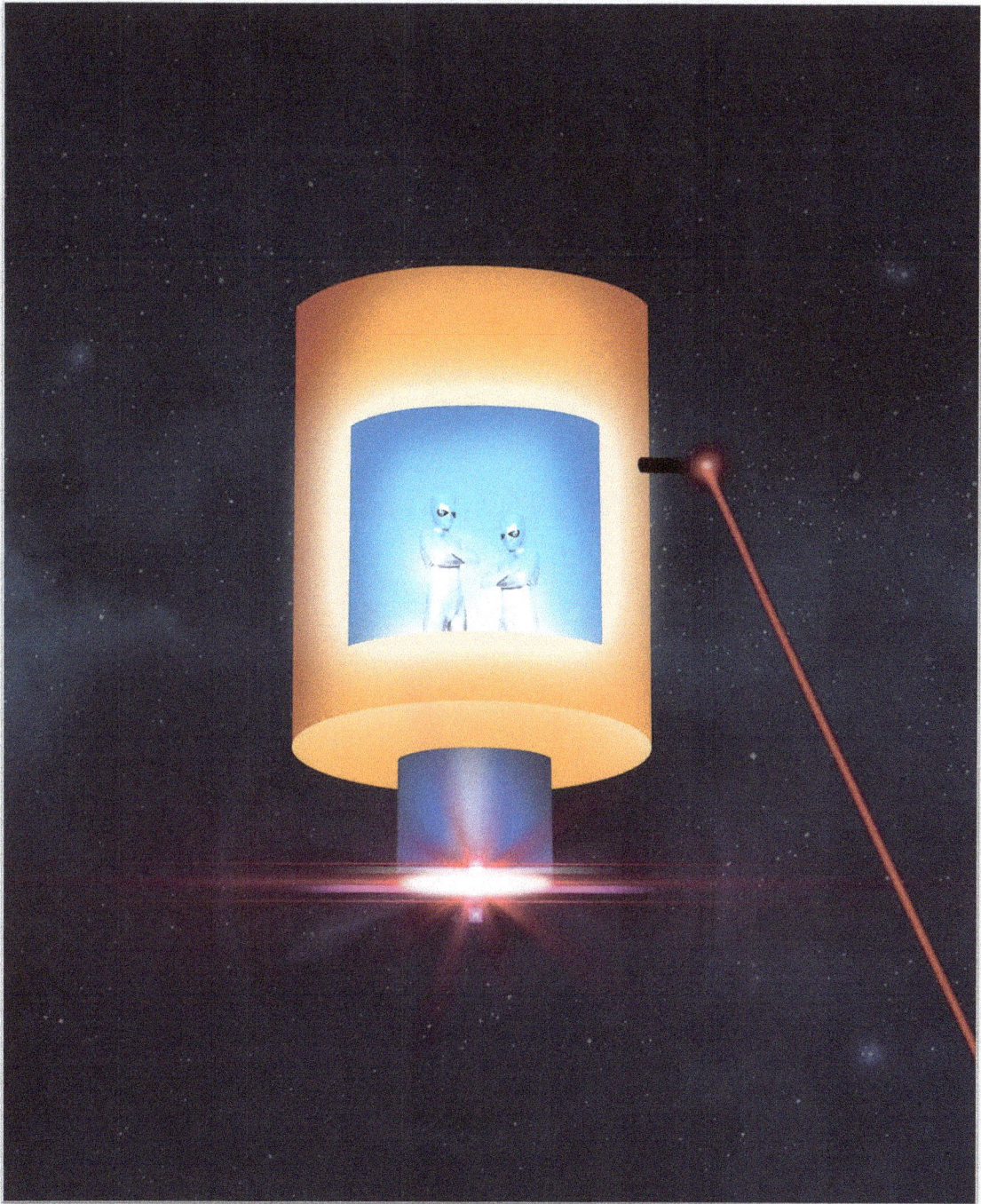

Figure 53. According to Manoel Espírito Santo, the UFO that attacked him had two human-looking occupants inside. (Credit: book "Guia da Tipologia dos UFOs". TICCHETTI, Thiago Luiz).

Chapter 14

Did the "Chupa_Chupa" prefer women?

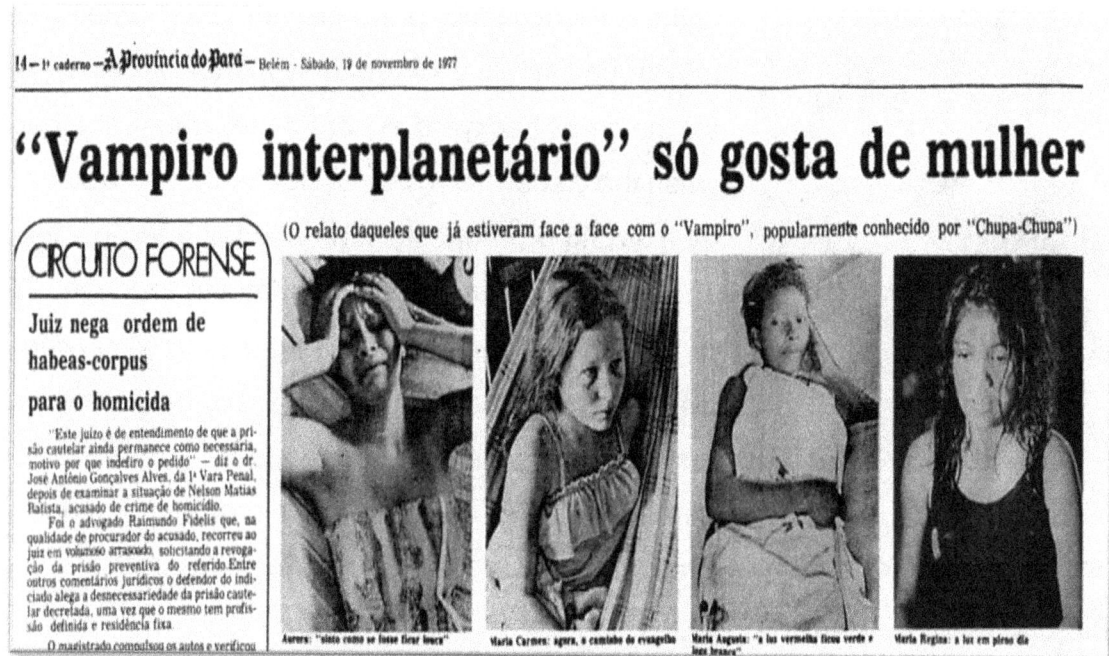

Figure 54. Cover of "A Província do Pará", **with the headline: "Interplanetary Vampire only likes women". The newspapers joked about the subject.**

Local news journal *A Província do Pará*, published a story with the headline "Interplanetary Vampire Only Likes Women". The newspaper said that the atmosphere of unrest among the inhabitants of several neighborhoods in Belém was growing every day, especially in Estrada Nova, Jurunas and Nova Marambaia, due to the appearance of the mysterious light, the "*interplanetary vampire*", popularly known as lollipop. The text said: "*Every 24 hours, we see an increase in the number of possible victims of this light that leaves purple marks on the body, small burns, in addition to a weakened physical state, attacks and causes fainting and immobilization of the limbs and severe headaches that can almost cause madness. One detail that intrigues those who have been attacked by this light, mostly women, are small marks as if they were injection points caused by the strange phenomenon, on the right breast of the victims through which a large amount of blood would be sucked*".

This symptom had not been clearly defined, and only the victims' versions persist. However, when 18-year-old Aurora do Nascimento Fernandes, was attacked, everything changed.

Figure 55. Thiago Ticchetti, Aurora Fernandes and George Knapp, in Colares, Pará, 2024. (Credit: Author's files).

A little above her right breast, the marks or spikes of a kind of wound made by a sharp object were clearly visible. Feeling a lot of pain, Aurora said that on Thursday night, around 9:00 pm, she arrived home from the playground set up on Avenida Roberto Camelier, heading for the small backyard in her house, where she began to wash the dishes from dinner. Suddenly, she was attacked by a strong draft of cold wind and felt a terrible fear as if she were surrounded by something strange. "*I was terrified. I called my mother, who was already lying down like the other people in the house. But before she arrived, a strong red light enveloped me, leaving me dazed. At the same time, I felt very fine punctures being made in my breast and fell to the ground unconscious.*"

Eunice Júlia Nascimento, the girl's mother, examined her daughter and noticed that a colorless liquid was coming out of her wounds: "*It was clear water and smelled like ether,*" reported Eunice. With the help of her brother, Waldir Nascimento, Aurora was put to bed and then taken by her father, Sinval Fernandes, to the Municipal Emergency Hospital. Aurora was not treated at the emergency room and had to return home. As her health condition worsened, she returned to the hospital again and was given a sedative. Next morning, Aurora was writhing in her bed due to severe headaches and was unable to walk because her legs were weak. "*The red light seems to torment me all the time. I feel like I'm going crazy,*" she said.

Maria Carmem do Socorro Lobo was another victim of the mysterious light. By the door of her home, in Passagem Japonesa, in the Estrada Nova neighborhood, she felt her entire body go numb when she was struck by a red light that changed hues, sometimes becoming brighter. She found the strength to run to bed, falling onto it and fainting. To the surprise of her mother, Neuci Ferreira Lobo, three streams of blood were running from the young girl's right breast through three orifices as if they were injection points. Maria Carmem was rushed to Santa Terezinha Hospital, where the doctors on duty confirmed that the girl had lost a liter and a half of blood, thus going into an anemia crisis.

After receiving medication, she returned home, where she continued to faint in short periods of 10 minutes. Due to the severity of her health condition, she was taken to the Adventist Hospital of Belém, where the doctors declared that nothing abnormal was happening to Maria Carmem. Her grandmother, Otília Ferreira Lima, was upset and asked the doctors for an X-ray of the young girl, but she was not granted permission. Skywatches were held there, with prayers and sacred songs, all asking God for the recovery of Maria Carmem, who on several occasions was overcome with violent crises.

"It was as if she was possessed by demons, such was the strength she demonstrated. Three men were needed to control her." The young woman had no memory of what had happened. She has already been cured by the *"power of God,"* according to her relatives.

The time when the mysterious light would attack was usually at night, but an incident that took place around 9:00 a.m. would be the first case of an attack by the light during the day.

Maria Regina Alves Freitas, a 13-year-old student at the Elementary School of the Federal University of Pará, was alone at home, when she was attacked by a bright light.

From the kitchen where she was cutting meat for lunch, she ran out into the street to seek help. *"As the light approached, my body temperature began to rise and in an attempt to escape the rays, I ran out into the street, but as I couldn't find anyone to help me, I returned home. I had the impression that I can see the light all the time. I'm very scared"*, she said to the journalists. She was not injured, but her left side was slightly swollen.

Maria Augusta Elizeu de Oliveira, single, 18 years old, was returning from the Imperial Esporte Club, when she found herself paralyzed by the red light. Terrified, her friends Francineide and Francilene ran away, leaving her alone on a street corner.

"*The red light suddenly turned green and then white. I realized when three friends came to my aid and, at home, I had several marks on my body, especially on my left arm, as if they were light burns,*" she said.

As she was not feeling well at home, Maria Augusta stayed at a neighbor's house, receiving home care. Her general condition is not a cause for concern. Like the aforementioned cases, reported *A Província do Pará*, other people claim to have been attacked by phantom rays, but no one has yet defined them, and the most disparate versions have emerged. Many seek an explanation in the supernatural. There is no knowledge of any measures being taken to at least reassure part of the population that is being victimized by autosuggestion and hysteria attacks, which have only caused social problems.

Chapter 15

The Operação.Prato **(Operation Saucer)**

As incidents began to pile up, the Brazilian Air Force started an operation to investigate the strange occurrences.

The first mission of the military would be to prepare a complete report on what was happening, respecting three aspects or guidelines, considered by the command as essential: the phenomena should be analyzed in depth and in an objective way, for their identification; the A-2 agents (Brazilian Air Force Secret Service) should seek every piece of information on the subject, investigate it and select it, according to the degree of importance and credibility; the investigation would be confidential and therefore public statements and comments on the subject should be avoided.

Figure 56. Captain Uyrangê of Hollanda, a true homeland hero (Credit: Revista UFO magazine).

In September 1977, along with the growing incidence in the region of Vigia de Nazaré, the first group of Brazilian Air Force (FAB) soldiers arrived in Colares Island. Soon after introducing themselves to the mayor, the military set up the first camp at Humaitá beach. By that time, many residents had already left their homes on the island and fled to the mainland. Of all the city authorities, only the mayor, priest Alfredo de La Ó, an American from Texas and Dr. Wellaide Cecim Carvalho, who had been treating the victims of the "Vampire Lights", had decided to stay.

The military, according to the mayor, even brought anti-aircraft guns, which were positioned in strategic areas. That reality showed that they did not really know what to expect from the mission. According to witnesses, the access to the base of operations at Humaitá beach was restricted. As they prepared to try and observe the phenomena and document them, in search of an understanding of what was happening, the military sought to listen to all who had experienced some contact, whether related to the attacks of the "Vampire Light" or even a "simple" sighting of flying objects.

Figure 57. Doctor Wellaide Cecim Carvalho (Credit: Revista UFO).

A few days after the mission started, 1stCOMAR intelligence agents were impressed. Contrary to what many people thought before, in addition to an aerial phenomenon of unknown origin and related to luminous objects that they themselves were beginning to observe, the fact was that, regardless of the picture of emotional unrest that the population actually lived, some of the cases involving the attacks of the "Chupa-Chupa" Phenomenon, whose victims were being treated by Dr. Wellaide, presented mysterious and difficult to explain signs as the result of a self-flagellation process.

According to the doctor, the first cases of attacks appeared soon after the first week of September, when she began to be sought by people, mostly women, who had mysterious burns. However, her first reaction was one of discredit. As the days went by, cases began to multiply and some were truly impressive. The Colares Sanitary Unit also received patients from other places that had the same burns and similar stories. The A-2 militiamen also sought out Dr. Wellaide daily to find out how many new cases had come up and then interviewed the victims of the phenomenon.

Another mission of fundamental importance to the military was the restoration of order. Those people had to be convinced that there was no reason for the emotional discontent that had taken over virtually the entire population of the island. At the time, part of the village met at night in a few houses and prayed for protection and sang religious songs. Bonfires were lit to scare the intruders and the bravest spent their meager resources to buy mortars and bullets for their weapons, to attack the enemy from the sky. There is no doubt that the presence of the military brought a progressive relief to that population, regardless of whether or not UFOs continued to be observed. The attitude of the agents of the A-2 in dealing with the popular ones was differentiated. In certain situations, they denied any confirmation of the presence of a phenomenon of extraterrestrial origin.

At other times, individually, when they realized that this would be the way to calm some people who saw in these manifestations something demonic, they went so far as to say that there was no reason to fear, "*because those things came from another planet and were not there to punish anyone*".

In relation to burns, the military had always sought alternatives that would disassociate them from any abnormal or unusual phenomena. Undoubtedly, they were the most problematic aspect of all that phenomenology. Dr. Wellaide herself told Revista UFO that the military, at the time, had requested their collaboration to omit the truth. She should say that everything was the result of a process of hysteria and self-flagellation. But as we know, she did not accept this kind of request.

According to the doctor, the burns were really impressive. "*It was as if something hot had been stuck in the skin of those people*", she said, "*the burns were long and one of the most surprising things was that they went into necrosis almost immediately after they had been produced. If it happened at night, in the morning they were already blackened. Any burn needs more than 90 hours to have this effect*". Dr. Wellaide says the wounds did not look like burns from hot water or fire. They did not form blisters. In fact, according to her, there was nothing like it in the medical books. The closest thing would be the effect caused by atomic bombs, that is, radioactive burns.

"At first, there was an intense redness in the place struck by the light. Then the hair began to fall and days later the skin began to fade. At this stage two small holes could be seen more easily, like those that would be caused by two needles", said Dr. Wellaide.

There was still a kind of elevation of the skin at the points that presented the small holes through which the blood was supposed to be withdrawn, at least in the eyes of the victims of the phenomena. *"Even though I took care of the huge number of inhabitants attacked by this unexplained 'being', I still doubted that it really was something I could see. What caught my attention was that when I was treating a person from a place called Airi, and another from a place called Candeúba, they were more than 100 km apart. In addition, the people told me the same story, without knowing each other, without ever having spoken to each other and on the same night and at the same time or even at different times. The burns were long, straight, extensive and wide, as if something had happened. There were always two parallel encounters, and if you pressed them, and they did not disappear, and they were raised as if two needles had been used to penetrate them. The population of Colares was taken by a crisis, perhaps of panic, and the city began to be empty. The fear was that we would reach a point where we would no longer have medicines or food, because we were heading for real chaos"*, said Dr. Wellaide.

Dr. Wellaide herself had a sighting of a "spaceship" in broad daylight, when she arrived at her house in Colares, during the wave of apparitions. The unidentified flying object was "something very beautiful" and moved low over her and another witness, her maid, who even noticed the presence of the device before, fainting soon afterwards. The artifact was cylindrical in shape and spiraled.

Figure 58. Doctor Wellaide Cecim saw a UFO in broad daylight over the village of Colares (illustrative image).

Also, according to the doctor, the people struck by the rays projected by UFOs were of several age groups, but there were no children among those affected by the phenomenon. She remembers perfectly that the mayor himself, even after the arrival of the military, went once a week to Belém to buy fireworks, which were distributed to the population. By the end of the afternoon everyone began to seal the holes in their houses, where light could enter. Even the keyhole openings were capped. In fact, no one could sleep at night in Colares. Cans were gathered during the days and used at night to make noise and, along with fires and fireworks, chase away the enemy that came from the sky. It was a truly unbelievable situation for those who did not believe in such things. Even afer the arrival of the military on the island, the events did not cease altogether.

At the time when the A-2 team was based in Colares, one of these objects almost landed on the town soccer field. Some folks rushed to the scene and started making noise and shooting fireworks. The military tried to stop the action, but without any result. In the end, that device ended up leaving and getting lost in the sky.

Also, according to Dr. Wellaide, *"Brazilian Air Force personnel spent the first week making people aware that UFOs did not exist, but if they appeared, they should not be scared away or attacked"*.

But what did the military really think of the whole story? Some of this material, including photos, was published in Revista UFO in 1985, published by its editor, A. J. Gevaerd. Subsequently, more documents were obtained from other sources. One of these concerns the situation found by the A-2 agents in Colares.

"The city of Colares lives in a state of collective hysteria. Its inhabitants, impressed by the appearance of the mysterious lights of unknown origin, do not sleep, they do not fish - the population's main activity - and, above all, they plunge in drinking, spending their meager resources in fires and beverages".

The secret report goes further by describing the events there: *"From nightfall to dawn, bonfires are lit in a daily procession. Fires and shots are constantly heard, as if to scare an 'enemy' who do not know when and where they will attack. Groups of 20 to 30 people, mostly men, travel the city in every way. The population lives in fear. Repeatedly, screams of dread are heard and the news announces then: 'The machine attacked that person.' The affected victims suffer from what we can classify as a strong nervous crisis, loss of better judgment, referring almost unanimously to partial or total immobilization, loss of speech, chills, dizziness, intense heat, hoarseness, tachycardia, tremors, headache with progressive swelling of the affected parts, in the great majority".*

The scenes of "terror" are meticulously described in the reports that the military mission prepared for the high command of the Brazilian Air Force:

"As we believe that the current situation, or its aggravation, will continue, we foresee problems of several orders, including the possibility of suicide on the part of the weaker ones as a consequence fear of the unknown. As a suggestion, the following preventive measures could be taken, such as prohibition of the sale of fireworks and alcoholic beverages; educate the population on how to keep watch, that is, in a more objective and rational way; divide and distribute - groups of no more than 10 men. The rest of the population would perform their

normal activities. In Imbituba, Campo Cerrado, Vila Nova and other smaller towns, the situation is almost identical, with one advantage: several families meet in a single residence - we counted 36 people in a shack. It is a matter of solidarity in the face of need. By bringing a word of comfort and sympathy to those humble people, our team has made them see they are not totally abandoned to their fate, and this was beneficial".

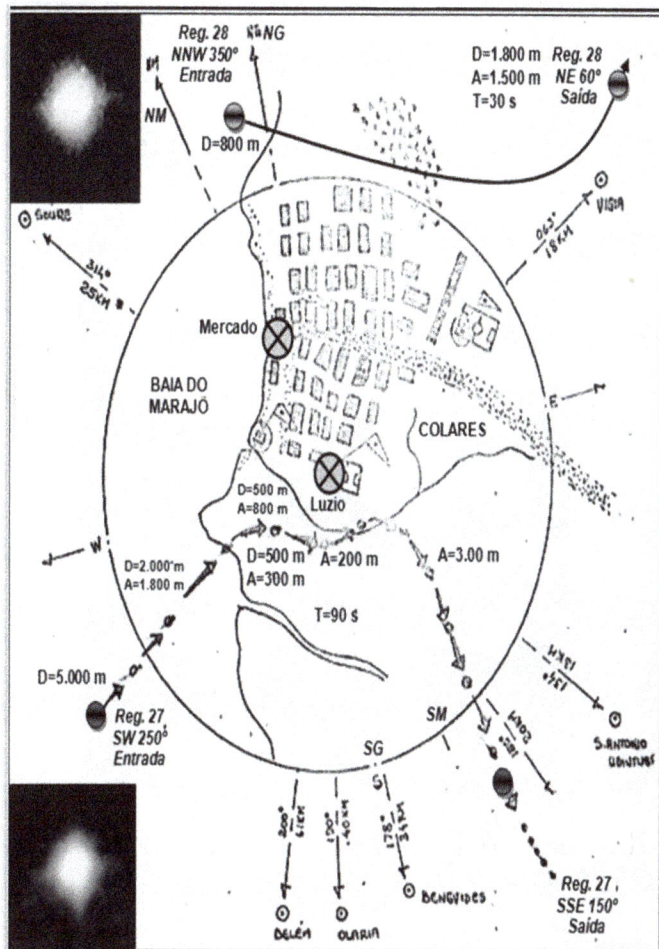

Figure 59. Records # 27 and 28, November 6th, 1977, of the "Records of UFO Observations" document. (Credit: Author's files)'

If anyone could still have any question about where things were going in Colares, the report clearly explains what the locals were experiencing. The same document reveals the situation after the performance by FAB, already in November of that year, 1977:

"The city of Colares, where we stayed, has a new atmosphere. Its residents learned to live with the problem. Maybe our lectures, contacts, slides contributed. The 'lights' continue to appear and what is amazing, as they seem to follow a schedule. The populace is not so frightened anymore. But there is still doubt: the 'monster' created by the press in its actions as a bloodsucker has marked in those minds the dread and a distorted and adverse image of reality".

This document ends with an emphasis on the reality in the existence and presence of unidentified flying objects in the area.

"We have seen several times these objects moving in varying altitudes and directions, performing complex maneuvers and indicating that these bodies and lights are intelligently directed. Our certainty is based on our personal observations and the reliable report of people in whom, by their actions and behavior, when properly analyzed, we can trust. Our

cinematographic records do not depict our convictions, since, lacking technical, material and personnel resources - only at the end of the period we use a type of film of high sensitivity -, they have left something to be desired. Sometimes we lose the opportunity, photographing with inappropriate material. We believe that with better resources we can reach reasonably satisfactory results". This impressive report is signed by the head of the A-2 agent team in Colares and region, the First Sergeant of the Brazilian Air Force, João Flávio de Freitas Costa.

Figure 60. Colares, Lusío soccer field - Place of observation record 28 - Records of UFO Observations (Credit: 1st COMAR).

Such a document serves to confirm what we have always said: an informed population, even on facts that may lead to unpredictable psychological behavior, will be more likely to find a balanced path. Disinformation will always be a greater risk.

A few weeks after the beginning of the A-2 missions in Pará, Captain Uyrangê Bolívar Soares Nogueira de Hollanda Lima returned to Belém, from Brasília, where he was taking a course.

Capitain Hollanda accumulated important positions within the structure of the 1stRegional Air Command (1stCOMAR), as the head of the Intendance Service of the Information Operations Service - the respected Brazilian Air Force Secret Service (A-2). In addition, he was coordinator of Special Jungle Operations and the Service of Search and Rescue with Parachutes (PARASAR). Hollanda was a man of direct confidence in the command of that division and was promptly summoned to undertake this task further.

Hollanda received the first reports from the A-2 teams and the mission to further investigations. Without detracting from the work that had already been done by his military until that date, Brigadier Protásio Lopes de Oliveira wanted to hear Hollanda's opinion. Once again, he would make a difference, with his ability to face challenges and perceive reality beyond appearances, however much it might be undercover. The Operation Saucer was born, with its name coined by the captain to designate the investigations from that moment.

The head of PARASAR reorganized the A-2 teams and refined the logistics of the missions, reiterating initial attitudes until they were understood. From that moment on the military would not leave the mission, even if they were in the jungle. Knowing what exactly was going on in his mind was not very easy, but he was willing to go every length to answer his commander. After all, what was happening?

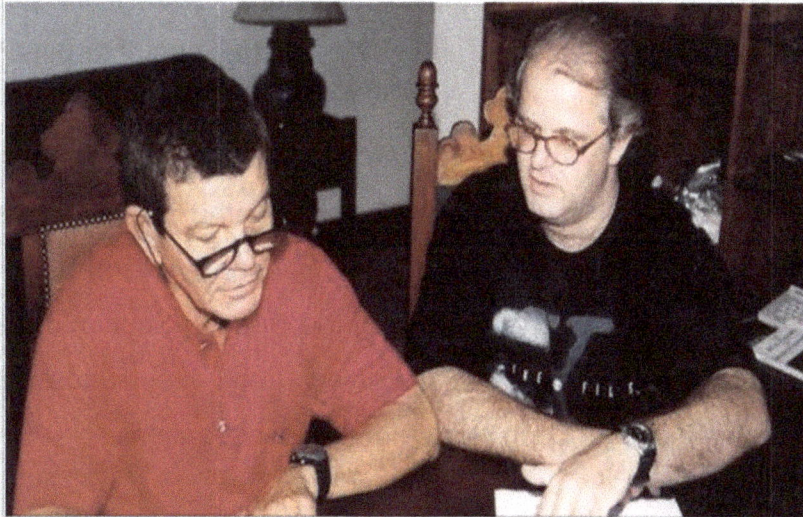

Figure 61. Uyrangê Hollanda and UFO Magazine editor, A.J. Gevaerd, during the historic interview, which took place in 1997. The Colonel told him everything about Operation Saucer and handed over official documents. (Credit: Revista UFO).

Despite acknowledging the ideas of the plurality of the inhabited worlds and having witnessed the appearance of a large UFO in the city of Belém when he was about 12 years old, Hollanda did not readily accept the current idea that the "Chupa-Chupa" Phenomenon was something extraterrestrial. Even after reading the initial reports, he was not convinced that he was in contact with objects of alien origin and was unwilling to let his personal belief in the reality of flying saucers interfere with his judgment.

His story would be told to the ufologists Marco Antônio Petit and A. J. Gevaerd, exactly 20 years later, in 1997, and a few months before his death. His decision to speak openly about what happened after taking charge of the investigations would change the history of Brazilian Ufology.

Ufologist Marco Antônio Petit, along with Gevaerd, participated in one of these rare and historical moments, whose importance, in its real dimension, was only perceived later. The story of Petit and Gevaerd's meeting with Hollanda began at the end of the first half of 1997, when Gevaerd received a first phone call from Colonel Hollanda. He was already willing to talk about everything that had happened in Pará. Hollanda, at that time, was living in the city of Cabo Frio, in Região dos Lagos neighborhoo, north of Rio de Janeiro state.

In his first interview, Hollanda described a bit of his love for military life, for the Brazilian Air Force and for the fact that from an early age he had no doubts about the existence of extraterrestrial life. He also spoke about his contacts with indigenous tribes during missions in the Amazon jungle, some related to the opening of airstrips, which were created in the past so that the military could have a more active presence and safeguard national interests in the region.

Continuing his testimony, Hollanda revealed how his involvement with the investigations, which were already being made by his fellow soldiers, came to be. Secondly, after returning from Brasilia and introducing himself to his superior and head of the Second Section of the 1st COMAR, Colonel Camilo Ferraz de Barros, he received from his hands the aforementioned initial reports and the mission to be in charge of the investigations.

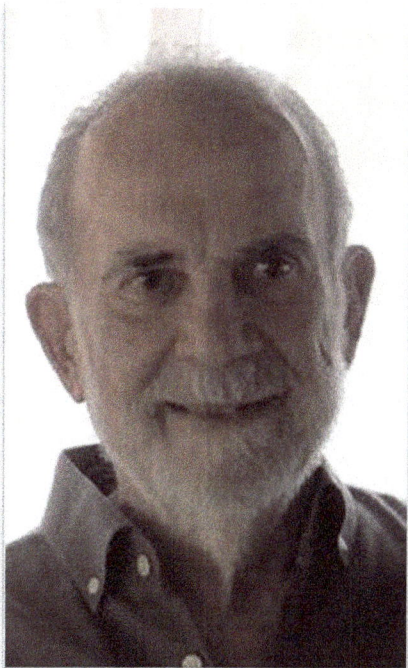

Figure 62. Colonel Camillo Ferraz de Barros was Chief of A2 of the 1st COMAR and Commander of the military team of the Operation Prato, in 1977. He passed away on January 26th, 2019. (Credit: Author's files).

"I went in as a devil's advocate. I went in to demystify that and to say that that thing (the "Chupa-Chupa" Phenomenon) did not exist, that it was a collective hallucination, which was nothing alien. I spent two months answering 'no' to my commander. When I came back from the jungle on weekends, he would ask and I would say no. I had seen nothing but lights", Hollanda said. But while Hollanda and his commanders were watching, he began to hear impressive testimonies that now also seemed to tell him that something serious was happening.

During one of the missions, after landing with a helicopter in Imbituba, another locality affected by these phenomena, Hollanda went to Colares to hear another witness. It was a lady who had been burned by a beam of light from one of those objects. It happened when she was in a hammock inside her house, helping a child to sleep. Suddenly she noticed that the temperature had begun to rise. Then one of the roof tiles was reddening, until it disappeared, as if it were glass, allowing her to see the sky in a surprising way. In the midst of that phenomenon, she was struck by a green light that left her half asleep and then by a red ray that struck her left breast, leaving a mark of a burn that Hollanda could examine.

As he stated in his interview, the wound had a brown color, like an iodine burn and had two small points or holes. While the captain heard her report in person, a child entered the house and warned everyone that another object was flying over the area. Hollanda had time to go out and observe a pulsating light heading north, which later changed its course and returned to fly over the region before disappearing. Other cases of the same type were reported directly to the military, including the "dematerialization of walls," and as he pointed out in his testimony, people seemed to be really telling the truth.

Documents		1977	1978
Synthetic-Chronological Summary	284	195	89
UFO Watch Records	130	82	48

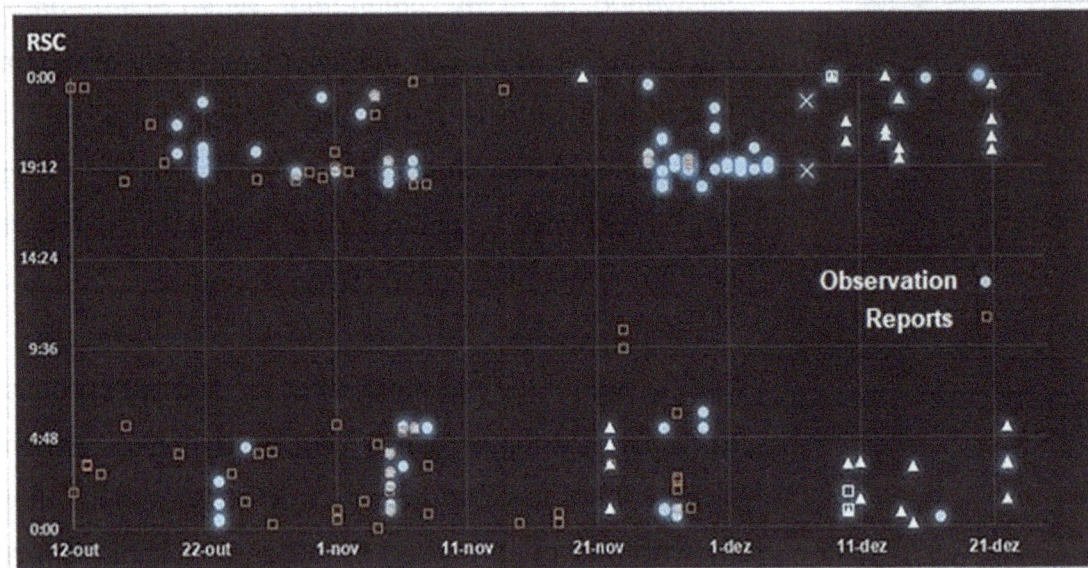

Figure 63. Graph with data from the Summary-Chronological document of I COMAR with reports and observations from October 12th to December 1977. The blue spheres indicate military observations, where at least one member of the A2 team was present, regardless of whether accompanied by civilians or not. The orange squares indicate the observations reported by civilians to the A2 agents. The triangles are a particular extract of the military observations, originating from the extra reports of Sergeant Flávio Costa. The "X" markings are military observations by Hollanda and civilians in Baia do Sol (Sun Bay). The empty white squares are the A2 observations on the Içuí-Guajará River. (Credit: UFO magazine).

In another case reported by the colonel, a resident of the area, seeing one of those objects near his residence, entered the house and took a gun, a shotgun, and went out and pointed it to the object. Then the UFO emitted a ray of light over him, leaving him totally paralyzed. According to Hollanda, the witness had his movements impaired for about 15 days, but as time went by he eventually regained his mobility. He was apparently targeted just so he would not go ahead with his goal of firing on the UFO.

At that time, there were also, according to Hollanda, the involvement of doctors from the Health Department of Pará, who were being sent by the state government to the regions affected by the phenomena and came to examine several of the witnesses struck by the light

162

projections from UFOs. Some of them had signs of anemia, which as the days went by would eventually disappear.

But it was his own later experiences with the phenomenon, increasingly objective, that led the soldier to the conviction that he was facing something surprising and unexplained by our limited knowledge. Manifestations of objects controlled by a high technology intelligence.

Despite the impact on Hollanda by the cases of the town dwellers in the affected neighborhood, and having personally observed many of those luminous phenomena, he still resisted and continued to respond negatively to his commander, mission after mission. The resistance was so great that one of his agents, already discontented, one day asked him if it would take one of those objects to go over his head for him to believe in its reality. Hollanda replied that if this were to happen, things would be more difficult. The surprising thing is that, a few days later, this was exactly what happened.

The commander of Operation Saucer was with his agents in the Baía do Sol (Sun Bay) in the Mosqueiro Island, when a light with the appearance of an electric welding arc came from the north and after pausing above him and his group of agents, turned around and disappeared. The device emitted an intense, bluish-colored light. The officers turned to the military man and asked: *"And now, chief, how are things?"*. Hollanda admits that the agents *"have always been right"*.

"They had arrived earlier in the operation. They had observed things I had not seen. Until that moment I had not observed anything that had convinced me", reported Hollanda.

From that time on, things would evolve rapidly. Wherever Hollanda and his commanders decided to set up their camp or stay overnight, the UFOs appeared. *"They knew where we were and went there. How, I do not know. It was strange"*, he said.

If the previous case had already had a strong impact on the investigating commander, his next experience would be even more surprising. Despite all the secrecy imposed on the missions and in relation to what the Brazilian Air Force was managing to investigate and document, it was common knowledge, even among the press, that the military of the I

Regional Air Command were involved in something important. UFOs were already being seen even over the state capital, Belém. This was the state of affairs when National Intelligence Service (SNI) officers contacted Hollanda and requested permission to attend one of the skywatches. Hollanda said that, as everyone was seeing those objects, there was no reason to stop military colleagues from participating. But the truth is that he could not assume such responsibility alone and requested that the participation be authorized also by the head of that agency in the neighborhood, Colonel Filemon Meneses - incidentally, also Chief of the Brazilian Air Force (FAB). Filemon had even been Hollanda's own flight instructor.

Figure 64. Officers from the Brazilian Air Force stood face to face with ships from other planets (Credit: Revista UFO).

With due authorization, all the details for the participation of SNI members were agreed. Once again, Hollanda and his commanders were in Sun Bay, now waiting for another group of military men who, running late, lost part of what was reserved for that memorable night. Around 6:30 p.m., three luminous objects in formation were observed passing at high speed over the region, at high altitude, going eastwards. As early as 7:00 p.m., two more pulsating objects were detected by Hollanda and his group. They arose to the north and began to move in the opposite direction, heading south. Soon after that sighting, the SNI's military finally arrived and while they were still talking and learning about the apparitions of the early evening, one of them drew attention to the presence of a dark, disc-shaped device with about 30m in diameter, which hovered exactly above the group, not more than 150m in height. The object's lower part had a light ranging from yellow to amber and making a sound like that of an air conditioner. At the same time, a noise like that made by a bicycle ratchet could be heard. Shortly thereafter, the device began to emit a pulsating yellow luminosity over the group at intervals of two to three seconds. According to Hollanda, the object projected that luminous emanation five times on the group, transforming the night into day. The yellow light at the bottom of the disc disappeared and was replaced by a blue one.

Then the UFO started at high speed eastward toward the Atlantic, leaving everyone under a heavy shock.

If Hollanda still had any doubts about what was really going on, the encounter with that ship was final. It was also evident, from that moment on, that the intelligences behind the demonstrations were already performing to the military and totally revealing themselves. There was an interaction between the members of Operation Saucer and those responsible for the phenomena. While the performance of the "Vampire Lights" diminished, a situation of contact with the military began to manifest itself. It was not very easy to understand what was really going on those days. The commander and his agents realized that something very special was happening.

However, if the contacts with the military took place in situations of total control, despite getting closer, the same thing cannot be said about the other experiences that were taking place.

According to Hollanda, there was a leak of information from the operation itself at the time, which allowed the former newspaper "O Estado do Pará" to identify one of the sites used by the group for the investigations. It was Buraco de Laura(Laura's Hole), a passage of water between the Mosqueiro islands and Colares. In that place, several journalists were surprised during a skywatch by the approach of a large luminous object. The device approached the car where the reporters were and projected a light on it, which turned the roof of the vehicle into something transparent. It was a similar effect to what Hollanda had found in other cases, when roofs and walls focused by an intense luminous radiation coming from those objects seemed to disappear. Despite the emotional imbalance that would have taken over some of the journalists, they managed to photograph the UFO, giving rise to a great story published days later. According to them, they would never want to get involved with such a mission again. Hollanda mentioned in his testimony various forms of objects that had been detected not only by the residents and journalists who covered the apparitions, but would also have been confirmed by the Air Force investigators. He spoke of a probe of cylindrical shape and the size of a 200-litre oil drum, which had been mentioned to him by Father Alfredo de La Ó. This has been observed several times on Colares. According to the military, the object was sighted by him and his team on several occasions, as well as photographed. It had an irregular flight and emitted an intense luminosity. There are records of the apparitions of this type of artifact in several locations, not only in Colares. There were devices of various shapes, some with the shape of a disc, others that resembled the body of a Boeing airplane, etc. In total, according to Hollanda, between probes and ships, about nine forms were detected.

From another military man, also an Air Force officer, Hollanda heard of another important case that had already occurred in the Guajará-Mirim River region, near Belém. The soldier liked to fish in the area and heard of a contact that had occurred a few days earlier with a young man who worked collecting clay for the Koaria Pottery, owned by a businessperson from Belém. Knowing what was happening and the Brazilian Air Force investigations, the officer sought the commander of the operation to report the story. Hollanda immediately sent one of his agents and a military vehicle to bring the witness to his presence.

The boy, Luiz Pereira Rodrigues, during his daily activities collecting clay with his boat on one of the banks of the river, had noticed signs left by an animal, probably a paca, (ground dwelling herbivorous rodent) which had been feeding on flowers present in that place. After taking the clay to the pottery and unloading a canoe of 8m in length with an engine, he returned to his house. Then he took a hammock, his weapon, a hunting rifle and returned by boat to the place. He set up a *jirau* (a hunting hideout) in one of the trees and set the net there, where he would be waiting for the animal. The boat this time was left lower in the river, so that it would not be trapped when the low tide came, whose influence is felt not only in the Guajará but, in all the rivers of that area. Luis had left a friend and a boy of 9 years taking care of the canoe and fishing for crabs. The hunter climbed up on the hideout and waited for the animal to appear. After dark, he suddenly noticed the passage of a luminous object above his position. The ship eventually returned and hovered exactly above him.

According to the witness, the object was not discoid and presented an ellipsoid shape, like the cabin of a Boeing. From the lower part of the craft appeared an opening, through which a creature of human form began to descend, amid a luminous projection. Luis left his lookout point and, already on the ground, hid quickly in the midst of vegetation. According to him, the creature possessed in one of his hands a red light, which was used to examine the point where he had been and his net, which remained there. Then, to his surprise and despair, the being focused that light exactly on the position where he was hiding.

Figure 65. Luiz Pereira sees a creature that came from the UFO (Credit: book "Luzes do Medo", MENDES, Carlos).

As soon as he realized that he had been discovered, the boy tried to run towards the boat, stumbling on trunks and roots of vegetation and jamming his feet in the mud of the riverbank. Then the creature returned to the ship. The object then began to chase or accompany the witness in his desperate escape. Slowly, the UFO followed the witness at low altitude, flying into the river's gulley, and close by the trees. As he approached the boat, Luis began to holler to his friends. The witness's idea was to get into the canoe and escape from that object. His friend had a battery radio. He was listening to a soccer game and was alerted by the boy. When both saw what was happening and noticed the approach of the ship, they ended up leaving the boat, jumping in the waters of the Guajará, to the witness's despair. Luis had no choice but to go back to hiding in the vegetation on one of the banks. That ship hovered over the boat as it had done before, the being descended again from the bottom of the object, inside the same spot of light. Then they watched the creature examine every detail of the boat with the same red light. After the procedure, the being returned to the interior of his machine, which rose quickly and disappeared. Luis and his friends returned to the boat and set off for safety.

Figure 66. Remains of the Keuffer Pottery, which went bankrupt in 1992. (Credit: www.operacaoprato.com).

After hearing this impressive account, Hollanda decided that the next day they would spend the night in that spot. He knew the owner of the pottery and managed not only to get his permission to use the property as a base, but also the boat. Earlier that evening he and his agents arrived and found the witness of the close encounter. The military, except for Hollanda, entered the house of the lookout while he stood attentively outside. Suddenly he noticed a very large and intensely illuminated object approaching at high speed and dived over where they were. As it was raining and the conditions for visibility were not the best, it was not possible for Hollanda to observe the shape of the device. But he revealed in his testimony that no pilot with the equipment we had would do the same thing, rushing over the trees during a night like that, because it would have meant certain death. The captain alerted his companions to what had just happened, and they went to the boat. They went up the river about two kilometers and anchored the boat. Then, using a smaller vessel, they drove to the point where the encounter with the ship and its crew had taken place days before. Everyone got out of the boat, and they were right under the tree where Luiz had made his hideout. But the tide was going up progressively and the team was forced to leave that lookout point. Hollanda and the agents got on the boat and returned to the main vessel. The officers thought the captain intended to end the search, but he had other plans and would not give up so easily.

Around 11p.m., about two kilometers away, Hollanda observed an object moving from north to south, crossing the Guajará River just at the point where there was the house of the pottery lookout, where at the beginning of that night another device had already been observed. It was a spherical object that glowed yellow, like the sun. According to Hollanda, the craft passed very low over the river - the light was so intense that it perfectly illuminated the water. The A-2 agent team, warned by his subordinates, was able to film and photograph that UFO. According to the military, the light of the object trembled, and the effect was documented in detail in the footage, considered by the commander of *Operação Prato* (Operation Saucer) as one of the most impressive.

After the sighting, Hollanda communicated to the rest of the team his intention to remain at that spot along the river, although they were without food, water and coffee. When they planned the mission that day, they had not planned to stay in the place for more than a few hours. Facing the situation, Luiz offered to go to his house by the river and bring something to the group. Soon after, once again, Hollanda noticed another luminous object, which, like the previous one, moved at low altitude over the point in the river where Luiz lived. However, this time the device was heading towards the boat, where everyone, already alert, was preparing to take new photographs and film the UFO. The object was indisputably moving towards the military, and it seemed to "know", as had already happened, exactly where they were.

Although it had the same yellow or amber appearance, it was much larger than the other one that had been spotted just moments earlier that night. Even with the gradual approach of that craft, all that was seen was light. But, as it drew closer, something surprising happened: all the light suddenly disappeared and the military were finally able to observe the shape of a gigantic object about 100m in diameter, as calculated.

The ship was shaped as a football, almost translucent and it had all around its extensions what appeared to be windows or portholes, behind which an inner light could be seen. The UFO, according to Hollanda, passed slowly above the boat, giving them time to photograph it and film it. Again, as had already been noticed in a previous contact, that air-conditioning noise was perceived as well as a sound like that of a bicycle ratchet. Hollanda even asked one of his agents, who was filming the passage of the UFO, to interrupt the footage for a few seconds so that he could be sure of the origin of the sound. He wanted to hear its emanation without overlapping the noise made by the camcorder. There was no doubt about it: that sound came from the UFO.

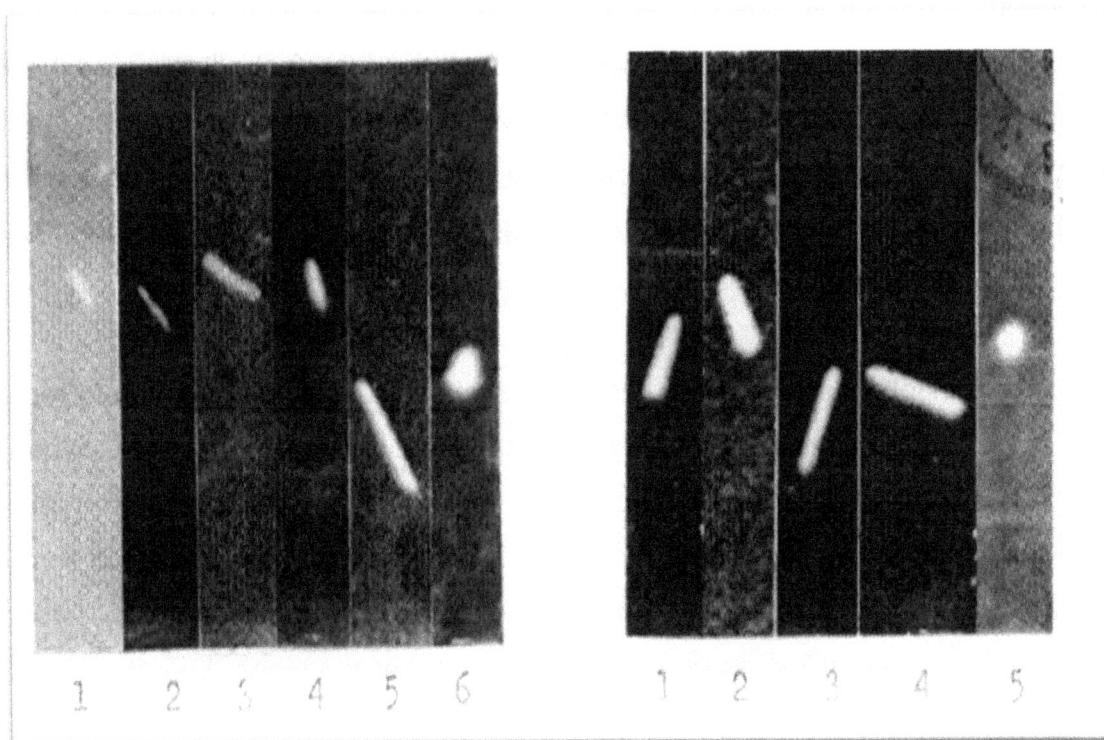

Figure 67. Photos of a UFO shaped like an American football. (Credit: www.operacaoprato.com).

That "American football" object, as Hollanda described it, passed over the ship and then followed the river toward Belém, before disappearing. The impact of that contact was visible to everyone present. Everything seemed obvious that this was a deliberate display. There was an interest in the operators of those objects, which had already been observed, photographed and filmed in the last few weeks, to an even greater extent. After midnight the group remained attentive. There was a very special feeling in everyone who was experiencing those events, a

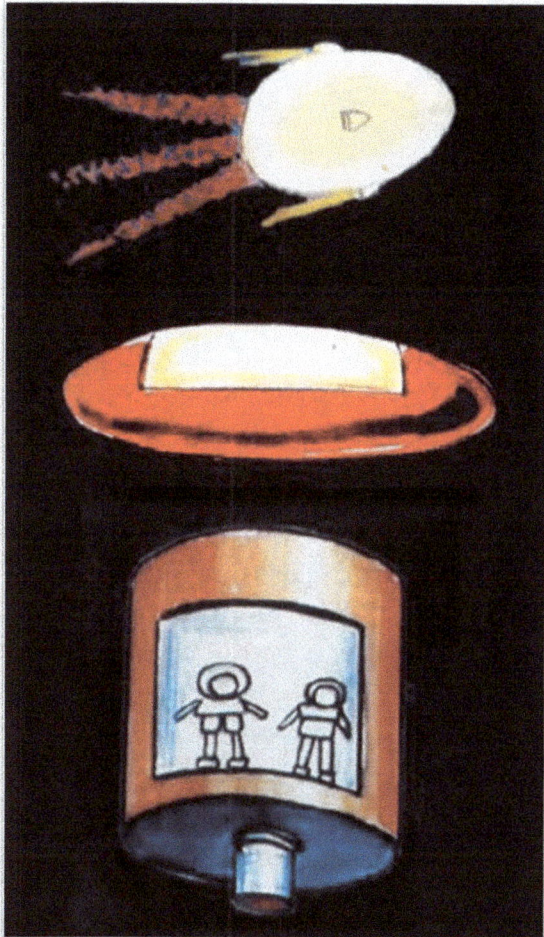

Figure 68. Various types of UFOs have been reported by residents and witnessed by military personnel. (Credit: UFO Magazine).

longing for something even more striking that night. Around 01.30 a.m., another object, this time with a blue glow, was sighted on the opposite bank of the Guajará River. The UFO flew even lower, at the trees' height. When it approached a small island, which the military boat had passed before reaching the skywatch point, it changed its course and headed back for Belém. Soon afterwards it reversed its course again. Everyone could see that strong luminosity. According to Hollanda, the device was actually very low. The UFO repositioned itself on the river and once again began to approach the men, eventually becoming stationary and completely immobilized near the opposite shore of the Guajará, right in front of the boat. Photos and videos were once again being made. The light was intense, but according to the commander of *Operação Prato* (Operation Saucer), it could be seen.

At that moment, for the first time, Hollanda was truly afraid. "*What will happen to me and my people now?*" He wondered. Later on, after three to five minutes, that entire luminous field that took the spherical form disappeared, revealing the structure of a gigantic ship. Apparently, it was the same device that had approached earlier horizontally, overflying the vessel. Now it was immobile, but upright. But this time there was not the slightest sign of the windows or portholes, nor was any sound that could emanate from the object being heard. The cameras continued to operate and document the encounter. If there was still any doubt about the interaction with unknown ships responsible for the strange phenomenon, it was dissipated. The military had been chosen to participate in something very special and there was no denying it. Nevertheless, the fear of what was happening, though fully controlled, was not something only Hollanda experienced.

Figure 69. Colonel Hollanda said that he would only believe what people were reporting if a spaceship flew over his head. And that happened. (illustrative image)

All the military present experienced in one way or another that feeling of perplexity and natural concern. What would happen next? After a few more minutes, the ship began to move slowly upwards, still not emitting any sound and now presenting a red light on the bottom and a blue one on the top. According to Hollanda, when it was about 1.500m high, it triggered some kind of propulsion, generating a sudden flash and a sound like thunder, shooting at high speed into the sky, being lost sight of in a few seconds. For the commander of Operation Saucer, the intelligence within that object had been careful.

To Colonel Uyrangê Hollanda, that wave of sightings did not seem to represent something negative. At the end of his testimony, he reached a clear and surprising conclusion. For him, everything was just another stage to be overcome before the day of contact, when a formal coexistence would begin between humanity and some of the extraterrestrial civilizations that visit us.

For the military, those beings responsible for the *"Chupa-Chupa"* phenomenon were collecting material, removing blood and even tissue samples, cells, etc. from humans. *"I do not know what that high-energy light could transport, something that could later be analyzed by them,"* said Hollanda. He remembered the time when he had contacts with indigenous tribes during his missions in the jungle and the care he took to avoid passing on our diseases to them, because they did and still do not have any effective immune defense. For Hollanda, those beings were taking biological material to develop an immune defense against our diseases. According to Hollanda, it would be necessary if they intended to seek direct contact with our species.

The commander of Operation Saucer's ideas were passed on by himself to several ufologists from Brazil and abroad, who found his considerations sensible and logical. One of them, scientist Jacques Vallée, would have declared it to be the most logical thing he had ever heard of. If we are to base ourselves on the official documents and the reports signed by the members of Hollanda's team, we will find no support for a negative view of those events, other than the psychological developments experienced by the population.

But the end of the military mission did not mean the closure of contact experiences, as Hollanda stated. Some of the officers involved were so impressed and interested in the phenomena that they continued to carry out skywatches on their days off. At least in the case of his commander, the most remarkable experience occurred within his own house, in the Military Village of the I Regional Air Command, in Belém.

"They only definitely activated the ship's propelling system when they were already far enough," he concluded. This was the last major contact maintained by the military during the jungle mission, which took place at the end of December. In a mysterious way, shortly after Hollanda reported the incident to his superiors, the order was issued for the investigations to be closed. In total, it was four months of research that produced more than 500 photos and 22 hours of footage documenting the presence of UFOs. However, what would be behind all those demonstrations witnessed by the military? And as far as the phenomenon is concerned, which peculiarities gave rise to so much perplexity and even panic among part of the population?

Captain Hollanda and his remarkable experience

One night, when he was lying in his bed doing meditation, there was an explosion of light and a loud click. Hollanda, who was lying on his side, immediately noticed the presence of two beings of short stature, five feet high. One of them was near the head of the bed and another gripped him from behind. The creatures were dressed in something like an astronaut suit, somewhat cute. The faces were dark gray, but he could not check details due to the presence of something like a diving mask. The being who held Hollanda spoke clearly in Portuguese, with a metalized voice, and told him that they would do him no harm. Soon there was another flash and the same kind of click and the beings disappeared. Hollanda does not remember anything else.

The next day, during an oath to the flag, already in the barracks, when he was saluting, the then captain began to feel a strong itch in the left arm. When the ceremony came to an end, the military noticed a red dot in that place, having around it a pinkish stain. A few days later, as he pressed the region where the itch had developed, Hollanda noticed for the first time two mysterious objects inside his left arm, one in the shape of a small plastic needle and the second in a rectangular shape.

Despite the existence of the objects, which were found by Petit and Gevaerd, still in 1997, there was no sign or scar that could explain how they had been introduced. Hollanda also revealed that one of his agents, who participated in the investigations of *Operação Prato*, also discovered two objects of the same type in one of his legs, more precisely in one of the thighs. The images of the first interview recorded with the colonel could not have been more impressive. Petit documented with his camcorder the mysterious implants - even in the video its existence is easily perceptible below the colonel's skin.

In the interviews that followed, Colonel Hollanda explained why he was revealing all those facts, after 20 years of silence. The colonel stated that even at the time of the events he and his commanders did not consider that the matter should be treated as a threat to the country's internal security or as a state secret. According to him, his superiors were determined that the secrecy be maintained, which he considered unjustified, but he was in the active service and was taking orders. He then described, in more detail than he had done in his initial statements, how his involvement in the operation had occurred.

He quoted Colonel Camilo Ferraz de Barros, his chief in the 2nd Section and how he had given him the task of commanding operations. According to Hollanda's statements, the main objective of the research group was "*to look at the phenomenon, to observe it carefully and, of course, to take testimony from witnesses*". He even said that even the US Air Force "*could do no more than that because of the technological disparity between our humanity and that of the planets from which the objects come*." Hollanda stated categorically that such artifacts were ships and probes under the control of some advanced extraterrestrial civilization. The colonel also emphasized that the presence of Brigadier Protásio at the head of 1st COMAR was fundamental for the implementation of Operation Saucer.

"*Without him in charge, despite the intense UFO activity in Pará at that time, the mission probably would not have continued. The brigadier was part of that group of senior officers who believed in the reality of the UFO Phenomenon and the importance of a serious investigation into the subject*", said Hollanda.

The retired colonel also spoke of the difficulties he and his team had in documenting the UFO manifestations at the outset of the investigations and the invitation he had made to a professional camera operator of his confidence working on TV Liberal to take part in some missions. His name was Milton Mendonça. According to Hollanda, with the participation and guidance of Mendonça, hundreds of photographs and several films were finally obtained, documenting the presence of UFOs. "*The whole population was seeing those objects, but the material obtained by the Brazilian Air Force would undergo a technical treatment*," said the military man, noting that he was not there to clap for anything.

Hollanda's statements pertaining to the military research procedures, the technical care for the photographic material being procured and all the treatment given to the information relating to such investigations serve as a response to those who today question the UFO Phenomenon, trying to minimize the importance of everything that was obtained amid the attempts of the already agonizing military cover-up. "*Altitude of observed objects, speed, path details, apparent size, luminous intensity, colors, time and location of each observation etc., were recorded in the reports. The photographs and films also received technical treatment. Information on the films and machines used, exposure time, everything was done within military rigor"*, he said.

In spreading his experiences openly, the colonel was struggling to make sense of his existence. But, unfortunately, on the night of October 2nd to 3rd, 1997 he committed suicide. When his death was revealed, totally irresponsible people, only interested in cheap sensationalism insisted that Hollanda had been murdered for divulging the truth about Operation Saucer, but according to his daughters, he had killed himself for the same reasons that had led him, years before, to another attempt: a depressive process related to family problems.

Colonel Uyrangê, Bolivar Soares Nogueira de Hollanda Lima, who thought of the possibility of a premature death, made a point of leaving before his final journey a proof of life: the certainty that somewhere among the stars in our sky there are beings who have already discovered us and if his conclusions are right, they are preparing the day for a definite contact.

However, would all the material from Operation Saucer be available to society? The answer is no. Where are the footages and the other hundreds of photos?

A few years ago, the government released new documents on Operation Saucer, this time having an unusual origin: the former and feared National Information Service (SNI), an agency then linked to the Office of Institutional Security of the Presidency of the Republic (GSI) today replaced by the Brazilian Intelligence Agency (ABIN). In other words, from the cellars of the Military Dictatorship (1964-1985) pages and more pages of documents about unidentified flying objects with the governmental seal were published. The documents are identified by the acronym ACE, of Chronological Entry File and have a numbering, ACE 3253/83 and ACE 3370/83. From Colonel Uyrangê Hollanda's historical interview to A. J. Gevaerd and Marco Petit, editor and co-editor of UFO Magazine, in 1997, SNI's involvement in the missions of Operation Saucer was known, but no document had ever been revealed about it.

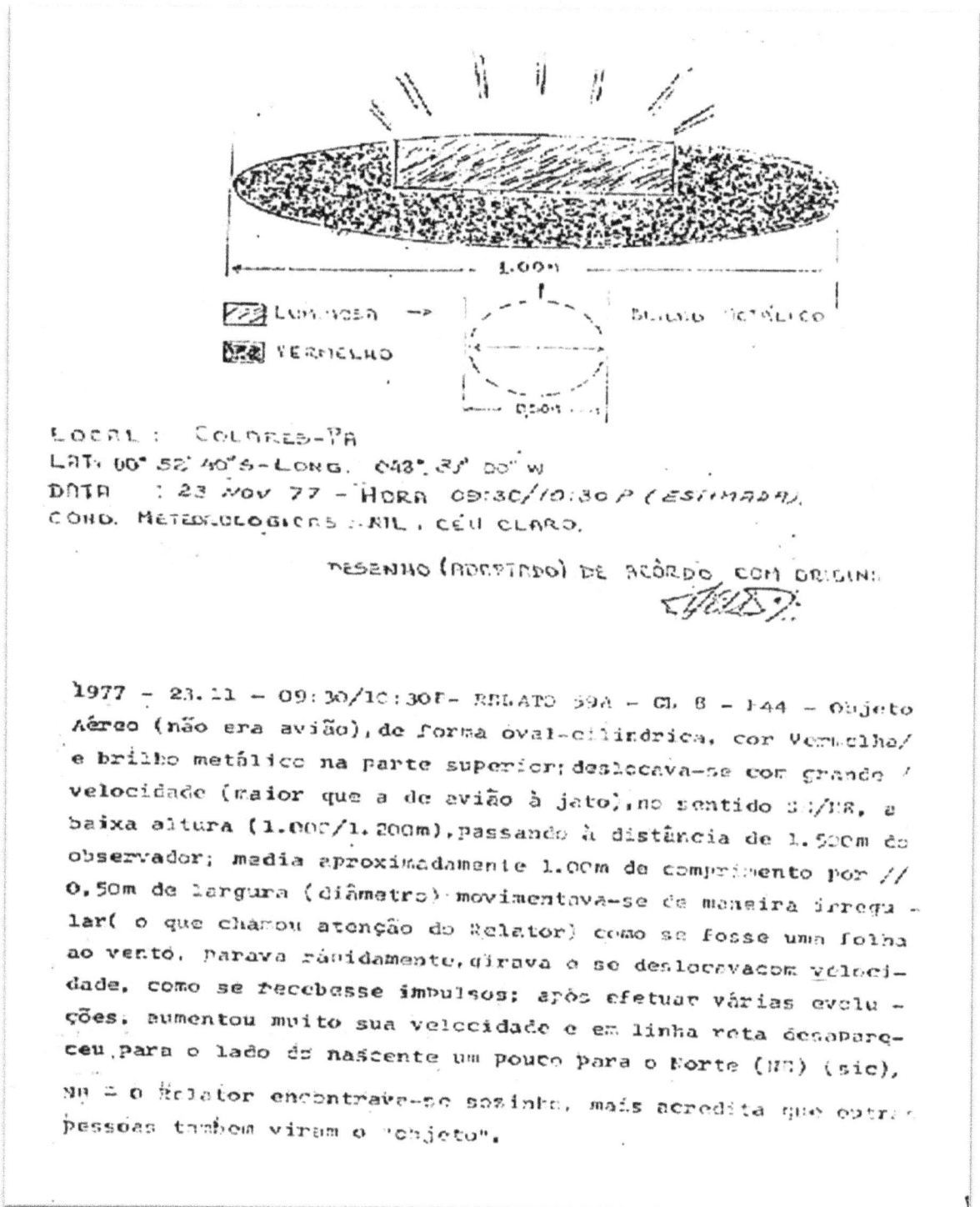

Figure 70. FAB document with report and drawing of a 100-meter-long cigar-shaped object seen in Colares on November 23, 1977. (Credit: Author's files).

Figure 71. Ufologist Marco Antônio Petit and Colonel Uyrangê Hollanda (Credit: Revista UFO).

Figure 72. Pilot spots UFO seen behind a cloud after takeoff. The object was bowl-shaped with landing gear. (Credit: Author's archives).

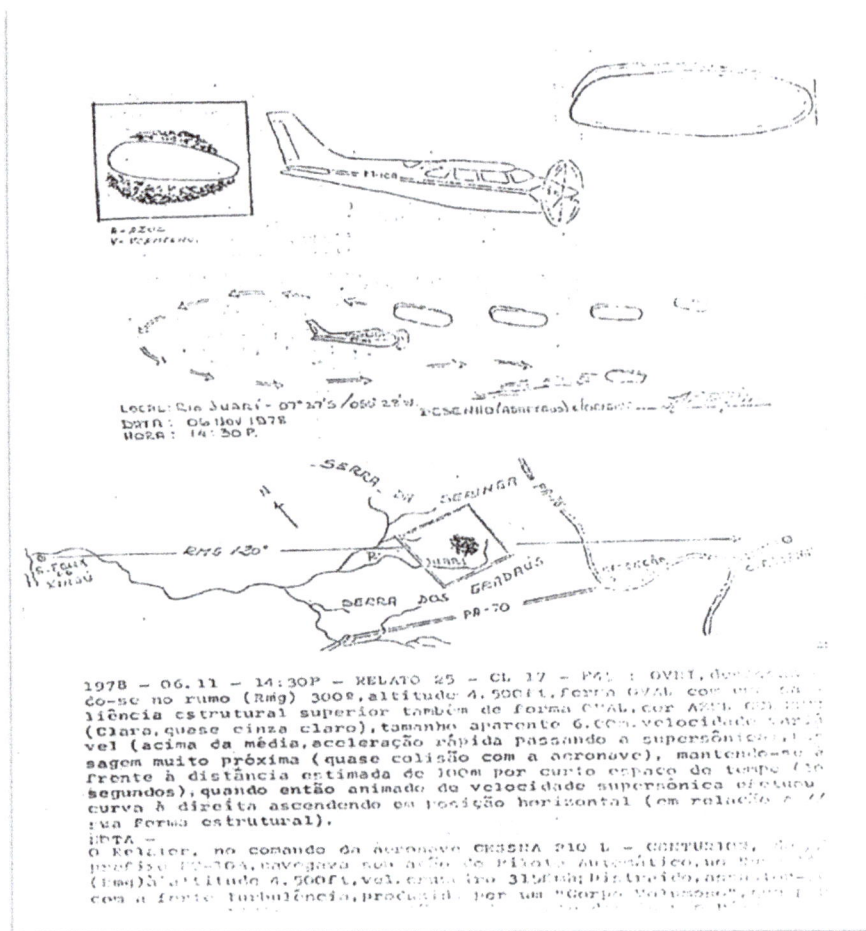

Figure 73. FAB document, with drawing, of the report by a pilot who observed a UFO circling his aircraft. Below is an illustration of the moment. (Credit: Author's archives).

In addition to the unofficial documentation leaked to ufologists, SNI's reports have their authenticity recognized and are available in the folders of the Regional Office of the National Archives in Brasilia, as well as on Revista UFO issues. They bring numerous reports of the UFO manifestation in the jungle and how and by whom it was investigated.

For example, in one of these files is the passage through the A2 of the I COMAR by the reporter of "*Homem*" magazine, Palmério Dória Vasconcelos, on January 25th, 1978. According to the document Information 0171/119 / ABE / 78, SNI, Vasconcelos' aim was to find out what the Brazilian Air Force (FAB) had actually discovered about the presence of UFOs in the skies over Pará. Obviously, the officer who attended to him did not give him any information, as it appears in the material: "*He tried to mislead me, not providing me with any of the requested data*".

Another document, entitled Information 1802/320 / ABE / 77, dated November 29th, 1977, although only five pages long, gives an overview of the military mission in the Colares region, what motivated it, what they concluded about it and some of the main witness reports. Among them, the doctor of the health unit of Colares at the time, Dr. Wellaide Cecim de Carvalho and the parish priest of the city, the Mexican American naturalized and later resident in Amazonas, Alfredo de La Ó. The document ends by indicating that a new contract is being prepared: "The Air Command I is organizing a new mission to continue the investigation." This passage indicates what was already known: that before Operation Saucer proper, which began in September 1977, another, smaller and more informative inquiry was conducted by I COMAR.

Figure 74. Photograph taken on December 10th, 1977, around 8:30 p.m., in Baía do Sol, in Mosqueiro, Belém (PA). The object was over Ponta do Machadinho, on Colares Island (PA). It was estimated to be 300 meters high, 3,000 meters away from the photographer, Sergeant Flávio and three other civilians. The UFO remained stationary for approximately 5 minutes. (Credit: Author's file).

Figure 75. Dozens of fishermen reported being chased and attacked by strange lights. (Illustrative image)

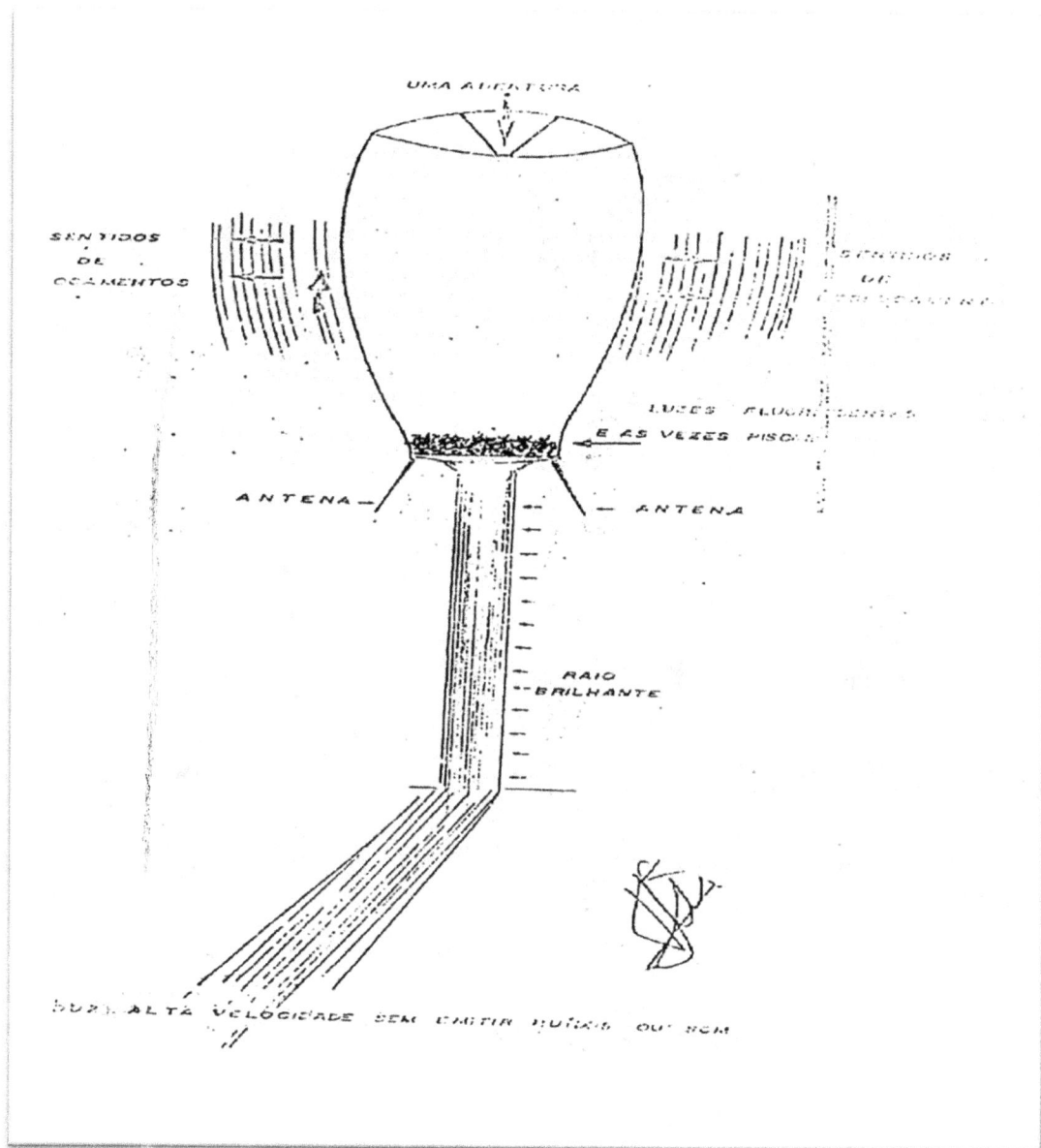

Figure 76. Official FAB document showing a UFO with an antenna and firing a ray that made a right-angle curve. (Credit: Author's files).

Associated with the registration of the UFO case in Pará, there is another important document from SNI, a request of councilwoman Dalgisa de Alcântara Garcia, from the city of Maracanã, which deals in an alarmed manner with the appearance of UFOs and their attacks. Taking the number 71/77, the proper authorities are asked to take the necessary "measures that the case requires, as a matter of urgency". As it turns out, the information service of the Dictatorship days kept a close watch on manifestations of all kinds about their interest - and flying saucers in the Amazon was one of them. And there were not many mayors, councilors, community leaders and authorities of all kinds who asked the State Government and the Air Force to resolve the issue.

Figure 77. UFO photographed by the A2 military team on December 17th, 1977, on Jeju Farm, in Pará. Twenty-six photographs were taken. The object had yellow-reddish color and bluish reflections (Credit: Author´s file).

Also included in the recently released SNI documentation are maps showing the locations where flying saucers and UFO probes would have appeared most frequently, as well as a UFO-shaped illustration depicted by several residents of different regions and various newspaper clippings which reported the apparitions of the "Chupa-Chupa" Phenomenon in Pará and Maranhão, highlighting the news of the death of a fisherman, after being struck by a strong light that penetrated the lower deck of the vessel where he spent the night, in Crab Island, which became famous in Brazilian Ufology.

One can find among the newspaper clippings of the time information that there was also an attempt on the part of the military to hide from the population the nature and dimension of the UFO manifestation. In an interview to the newspaper Estado do Pará, on November 5th, 1977, Lieutenant Colonel Camilo Ferraz, an assistant commander of the 1stCOMAR, says that *"everything was nothing more than a mere illusion from the population's perspective, which is of low intellectual level. Residents confused the existing artificial satellites in the area and the meteorites that blink in the sky with aliens"*. Analyzing the reports of *Operação Prato* it is evident that this statement does not demonstrate the reality of the facts. It was another blow to the policy of cover-up, in vain.

Figure 78. Expedito showing the hole made by the UFO landing at the Jejú farm (Credit: Author's file). Bellow, drawing by researchers of the Brazilian Air Force of the holes left by the UFO at the Jejú farm (Credit: Author's files).

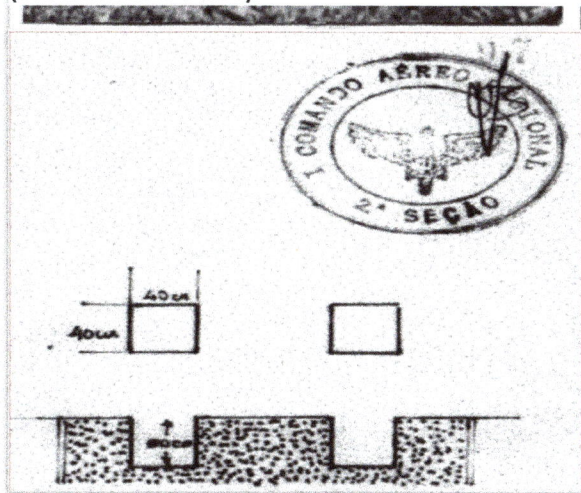

Fortunately, the Information Disclosure initiated in Brazil and in many other countries begins to change that reality. Specifically about Operation Saucer itself and not the general UFO incendence in Pará, the documentation originating from SNI contains a report restricted to the events in the town of Colares. But, its 18-page content, comprising cases from October 20th to November 10th, 1977, is similar to the one found in unofficial documentation previously achieved by Revista UFO.

The dates and times of events, as well as the names of witnesses, are practically the same in both materials. The difference is due to the way in which the facts are described and the text of the report that was officially released appears to be more informal. For example, the word "light" in the SNI document is replaced by "luminous body" in the unofficial material.

Another detail that sets one document apart from the other is that not all the occurrences recorded in the SNI report are in the unofficial documentation, which seems to have only the most relevant events. It seems that this is a preliminary text and another end to the same mission. Still in the SNI docs, there are illustrations of UFOs made by the military of Operation Saucer, based on their reports, in which it is possible to see even UFO crews and the evolutions made by the objects. Still, the expression "documenting to perpetuate", a slogan of the Center for Documentation and History of Aeronautics (CENDOC), a key organ in the UFO Disclosure process bythe Brazilian Government, has much to do with Operation Saucer's material.

As a result of the campaign "UFOs: Freedom of Information Now", created by the Brazilian Committee of Ufologists (CBU) and Revista UFO in 2005, which has triggered the main official releases, in 2009, the Cendoc issued a letter dated February 14th, 1979 from Major Protasio Lopes de Oliveira, at that time commander of the First Regional Air Command (1stCOMAR), in which the officer, at the request by his General Staff, sent to Brasília a file with 159 pages referring to UFO records in Pará , ranging from September 1977 to November 1978, all occurring in 1st COMAR's jurisdiction area. This is further evidence that the end of Operation Saucer in December 1977 did not determine the closure of official UFO surveys in the north of the country.

Protasio Lopes de Oliveira is a key figure in this whole story. It was he who determined the creation of the operation, which appointed the then captain Uyrangê Hollanda to command it and determined its closure - at least in the format in which it was conceived - shortly after Hollanda revealed to him a direct contact with a crew member of a cylindrical UFO of large proportions on the banks of the Guajará-Mirim River, which runs along one side of the island of Colares in mid-December 1977. His daughter, pedagogue Nahima Lopes de Oliveira Gonçalves, revealed to Revista UFO that her father eventually brought home, for family and friends to watch, footages of unidentified flying objects on the Amazonian rivers, obtained during Operação Prato. "You could see lights in every direction," she said. She also said that her father always believed in extraterrestrial life.

Figure 79. Another photo of the marks left by the landing of a UFO at the Jejú farm (Credit: Author's files).

The document sent by Oliveira is a compilation of 130 records of observations contained in 139 pages that were also released, along with other papers. They are duly registered cases with a registration number, description of place of occurrence, including geographic coordinates, date and time, as well as details of observed objects such as color, shape, size, movement, speed, altitude, distance, etc. The names of the observers are also included in the material, as well as illustrations of scenarios and UFOs, maps with the routes of the objects searched and dozens of photos showing spheres of light and disk-shaped objects. Another 20 pages with newspaper clippings of the time are in the archives of CENDOC, which is one of the major UFO cases of Brazil and the world.

The documentation available today on UFO research developed by 1stCOMAR, which includes Operation Saucer, came from various sources, official and unofficial. Due to its volume, it is a real puzzle to classify all these documents, identify duplicate pages and organize all the documentation, so that it is possible to reconstitute the steps of this historic military operation carried out in the Amazon, considered the greatest of all the military initiatives to do such work, all over the planet. The information is still fragmented because not all documents are believed to have been officially released, not even the film records cited by Hollanda in his interview to Revista UFO, which would amount to 22 hours in super 8mm and 16mm format films. In addition, unfortunately, some pages already released or leaked have virtually illegible parts - they are photocopies that have been decayed with the passing years and were later digitized.

What we have from the Brazilian Air Force (FAB) regarding Operation Saucer, through the CENDOC, are mission reports that occurred between September and December 1977, and its premature closure is still a mystery. After that period, surveys continued throughout 1978, but through smaller incursions, with fewer agents involved and for shorter periods. From this point, the records cease and in early 1979 a compilation of the main cases was made, which resulted in a folder that was sent to the Aeronautical General Staff, specifically on February 14th of that year. Assembling this puzzle is hard work, but in the end, it will be rewarding.

Figure 80. Humaitá Beach, site of the first military skywatches in Operation Saucer. (Credit: Author's files)

Figure 81. Machadinho Beach, Colares (Credit: Author's file).

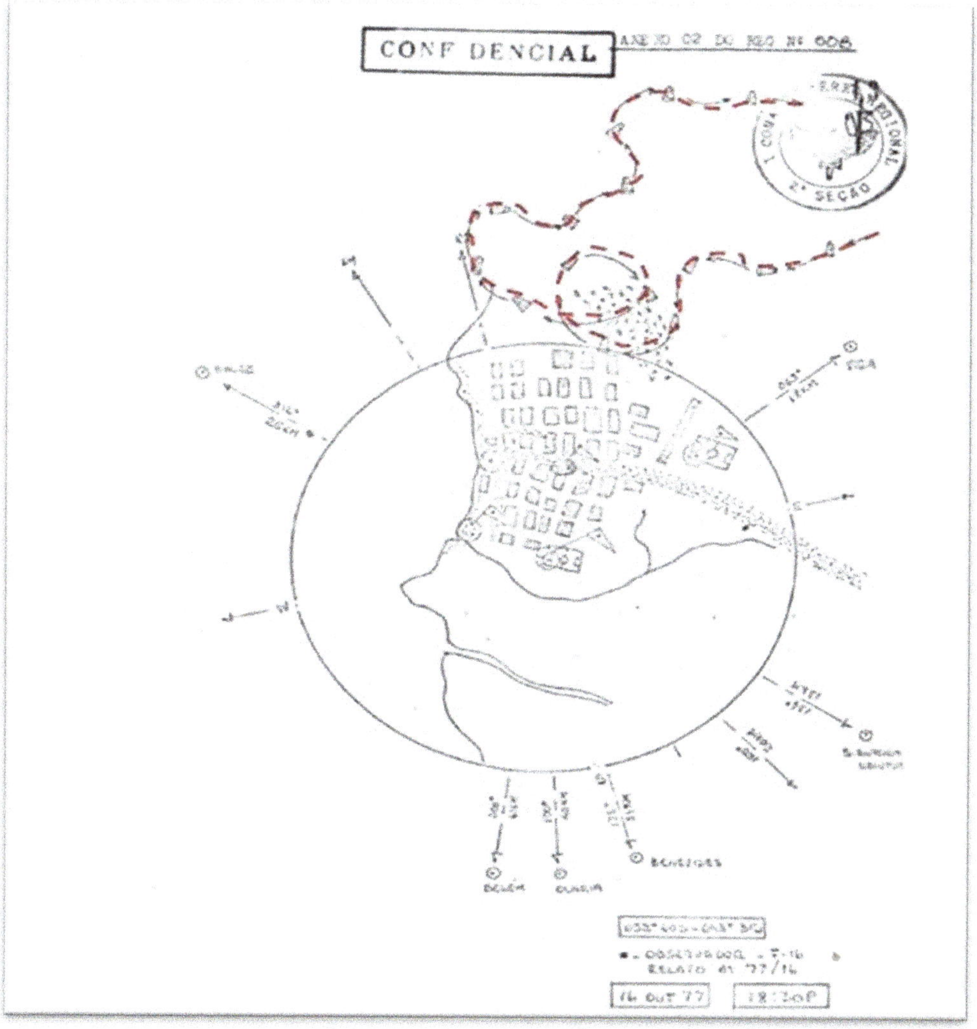

Figure 82. Map made by the Operation Saucer team showing, in red, the trajectory of the UFO seen on July 16, 1977. (Credit: Author's file).

Figure 83. Based on the information in the previous document, it was possible, through recent aerial photographs (Google Maps), to reproduce the trajectory of the object and the position of the observers at the time. In red, the camping sites of the Operation Saucer soldiers who witnessed the apparition are highlighted. In orange, other observers. (Credit: Author's file)

Chapter 16

The Discovery of a Peruvian CIA Spy

This incredible episode shows that there were international interests, mainly from the US, in the events that were happening in Pará state. It strengthens the claim by Brazilian ufologists that US intelligence agents and military personnel, were present in Brazilian territory to steal information and documents, threaten witnesses, influence the actions of the Brazilian military and try to discover, and who knows, capture the "*Chupa-Chupa*" technology.

According to journalist Carlos Mendes, who was the protagonist of this episode, a Peruvian CIA agent, disguised as a journalist, infiltrated the editorial office of the newspaper *O Estado do Pará*.

The first time he (the secret agent) showed up at the *O Estado do Pará* newsroom was not to give an interview, but to ask for a job. He said he was from Lima, the capital of Peru, and needed to work because he had come to Belém to study at the Federal University of Pará (UFPA). He said his name was Juan Mendoza Ibardo and that he had lived in Washington, in the United States. Carlos asked if he had any experience in reporting and he said he liked the police department. Carlos told him to talk to Biamir Siqueira about it. From then on, Juan started to stick with Siqueira and "Riba", but although he spoke decent Portuguese, he would have to get to know the city, the ins and outs of covering police stations, the Forensic Medicine Bureau (IML) and hospitals.

Carlos did not pay any more attention to Juan, who seemed to be just another applicant for a job at the newspaper, like so many others who were looking for a place in the sun in the profession. Siqueira, however, commented to Carlos three times about the strange movements of the Peruvian, who would insinuate himself, brazenly get into the newspaper's car and, without being an intern, went on patrol with the police reporters through the neighborhoods. So far, nothing out of the ordinary. He wanted to familiarize himself with the news coverage. The problem was that he seemed to be obsessed by the newspaper's film development lab.

It was November and the newspaper photographers, including Riba, already had unpublished material about the "Vampire Lights" that appeared in the countryside and in the capital. If Juan had wanted to be a police reporter before, he would now ask to develop films and work in the studio. Riba began to suspect. He decided to put a sign on the door, saying that "strangers are not allowed" to enter the premises.

"Walmir Botelho needs to take action and issue an order also prohibiting Juan from entering the newsroom and even the newspaper building," said Siqueira to Carlos Mendes. So, Carlos started paying more attention to Juan. One day, Carlos was going to a restaurant in the square behind the newspaper, when he noticed Juan talking with Captain Hollanda, who was walking down the street, as he was heading toward the newspaper building. When Juan saw Carlos, the Peruvian looked up and walked away from Hollanda and headed toward a bus stop. Hours later, Carlos told Siqueira about that strange encounter between Hollanda and Juan. *"That's why Walmir should keep an eye on him and kick this guy out of the newspaper. He won't be allowed in the reporter's car anymore; I've already cut him off."* To make matters worse, Riba had also seen Juan talking in a bar with a person who was from the SNI.

According to Carlos, Juan was a bit of a strange guy, although he was sometimes playful. He imitated the way his colleagues at the newspaper talked and the way they walked. A short, stocky, dark-skinned man in his thirties, Juan wore a faded cap that he took off when he came into the newspaper. He said he had worked in Cuzco, Peru, for a newspaper and that he wrote police reports. One day Carlos asked him if he had any written and signed reports to support his intention to work in Belém. He changed the subject, saying he would bring some newspaper clippings with articles he had written. He never did.

Walmir asked several times who was that person with the *"Bolivian Indian face"* who, from time to time, would show up at the newsroom, work on the typewriters or take the daily newspapers out of the collection. In fact, no one knew who he was or where he lived. When asked, the Peruvian replied that he lived in the house of a Venezuelan who had come to Belém to study and take an exam to the Federal University of Pará. He never gave the name of the street but said only that his address was in the Guamá neighborhood, one of the most populous in Belém.

Figure 84. Photographer José Ribamar dos Prazeres, "Riba" [left], a colleague of journalist Carlos Mendes, had a terrifying experience with the "Chupa-Chupa" phenomenon. (Credit: book "Luzes do Medo". MENDES, Carlos).

Siqueira, Riba and Carlos already had strong reasons to suspect that Juan was a spy in the newspaper and was working for some person or entity. He was probably someone "planted" by the SNI, or even by the US Central Intelligence Agency (CIA). "*I say this because another strange fact sharpened my perception regarding Juan. Riba told me that after he and Siqueira had come from Bahia do Sol to develop the films with the frightening appearances of the unusual lights, Juan, who had been missing from the newspaper's offices for about five months, reappeared with a different appearance. He was shaved, his hair was cut short and without the usual cap with the inscription "Machu Picchu"*", said Carlos. One day, he tried twice to enter the photo development studio while Carlos was busy examining films, using the excuse that he needed to talk to me about something. "*I told him to wait for me in the office, because no one was allowed to enter the laboratory except those who worked there*", said Carlos.

But Juan seemed uncomfortable and curious when he heard the newspaper reporters talking about Biamir Siqueira and Riba finally having proof of the existence of flying saucers in *Baía do Sol* and *Mosqueiro* and that this was an exclusive "bombshell" that the newspaper was going to publish.

On the same day that Juan tried to enter the laboratory and had been warned by Carlos Mendes that people outside the newspaper were not allowed in that area (*"If Walmir finds out about this, he won't like it. This Juan is going too far"*, the reports said), Carlos, Biamir and Siqueira took a taxi in front of the newspaper and went to Pagode Chinês, an entertainment venue in the Cremação neighborhood. Osvaldo Rodrigues, the owner of that famous nightclub, which is very popular with tourists visiting Belém, welcomed them kindly as always. They sat at a table a little away from the dance floor and the noisy music. That night, they decided they would find out who Juan was, what he really did in Belém, "where he lived and why he always appeared in our path while we were writing about the mysterious lights that were terrifying the entire northeast of Pará". The friends came up with a strategy that seemed foolproof. As a good police reporter that he was, Biamir had friends in the Civil Police. One of them, a trusted investigator and the reporter's source, would be called in to "stick it to" the Peruvian and find out, after all, who that mysterious guy was.

Two days after that conversation, the person who appeared in the newspaper that afternoon, during Carlos' Saturday shift, was Captain Hollanda. That day, he showed up in military attire, unlike other times when he had dressed as a civilian. Hollanda came to Carlos and asked for Walmir Botelho. Walmir's office, surrounded by glass walls, was at the back of the newsroom. Walmir was off duty that Saturday and had traveled to his hometown, Maracanã.

Since the newspaper editor was not in the office, Captain Hollanda saw another "target": journalist Carlos Mendes.

The captain pulled up a chair and sat next to Carlos while he wrote some news, looking at the notes on the table. *"These reports of yours are taking a direction that is not good for anyone. People are confused,"* Hollanda pondered, in a low voice, without showing arrogance. His manner was always the same. Polite, but firm and direct in his words. He told Carlos that Brigadier Protásio Lopes de Oliveira, the commander of the Brazilian Air Force in Pará, had deep respect for the newspaper *O Estado do Pará* and its reporters — even those who were identified by P2 and SNI as "communists". According to Hollanda, the brigadier was feeling a bit sad about the sensationalism from the press. He complained that the "men" in Brasília were demanding that the staff at the 1st COMAR in Belém take a more incisive stance towards the newspapers' conduct in Pará. The captain used an argument typical of someone who would like nothing to be published other than what had already been published. Hollanda said according to the information he had received only our newspaper treated the apparitions differently, trying to influence readers. He was also ironic when he said that Carlos wrote better about land and street protests than about the "Chupa-Chupa". That was yet another provocation to the tense relationship we had always had. Despite everything, Carlos respected the captain, but did not fear him. Once, in Colares, during a conversation with Carlos in the office of Mayor Alfredo Ribeiro Bastos, he classified the three daily newspapers in Belém in the following way: "*The Province of Pará is right-wing, says it is ours, that is, sympathetic to the military regime, but sometimes it writes nonsense about the Chupa-Chupa. The Liberal newspaper supports us and has more balanced news. As for you, from O Estado do Pará, besides being socialists, I don't know, you also like to create sensationalism*". For Hollanda, of the three newspapers, the only one that used the expression "flying saucer" when talking about the lights in Baía do Sol, Colares and Vigia was O Estado do Pará. The language, he defined, was "*scandalous and created panic*" among the population. "*You want to sell newspapers, and you sell very well, but not even you know if what you write is the truth about what is really happening*," said the captain. "*Long live the flying saucers*," he said ironically.

Carlos asked the captain what the truth was. Hollanda did not answer and, without saying anything else, got up and left. Carlos did not ask about his meeting with Juan a few days ago, on the street next to the newspaper. The Peruvian, however, either because he wanted to make an excuse or because he suspected that Carlos was suspicious, had told him on another occasion, when asked if he knew the captain, that he had approached him "just to ask for a cigarette".

The three reporters, with the information given by the police agent, "squeezed" Juan. He said he was not Juan Mendoza Ibardo, but Alejandro Montoya. He denied being a CIA agent but confessed to having worked as a private detective in a city near Washington. What's more, he had come to the state of Pará looking for an ex-wife who had fled Goiânia for Belém with their 5-year-old son. He told another story about the encounter with Hollanda on the street next to the newspaper. He had met Hollanda at the *Bar do Parque*, but he did not know he was in the Brazilian Air Force or that he was investigating the mysterious lights.

Regarding the various attempts to enter the newspaper's photo development studio, he admitted that he wanted Riba's material taken in Baía do Sol but claimed that he intended to negotiate with Riba himself to sell the photos to Europe and the United States, because he was sure that the photojournalist would become rich and his name would be mentioned all over the world.

In any case, after the episode in the nightclub and the confession of ideological falsehood, on the following days, "Juan", "Alejandro" or any other name he had, did not appear again in *O Estado do Pará*, let alone in the police pages of the newspaper where he supposedly intended to work. He disappeared from Belém.

At for his real address in Marambaia, a neighbor informed Siqueira that the Peruvian paid the rent, handed over the keys to the owner of the property, and left "*with only his clothes in a bag*". His incredible adventure in Belém turned into a bitter nightmare. He left without much ado. Actually, without the photographs of the bizarre lights.

Figure 85. Journalist Carlos Mendes with Thiago Ticchetti, holding the book "Luzes do Medo", by Carlos Mendes. (Credit: Author's file).

Chapter 17

The Patterns and Characteristics of the
"Chupa_Chupa"

The *"Chupa-Chupa"* Phenomenon presents peculiar characteristics that impress both researchers who are already accustomed to UFO cases and skeptics of the phenomenon, who avoid addressing facts related to incidents that occurred in northern and northeastern states.

These peculiar features that surround these occurrences demonstrate that something unusual, intelligent and whose origins are beyond our knowledge were operating in populated areas of the Amazon.

Evolution of the Phenomenon

The first interesting aspect that we can mention about the *"Chupa-Chupa"* Phenomenon would be concerning the evolution of the phenomenon. The attacks occurred very sporadically in 1976, gradually increasing until April 1977, with the occurrence of Crabs Island, spreading throughout the state of Maranhão, later reaching Pará and then Amazonas, in Brazilian territory. Unconfirmed information suggests that the phenomenon also occurred in nearby countries, Guiana and Venezuela. This evolution by region occurred in a standardized way, as in a scientific mapping. This already gives the phenomenon an intelligent activity.

Recurring Details

When checking the *"Chupa-Chupa"* cases, several recurring details are observed.

Almost all incidents occur at night, and refer to lights that suddenly appear, illuminating the entire place near the encounter spot. Practically all cases reported by riverside dwellers, whether in Maranhão, Amazonas or Pará, contain a description of the intensely illuminated object. The light is usually so strong that it prevents the source of the light from being seen, that is, the object itself.

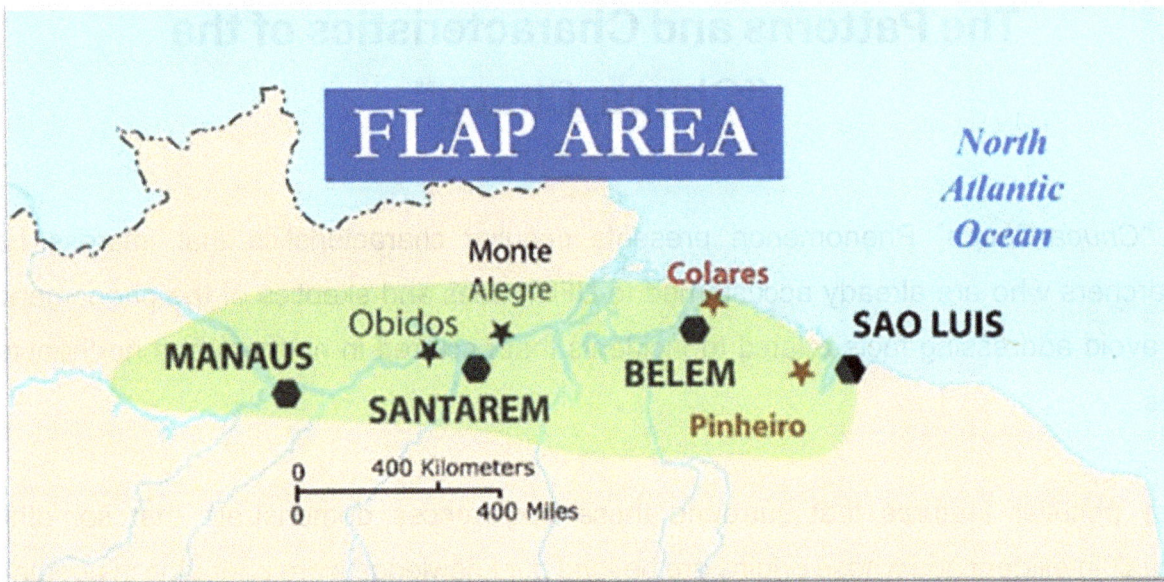

Figure 86. The map above shows the vast area that was the scene of the "Chupa-Chupa" phenomenon. The phenomenon occurred in waves or bands, exactly as it occurs in aerial mapping. (Credit: www.fenomenon.com.br).

When the object suddenly approaches, the "attack" occurs, which is almost always described in the same way: the object projects a beam of light that paralyzes its victims, stopping them from moving any part of their body, or even screaming for help. In addition to the paralysis, the victims describe that they felt unbearable pain and were soon overcome by profound weakness, as if they were being drugged. Under these conditions, most of them were still able to observe a second beam of light that hit the women above the left breast (in most cases) or the men, at the height of the neck. The beam of light produces a long, straight, extensive and wide mark, as if something had happened to the the victims' skin. In all documented cases, between two and three holes were found, as if needles had penetrated the skin at that location. The difference is that, when pressed, they did not disappear.

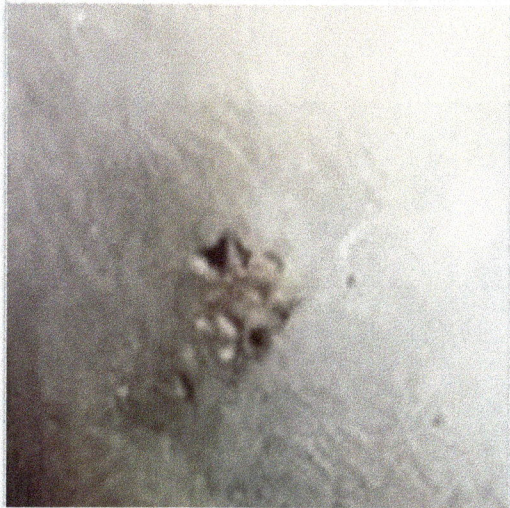

Figure 87. A mark by the "Chupa-Chupa" attack on Aurora Fernandes. (Credit: www.fenomenon.com.br).

The burns resulting from the attacks were very different from conventional ones in practically all cases. Whereas conventional burns take around 96 hours to heal, the burns caused by the "Chupa-Chupa" healed immediately.

After the incident, the victims of the "Chupa-Chupa" complained of dizziness, body pains, tremors, lack of energy, destruction, defeat, hoarseness, hair loss, peeling of the damaged skin and frequent headaches. These symptoms were confirmed by doctors.

In general, the affected area was permanently devoid of hair. In addition, the affected individuals had permanently low immunity and all of them became ill easily.

Flying Objects Observed

As for the objects observed, a characteristic pattern was observed in all cases. In general, the objects were observed approaching the affected areas from either the sky or the ocean, or from the river.

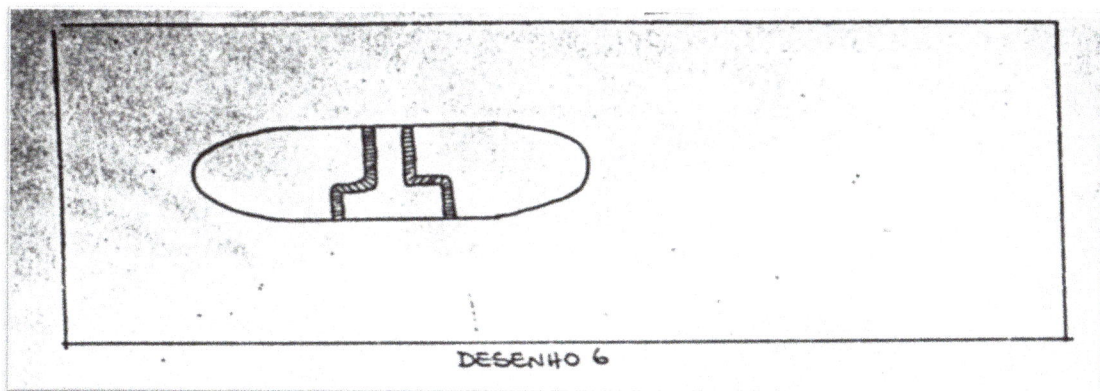

Figure 88. Militaries conducted nighttime vigils, documenting "lights" with a "cylindrical, almost conical shape," moving at high speed. (Credit: Author's files).

The cases were predominantly nocturnal, with rare daytime exceptions. Eight forms of objects commonly observed in different regions were identified. Some of these objects had portholes and in some specific cases, the presence of crew members approximately 1.5m tall was seen.

Figure 89. Eight types of objects were officially recorded by the FAB during the "Chupa-Chupa" phenomenon. (Credit: Author archives).

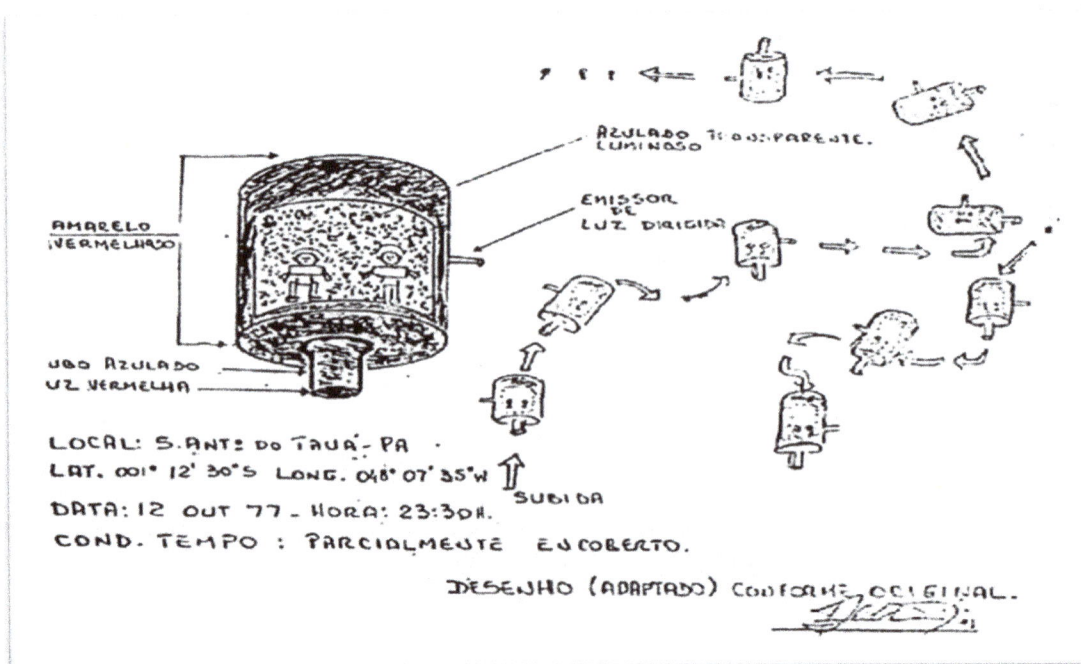

Figure 90. Cylindrical UFO with humanoid beings inside it. (Credit: 1st COMAR files).

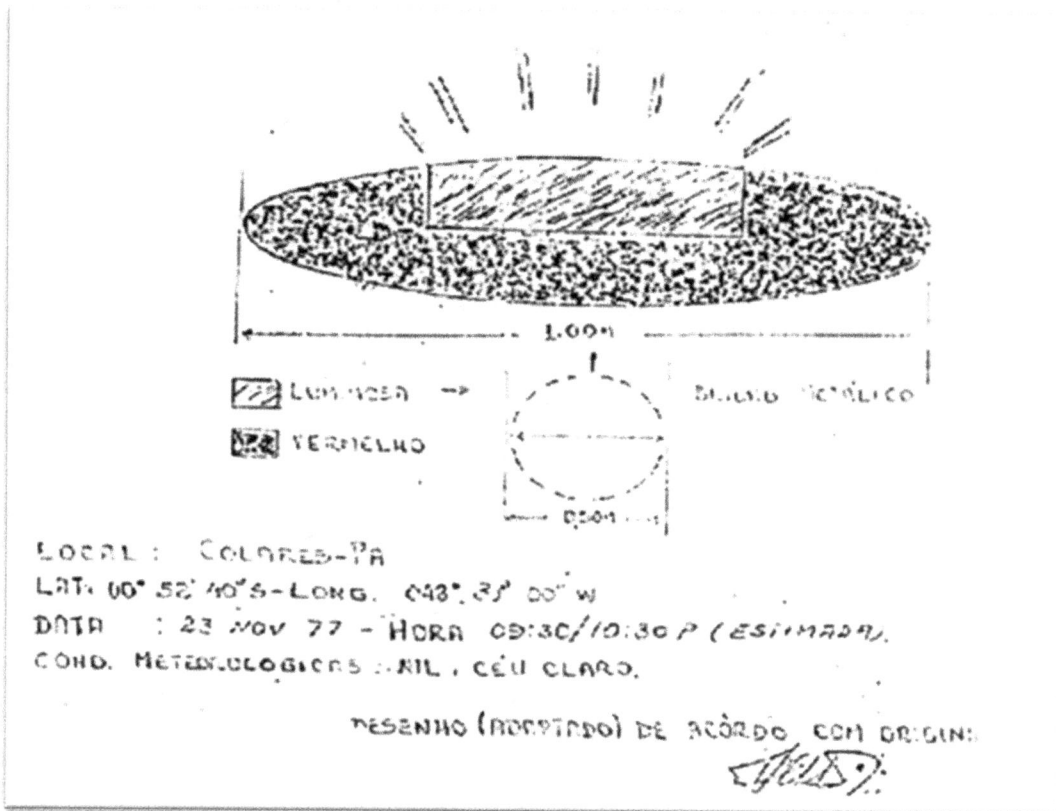

Figure 91. A large UFO, 100 meters long, which had a yellow light on top. (Credit: 1st COMAR files).

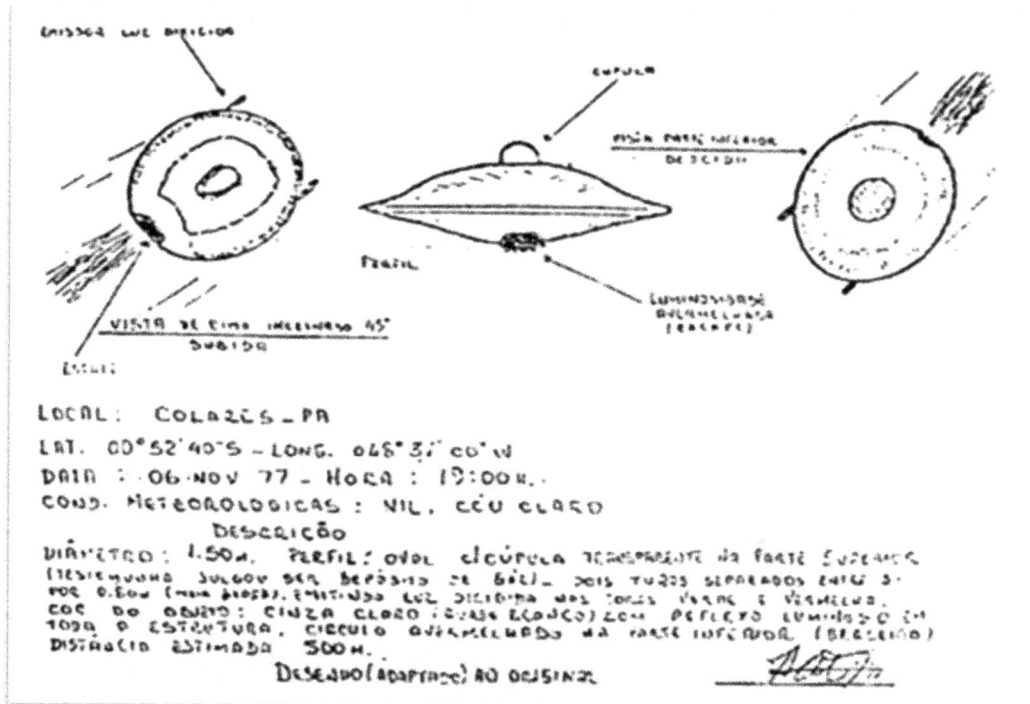

Figure 92. A classic disc-shaped UFO, with two antennas on the front that emitted the rays. (Credit: 1st COMAR files).

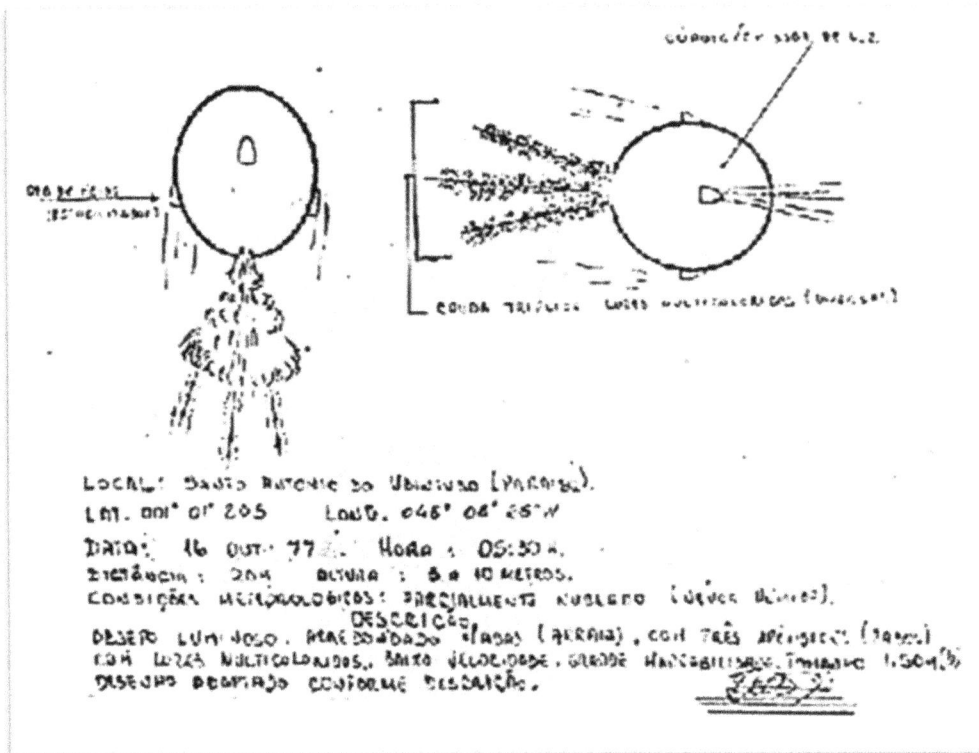

Figure 93. A UFO shaped like a stingray because it had three colored tails. (Credit: 1st COMAR files).

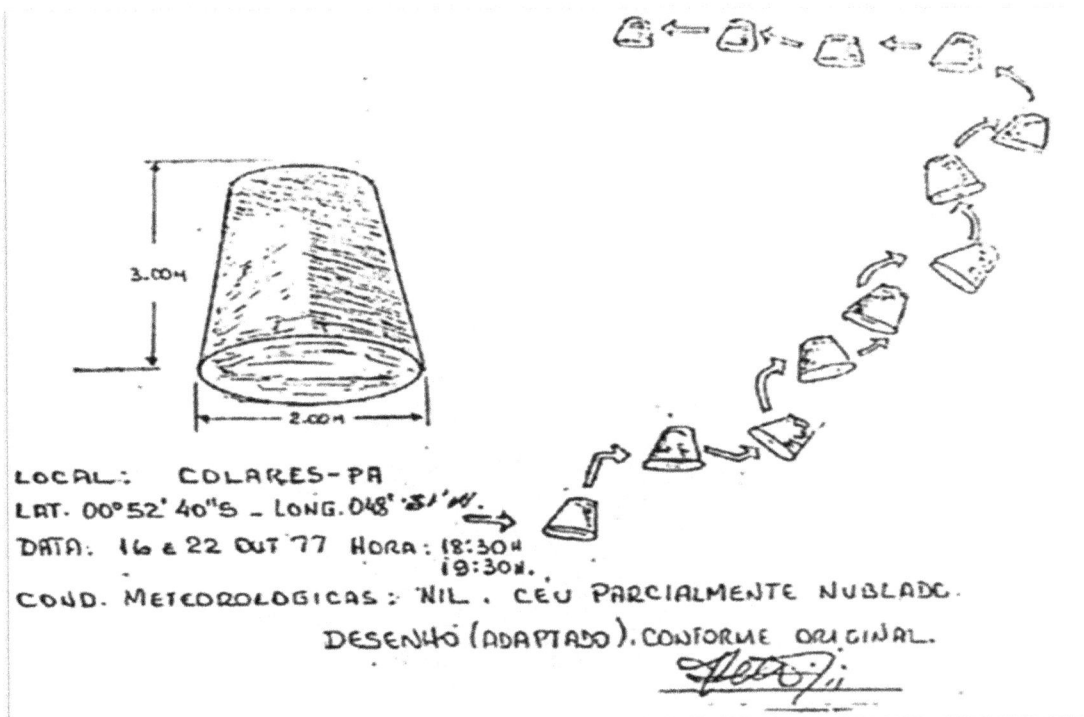

Figure 94. A cup-shaped UFO was flying in irregular movements. (Credit: 1st COMAR files).

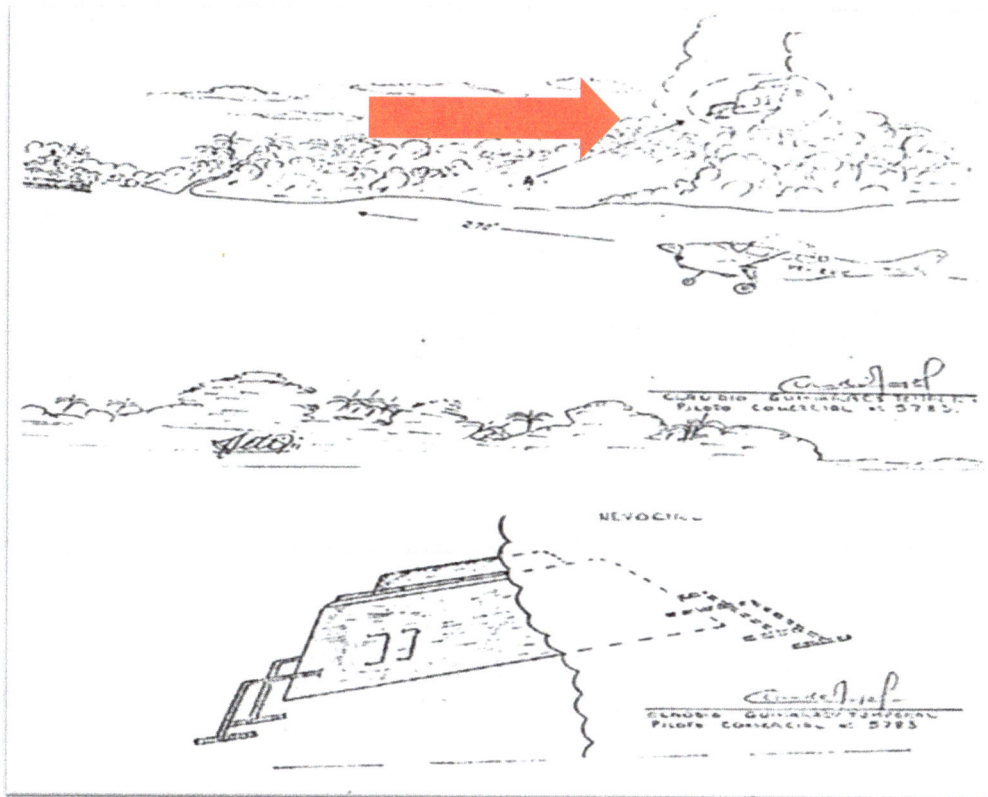

Figure 95. A UFO was seen, hidden behind a cloud, by a pilot. The object had landing gear and was shaped like an inverted bowl. (Credit: 1st COMAR files).

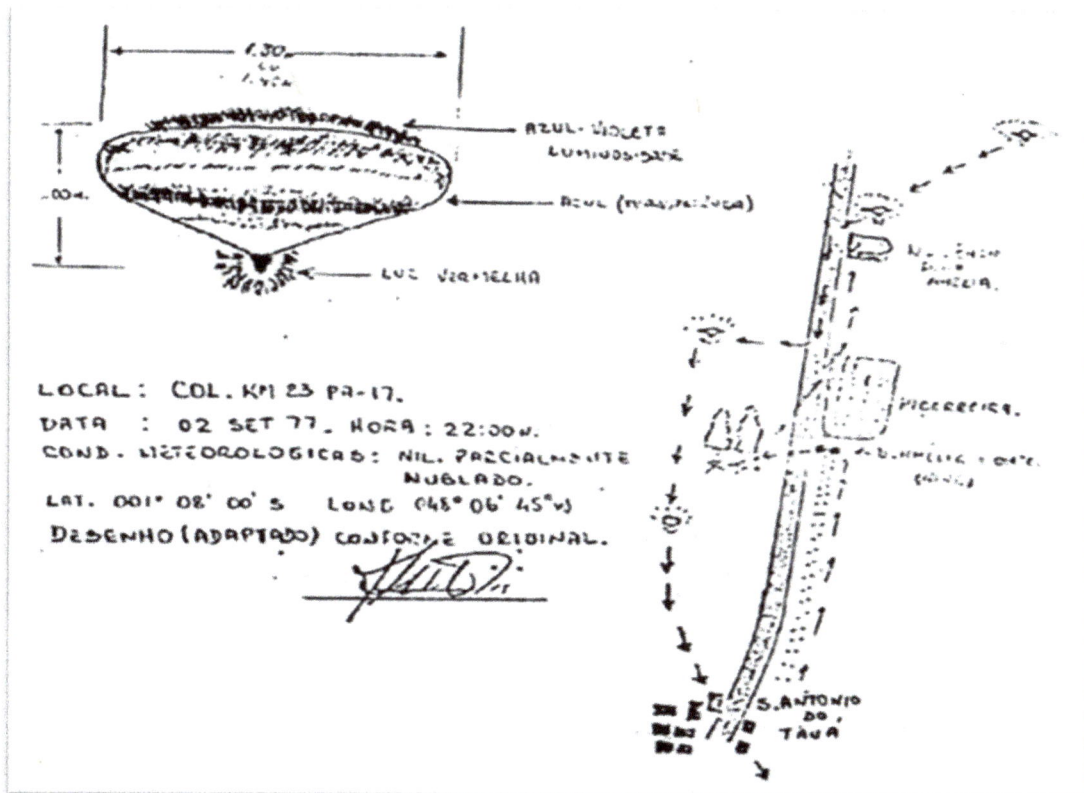

Figure 96. A UFO in the shape of a top was seen by an entire family. (Credit: 1st COMAR files).

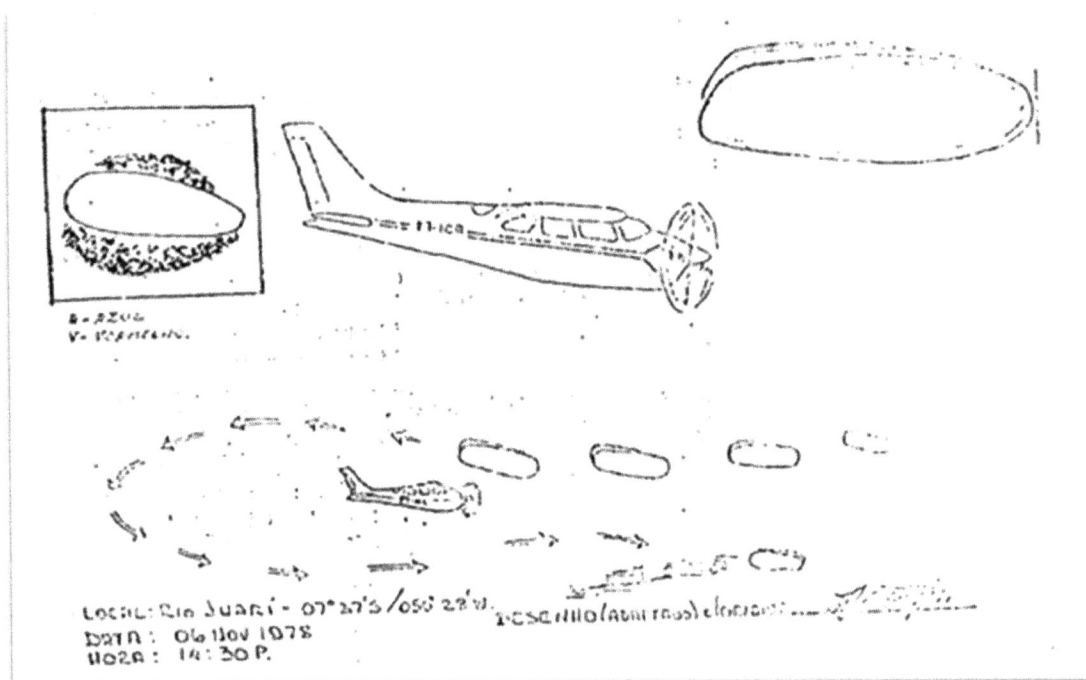

Figure 97. A "Tic-Tac" UFO was reported by a pilot during the "Chupa-Chupa" phenomenon, but an object of this shape was also reported by another pilot (Credit: 1st COMAR files).

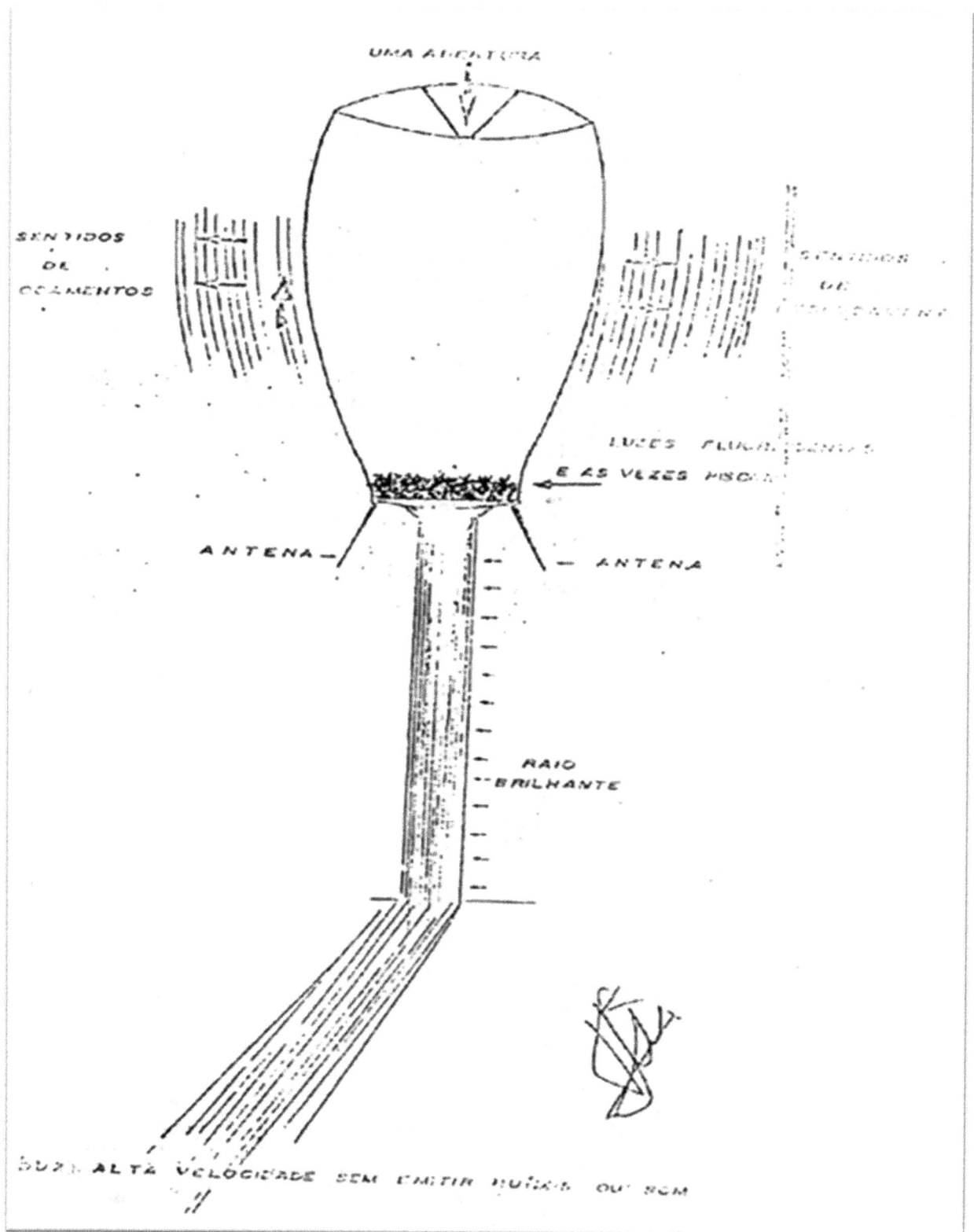

Figure 98. A UFO that emitted a ray that made a "straight curve" until it hit one of its victims. (Credit: 1st COMAR files).

Chapter 18

Main characters of the "Chupa-Chupa" phenomenon

There were more than 250,000 witnesses to the "Chupa-Chupa" phenomenon. There were more than 2,000 victims of these objects. There were two missions by the Brazilian Air Force (FAB) to investigate the incidents. And thousands of other people who witnessed what happened in at least three Brazilian states. Here I present some of these people.

MILITARY

Sgt. João Flávio de Freitas Costa

Born in Belém. At the time of the Operation, he was a first sergeant and served in the second section (A2) of the 1st COMAR. A meteorologist graduated from the Air Force Specialist School. During the "golden age" of the operation, he was second in command of the military team, which was headed in the field by then Captain Hollanda. The officer arrived as a skeptic and was gradually convinced by Sergeant Flávio to change his mind. As the Operation progressed, not only did Holland begin to believe, but also reported possible experiences. Flávio had versatile skills, including drawing and photography. He created the vast majority of the sketches, maps and drawings produced during the operation, originating from his own observations or through reports from natives interviewed in the places where the phenomenon occurred. Flávio read and spoke English reasonably well, as well as basic Spanish. In June 1967, he was sent to take a counter-espionage course at SOA (School of the Americas), in one of its best-known headquarters, Fort Gulick, located in the Panama Canal zone. During the Cold War, the SOA trained almost all of the military personnel in Latin America who participated in coups d'état, in the face of any possibility of left-wing politicians coming to power in most countries on the continent.

Cel. Uyrangê Bolivar Soares Nogueira de Hollanda Lima

In 1958, at the age of 17, Uyrangê Hollanda joined the Brazilian Air Force (FAB), where he remained until March 10th, 1992, when he was retired. For 24 years, he was a member of the A-2, the Brazilian Air Force intelligence service. He was fluent in English and French. He was a finance officer, pilot, paratrooper, and a jungle operations specialist. For 5 years, he was in charge of an anti-guerrilla unit in southern Pará. However, it was in 1977 that Capt. Hollanda experienced one of his greatest adventures, which brought him international renown. Cap. Hollanda received orders from Brigadier Protásio Lopes de Oliveira, commander of the 1st COMAR in Belém, and from the Chief of A-2 Col. Camilo Ferraz de Barros to command an operation that was already underway, but still unnamed, regarding strange lights that invaded our airspace and disturbed the residents of some areas of Pará, where beams of light were fired, causing burns to their victims and terror in the population. It was he who named the operation "*Operação Prato*" (Operation Saucer). According to his statements, he led a team of five agents, all sergeants, and went to the affected areas to investigate. Although he believed in the possibility of extraterrestrial life, at first, Cap. Hollanda believed that they could be guerrillas or even smugglers in search of the local natural resources. The military team interviewed witnesses, produced reports, photos, films and had their own experiences. Although Operation Saucer officially took place between October and December 1977, we know that Capt. Hollanda, along with other military and civilian personnel, continued to investigate the strange phenomena on their own. In 1997, already retired, the then Col. Hollanda decided it was time to share his experiences during that period. He gave an interview to ufologists A.J. Gevaerd and Marco Petit where he recounted some fantastic events that he had experienced and learned about. He believed that he had a kind of "chip" implanted under the skin of his arm and that he was visited by extraterrestrial beings in his home, even years after the end of his military mission. On October 2nd, 1997, shortly after the interviews, Uyrangê Hollanda hanged himself in his bedroom, located in a condominium in Iguaba, a city on the coast of Rio de Janeiro. He left behind a wife and children from his two marriages.

Brigadier Protasio Lopes de Oliveira

He was the commander of the 1st COMAR in 1977. He is said to have given the order for the military to investigate the strange phenomena that were occurring in the municipalities of Pará, where strange lights were attacking residents and causing panic among the population. According to statements by Col. Hollanda, Brigadier Protásio was interested in Ufology and believed that Operation Saucer was carried out for this sole reason. According to statements by Brigadier Protásio's daughter, Nahima Lopes de Oliveira, he would take the videos produced by the military team home to watch them. On one of these occasions, she herself had the opportunity of seeing them. Brigadier Protásio ordered the official closure of Operation Saucer on December 5th, 1977, for reasons that remain unknown to this day.

Colonel Camillo Ferraz de Barros

Col. Camilo Ferraz was the Chief of the 2nd Section (A-2) of the 1st COMAR. Col. Camilo was, therefore, Cap. Uyrangê Hollanda's the immediate hierarchical superior. On some occasions, he participated directly in investigations and surveillances together with the military team.

Sgt. Alvaro Pinto dos Santos

He was one of the first military investigators to go to the city of Colares in 1977. He was one of the intelligence agents who participated in Operation Saucer, but every now and then he joined in the investigations along with Col. Hollanda's team. He was 42 years old at the time of the investigations. At the age of 64 in 1999, he is said to have told foreign researchers about the population's mental state: "*People were really scared. They were so terrified that they didn't fish. They wanted guns for shooting. We had to explain to them that they couldn't shoot at the UFOs or things could get worse*".

Subofficial Moacir Neves de Almeida

At the age of 34 and with the rank of Second Sergeant at the time of Operation Saucer, he was part of the military team that carried out the first raid towards Colares, which began in the second half of October 1977. In military reports he was identified under the codename "Luciano". The first Operation Saucer mission lasted 21 days and was divided into two periods, the first began on October 20th, 1977 and lasted until October 27th, 1977. On that last day, the team returned to COMAR, going back to Colares on October 29th, 1977 and staying there until October 11th, 1977.

Lieutenant Colonel Isaac Samuel Benchimol

In 1977, as a Major Doctor in the Brazilian Air Force, he was head of the health section of the 1st Air Zone. He participated in one of the Operation Saucer skywatches in January 1978, being invited by then Captain Hollanda.

Carlos Eduardo Gomes de Sá Kuster

When he joined the Air Force, he was accepted into the Mechanical Engineering course, which unfortunately he was unable to complete due to his duties as a lieutenant in the Brazilian Air Force (FAB). He took a pilot course at the FAB in the city of Natal – RN, flying a T-25 in 1975. He later took a course to fly a helicopter in Santos and served another two years in the city of Belém (PA), participating in the RADAM Project in the Amazon, and flying a helicopter (UH-1H) that he called "sapão" (big frog). He chose to become a reserve officer and began his career in commercial aviation by flying a TAM Bandeirante, later going on to fly a Boeing 737-200 for the former Vasp airline. His participation in Operation Saucer occurred on November 1st, 1977, when he piloted a helicopter (UH-1H) from Belém to Colares, carrying a military team made up of five agents, among them Colonel Camilo Ferraz de Bastos and then Captain Uyrangê Hollanda. He died as a passenger in a airplane crash in 1982.

CIVILIANS

Doctor Wellaide Cecim Carvalho

Born in the city of Nova Olinda do Norte, she moved to the city of Santarém when she was 12. At the age of 21, right after graduating, she was appointed on December 10th, 1976, to be responsible for the health unit in the city of **Colares**. She specialized in psychiatry. She had contacts with the first people 'attacked' by the *"Chupa-Chupa"* phenomenon. According to her statements, she was extremely skeptical at first, but began to notice changes in the victims that were inexplicable to medicine. She was pressured by the military personnel of Operation Saucer to convince the people affected by the lights that they were victims of a mass hallucination and to prescribe tranquilizers. She refused. She stated that in October 1977 she had seen a cylindrical metal object (slightly conical), apparently measuring 3 x 2 m, which was making elliptical movements. This sighting left a deep impression on her.

Father Alfredo de La Ó

Alfredo de La Ó was an American, the son of Mexican immigrants. He was born in El Paso, Texas on October 18th, 1932. He was the first parish priest of the city of Colares and took over the parish of N. S. do Rosário on March 14th, 1976. He was also an otorhinolaryngologist. Father Alfredo was an eyewitness to the attacks of the lights suffered by the population and to the military presence on the island during Operation Saucer. He was also the protagonist of some sightings. According to information, the priest had a very strong temperament and walked around the city of Colares carrying a gun. He remained as priest in Colares at least until December 8th, 1978. He died on an unknown date and from unknown causes. He was buried in the cemetery of Castanhal, in Pará sate.

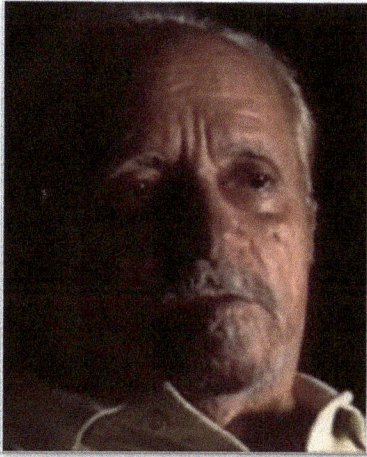

Ubiratan Pinón Frias

He was a commercial pilot for Belém Aeroclub. Thanks to his friendship with Capitain Hollanda and Sgt. Flávio, Pinón participated in several skywatches with the military during and after the end of Operation Saucer. In addition, he provided logistical support to the military, as they used his private plane for reconnaissance flights in the locations where the camps would be set up. He died at the age of 74 on August 5th, 2014 after being hospitalized for more than two months in a hospital in Santarém, Pará, due to respiratory failure and multiple organ failure.

Bob Pratt

Robert Vance Pratt was born on August 12th, 1926. A journalist and UFO researcher, he retired in April 1999 after 48 years as a reporter and editor for various newspapers and magazines. He attended the University of California, Berkeley, an American University. He was a skeptic on the subject of UFOs, but in May 1975, as a reporter for the National Enquirer, he came to believe that they could be real after interviewing more than 60 people in one week who had seen UFOs. From that moment on, the subject became a major interest in his life, and he has since interviewed more than 2,000 people who have had experiences. He traveled throughout the United States and Canada, Mexico, Argentina, Bolivia, Brazil, Chile, Puerto Rico, Peru, Uruguay, the Philippines, and Japan, looking into reports of UFO sightings. During the six years he worked as a UFO reporter for the Enquirer, he traveled to Brazil four times. He discovered that UFO encounters were often more hostile and harmful here than anywhere else in the world. As a result, after leaving the Enquirer in 1981, he began investigating reports on his own and traveled to Brazil 10 more times, most recently in 2003. He wrote the book "UFO Danger Zone." He personally investigated the incident that occurred on Crabs Island in Maranhão and, together with Capt. Hollanda, the events of 1977 and 78 in the regions affected by the strange phenomenon of the "*Chupa-*

Chupa". In recognition of his research in Brazil, on May 3rd, 2003, at a conference in the city of Curitiba, he received a diploma as "Honorary Brazilian Ufologist." In more than 28 years as a UFO researcher, he never saw a UFO. He died on November 20th, 2005. He was 79 years old.

Rosil de Oliveira

Rosil de Oliveira was the owner and pilot of the vessel that was used by Captain Hollanda for all military incursions carried out in the waters of Marajó Bay and the rivers of the Amazon region. He participated in all military actions at sea and became a great friend of Captain Hollanda. At the request of Captain Hollanda himself, he kept quiet about it for almost 40 years after the end of Operation Saucer.

Carlos Mendes

Carlos Augusto Serra Mendes was born in Pará on December 19th, 1949. He has been a professional journalist for over 40 years. He is a correspondent for "O Estado de São Paulo" in Pará and for "O Correio", the newspaper in Carajás. In 1977, he worked for the newspaper *O Estado do Pará* and was assigned to investigate the sightings of strange lights that were attacking the inhabitants of some towns in Pará. Despite having had direct contact with dozens of witnesses and with the military personnel of Operação Prato, he stated that he had not observed any of the phenomena at that time. He is the author of the book "*Luzes do Medo*" (Lights of Fear).

Figure 99. Thiago Ticchetti, George Knapp and Carlos Mendes in Belém, in 2024. (Credit: Author's files).

MOST KNOWN VICTIMS
Aurora Fernandes

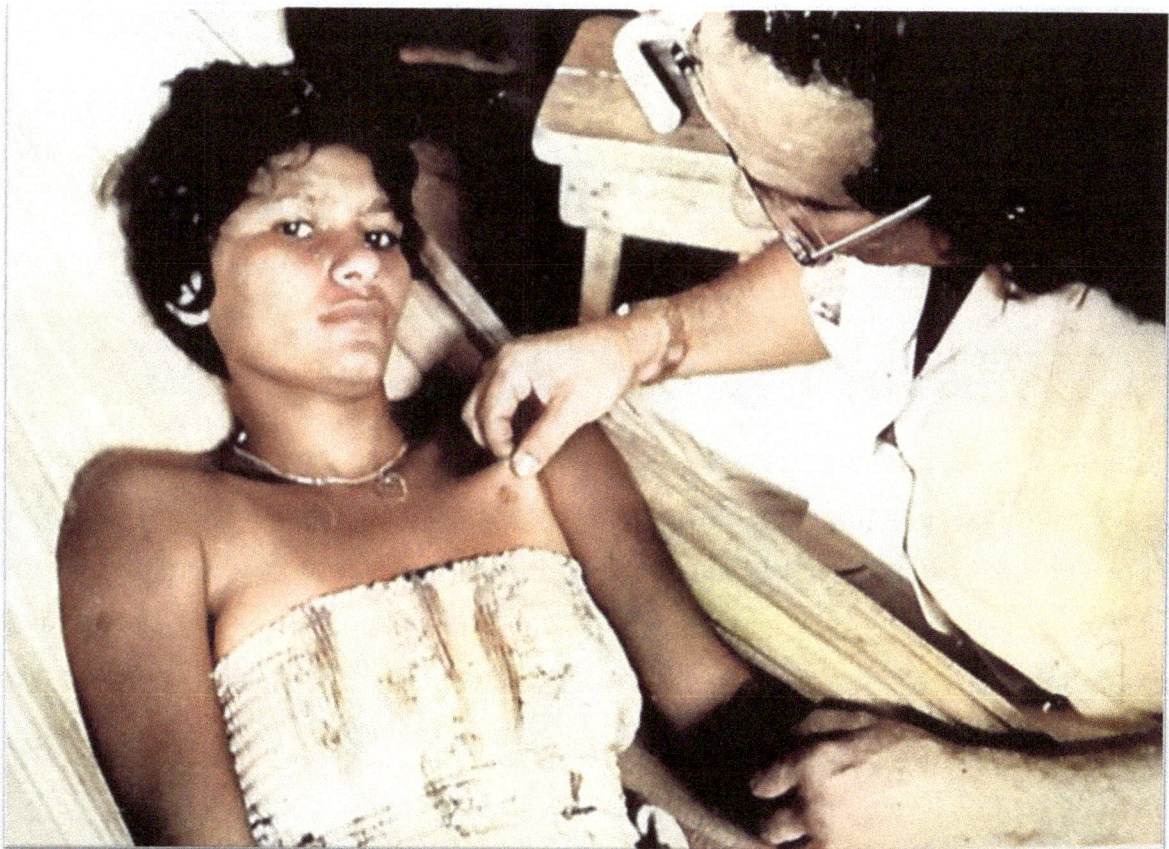

Aurora do Nascimento Fernandes, then 18 years old and living in the Jurunas neighborhood, reported that on November 17th, 1977, at around 9:00 p.m., while washing dishes in the backyard of her house, she was attacked by a strong current of cold wind and felt a terrible fear as if she were being enveloped by something strange. She said: "*I was terrified. I called my mother, who was already lying down like the other residents in the house, but before she arrived, a strong red light enveloped me, leaving me dazed. At the same time, I felt very fine punctures being made in my chest, and I fell to the ground, unconscious.*" Her mother, who attended to her, noticed that a colorless liquid was coming out of the wounds and said: "*It was clear water and smelled like ether*." After being treated at the hospital, she returned home and had severe headaches and was unable to walk, because her legs were weak.

Newton de Oliveira Cardoso, "Lieutenant"

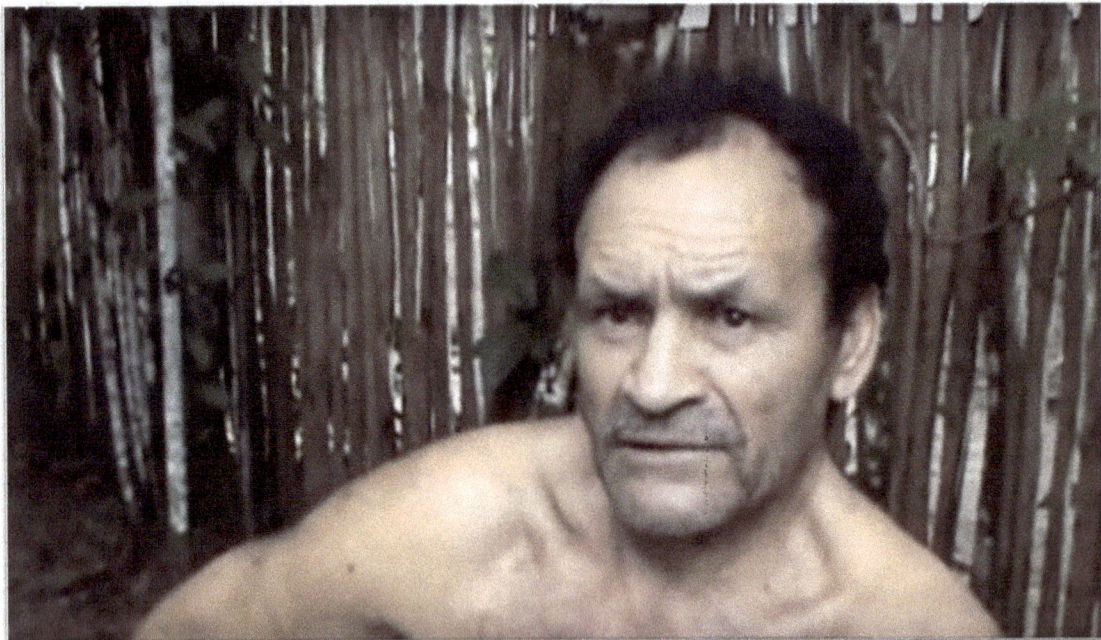

At the age of 18 in 1977, Newton was sleeping in a hammock in the room of his girlfriend's house and in the middle of the night he was struck by a mysterious light. Unaware of what was happening, he became agitated. His girlfriend, upon hearing the noise and seeing the scene, began to scream and ask for help. That's when he woke up without understanding what was happening, feeling sick, weak and his body ached a little. He said: "*I ended up with a mark on my neck, like a burn. I had several tests done at the time*".

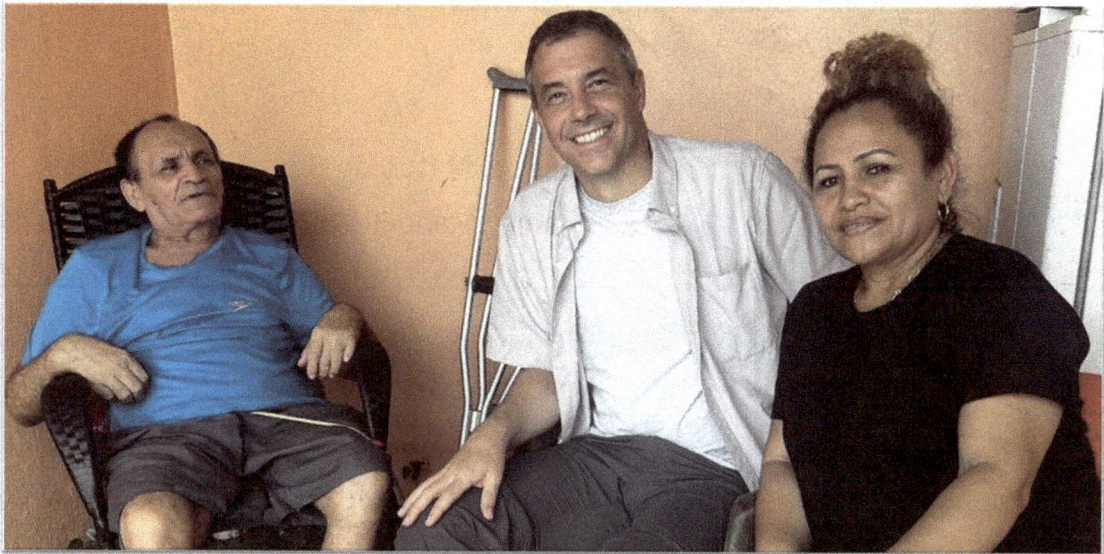

Figure 100. Mr. Newton Cardoso and Cardoso's daughter, Gleice, in Belém, in 2024. (Credit: Author's files).

Claudomira Rodrigues da Paixão

Born in Colares, this farming lady was 35 years old at the time, literate, and was with her children at a cousin's house. On October 18th, 1977, at 11 p.m., she was sleeping in a hammock in one of the rooms of the house when she felt a light travel through her body, like a flashlight, and attach itself to her left breast, sucking it. Then, the light went down to her right hand, at which point she felt the sensation of being pricked by a needle. The victim reports that she tried to scream for help, but her body was partially paralyzed and she felt a strange numbness, followed by pain in her head and right hand, numbness on the left side of her body, and a great localized heat in her breast. The military report recorded her 18/19 words immediately after the attack: *I'm already ruined. The animal sucked me!*" Years later, she added new elements to her story, such as the presence of a being with light skin, eyes like an Asina person, large ears, and wearing a diving suit, and it was holding a kind of "pistol" in its hand that emitted a greenish light towards

219

her left breast and arm, and at the time she felt a heat in the area. Claudomira died in the mid-1990s.

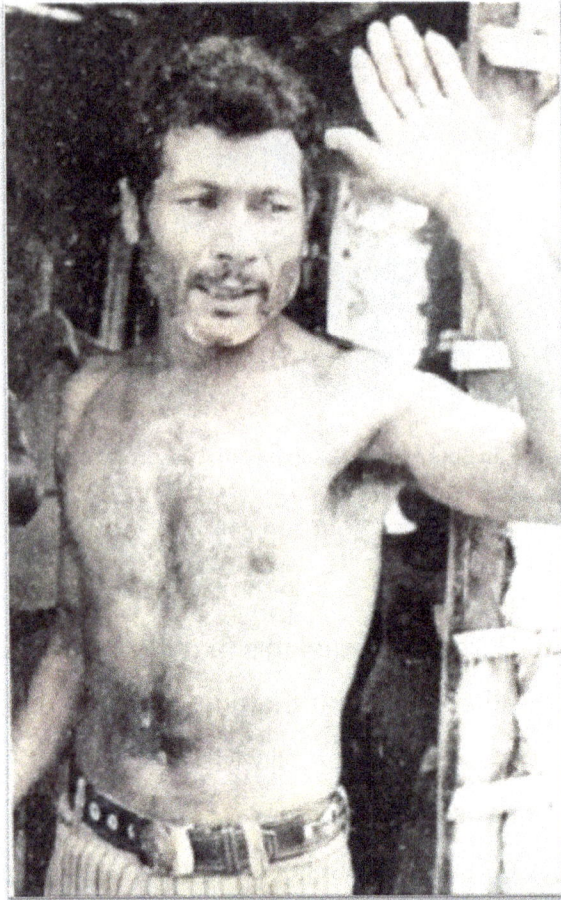

Manoel Matos de Souza, "Coronha"

Farmer Manoel Matos de Souza, better known as "Coronha", 44 years old at the time, living in the locality known as Monte Serrado, in Santo Antônio do Tauá, Pará, experienced the following on October 18th, 1977: "*I was sleeping, it was about two to three in the morning on Tuesday, when through the gaps in the wall of the shack, above the door, I saw rays of bright light penetrating with great intensity. I got up and, opening the door, I saw a strange object standing about 5 meters away from me and, inside, a couple laughing out loud as if they were joking. Even though I was blinded by the red rays that were reaching me, I quickly went into the house and grabbed my cartridge rifle. I pointed it towards the "ship", pulled the trigger,* but to my surprise the gun did not fire. At this point, the right side of my body began to go numb. That's when I screamed for my wife and children, causing panic inside the house. Meanwhile, the device vanished without making a sound, disappearing into the sky,*" said "Coronha".

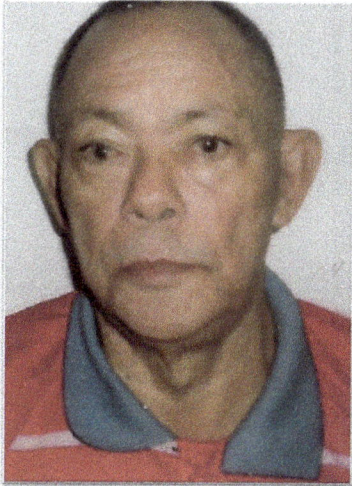

Luiz Pereira Rodrigues

Luiz worked collecting clay for the Keuffer Pottery Factory and on November 2nd, 1977, he was allegedly chased by a luminous object and a humanoid being that emerged from the object. The episode was narrated in 1997 by Colonel Uyrangê Hollanda in an interview with ufologist Ademar Gevaerd, and was also recorded in an official Brazilian Air Force report. He was 26 years old in 1977.

Chapter 19

Unpublished Military Document: Super 8 Films

Although the knowledge that footage was taken during the military mission is not new, since Colonel Uyrangê Hollanda himself mentioned it in his historic interview in 1997, this document provides valuable information such as: number of footage, type of equipment used, days, locations, duration, among other valuable pieces of information for those researching the subject. Unfortunately, no film has been officially released or even leaked to date.

The document was prepared in the late 1970s and the the document does not have the institution's letterhead, numbering or signature, however, it does have elements that link it to the 2nd Section of the 1st Comar. In this case, although the document does not contain important elements such as a signature, for example, we can count on its logical relevance when compared to official documentation, a fact that will be addressed in detail in this analysis, and on the reliability of the source itself, which we guarantee is absolutely reliable.

If we look at the military reports below, we will see that December 9th and 10th, 1977 were particularly intense in terms of observations of lights exactly on the Guajará River.

Date / Time: September, 12th, 1977 - 11:50 p.m.

> .BRG770912/01A- Corpo luminoso,cor amarela avermelhada,rumo N/S,baixa altitude 150ft,distância 300m,sobre a calha do rio,passagem muito rápida,ausência de ruído.(23:50P).

Type: Military observation

Location: Guajará River, Ananindeua, Pará

"Bright body, reddish yellow color, heading N/S, low altitude 150ft, distance 300m over the river channel, very fast passage, absence of noise".

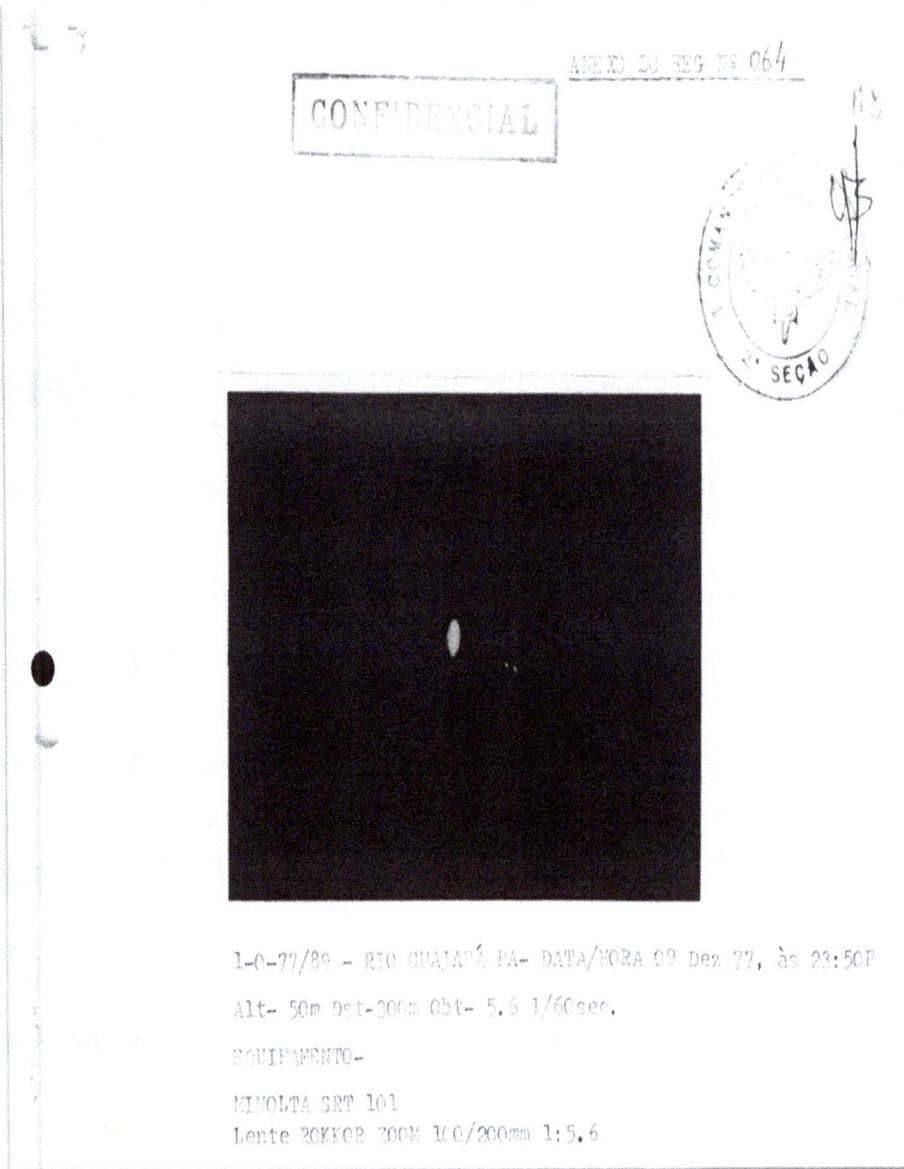

ANEXO AO REG Nº 064

CONFIDENCIAL

1-0-77/80 - RIO GUAJAPÁ PA- DATA/HORA 09 Dez 77, às 23:50P

Alt- 50m Dst-200m Obt- 5.6 1/60sec.

EQUIPAMENTO-

MINOLTA SRT 101

Lente ROKKOR ZOOM 100/200mm 1:5.6

BRO771012/02A- 00:50P-Corpo Luminoso,cor amarela avermelhada,rumo SE/
NW,baixa altitude (abx 150ft),distância 2.000m,tamanho aparente 25cm;
velocidade variável, efetuou várias evoluções,intensidade váriavel.

Date/Time: December, 10th, 1977 – 00:50 a.m.

Type: Military observation

Location: Guajará River, Ananindeua, Pará

"Luminous body, reddish yellow color, heading SE/NW, low altitude (below 150ft), distance 2,000, apparent size 25cm, variable speed, made several evolutions, variable intensity."

ANEXO 01 DO REG Nº 065

CONFIDENCIAL

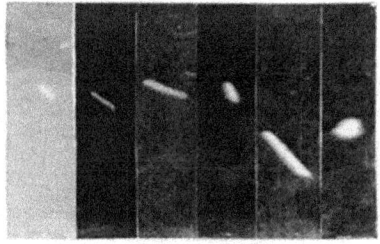

1 2 3 4 5 6

O-77/91 — DETALHES

EQUIPAMENTO—

MINOLTA SRT 101
Lente ROKKOR ZOOM 100/200mm 1:5.6
Filme RB-2475 (1000asa)

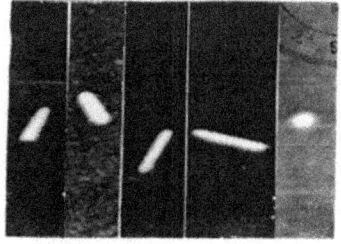

1 2 3 4 5

O-77/92 — DETALHES

FILMADORA—

CANON 514 (x3)
Filme C160asa

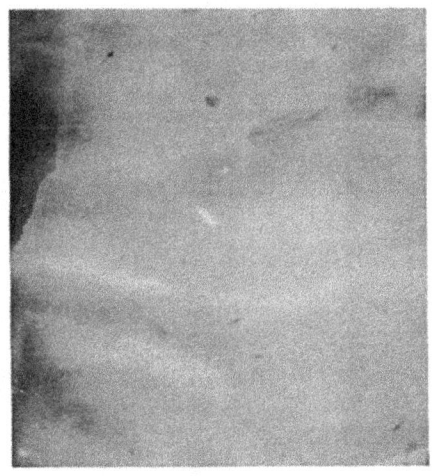

1-O-77/91 — RIO GUAJARÁ (GLÓRIA)

Alt-50m Dst-2000m Obt 5.6 1/30seg.
DATA/HORA — 10 Dez 77, às 00:50F

CONFIDENCIAL

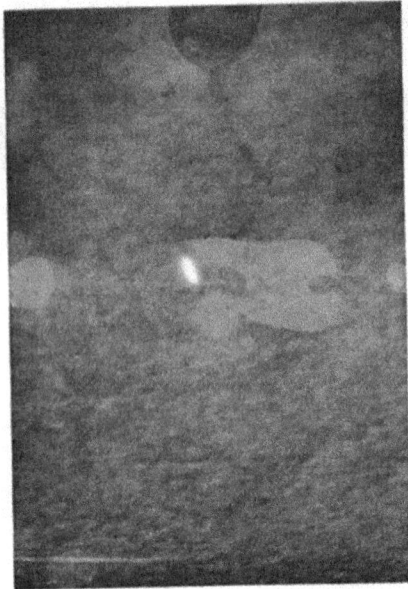

2-0-77/91

Alt- 50m Dst- 1.500m Obt 5.6 1/30seg.

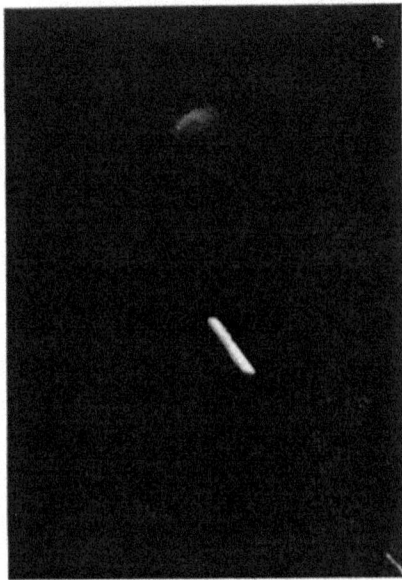

Date/Time: December 10th, 1977 – 01:50 a.m.

Type: Military observation

Location: Guajará River, Ananindeua, Pará.

"Luminous body, above the observation point, low altitude (300ft), distance 500m, apparent size 30cm, reddish color with blue, violet reflections. Several evolutions. Variable speed."

227

CONFIDENCIAL

4-0-77/92

Alt- 90m Dst- 1.500m Obt 5.6 1.40 seg.

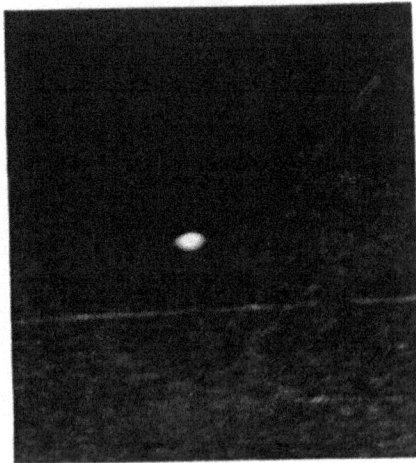

5-0-77/92

ANEXO 02 DO HC Nº 066

CONFIDENCIAL

2-0-77/92

Alt- 100m Det- 800m Obt 5.6 1/30 seg.

3-0-77/92

According to the document, the shooting of this film started in the city of Ananindeua, Pará, in the Guajará River, at the Sawmill Volkswagen facilities at two different times. The first moment would have been at 11:32 p.m. on January, 27th, 1978, with the UFO having an estimated altitude of 900 ft (300 m); the shortest distance to the observer being 3,000 ft (1,000 m); an apparent size of 5 cm, and heading in a SE/N direction. The second moment would have been at 1:08 a.m. on January, 28th, 1978 in conditions almost identical to the previous one. The UFO was heading north and made a 175° turn, disappearing at a low altitude.

Thirty feet of film would have been used to shoot these two passages. This information is important, as it allows us to calculate the approximate time of the footage.

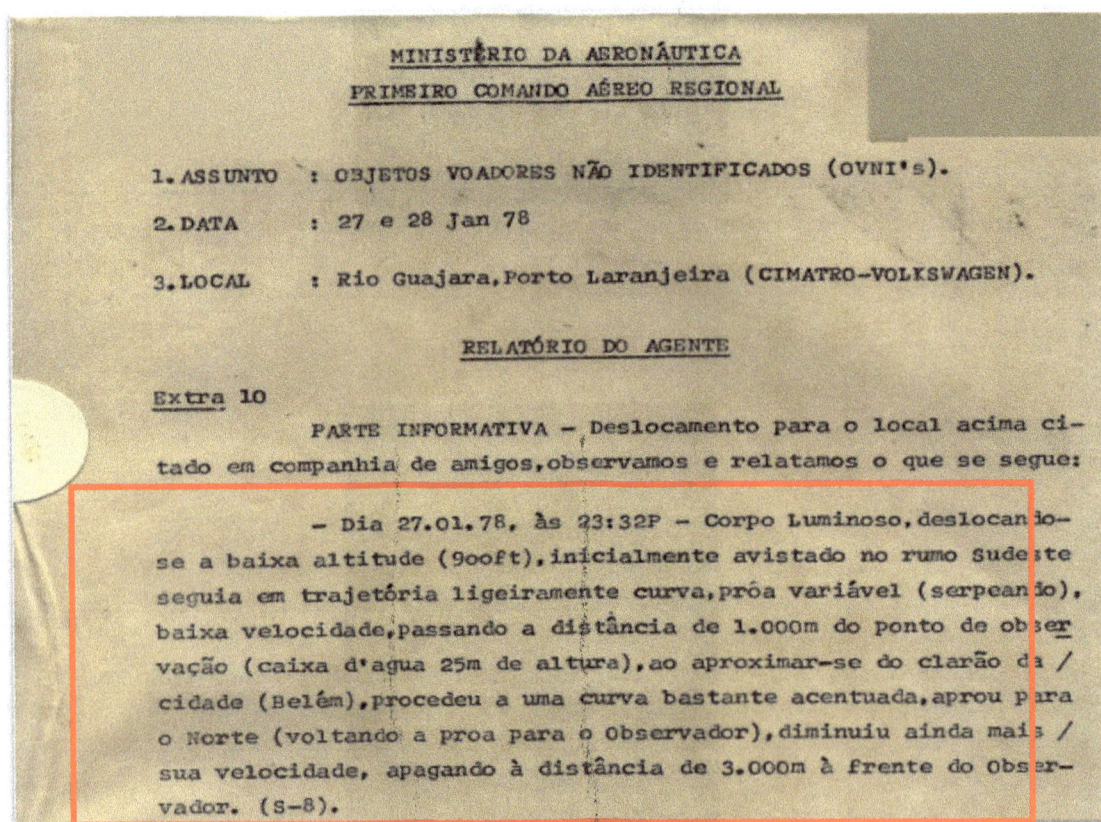

Date: January, 27th, 1978

"Luminous body, moving at low altitude (900ft), initially sighted in a southeast direction, followed a slightly curved trajectory, variable heading (snaking), low speed, passing at a distance of 1,000m from the observation point (water tank 25m high), when approaching the city lights (Belém), it proceeded to a very sharp curve, headed north (turning its heading

towards the observer), decreased its speed even further, disappearing at a distance of 3,000m in front of the observer."

> – Dia 28.01.78, às 01:08P – Corpo Luminoso,deslocando-se a baixa altitude (900ft),no sentido Norte para Sul,passando à distância de 1.500m do Ponto de Observação,afastou-se em trajetória curva à direita,procedendo exatamente como no relato anterior, evitou o clarão da cidade efetuando curva aproou para o rumo Nordeste,retornou e no mesmo ponto onde anteriormente havia "apagado" o outro Objeto; sumiu.

Date: January, 28th, 1978

"Luminous body, moving at low altitude (900ft), from north to south, passing at a distance of 1,500m from the observation point, moved away in a curved trajectory to the right, proceeding exactly as in the previous report, avoided the glare of the city by making a curve, headed northeast, returned, and at the same point where it had previously "erased" the other object, it disappeared".

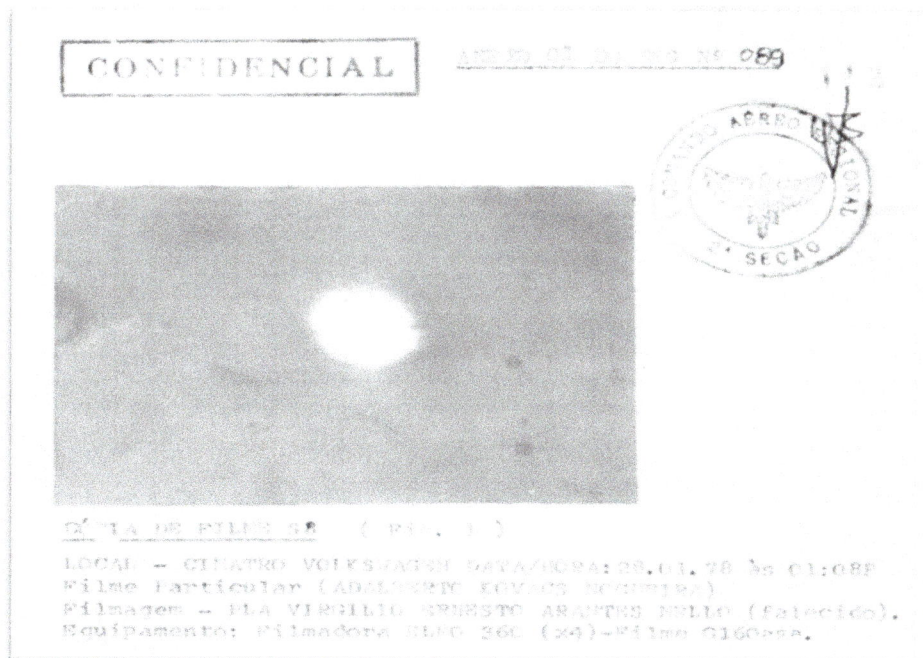

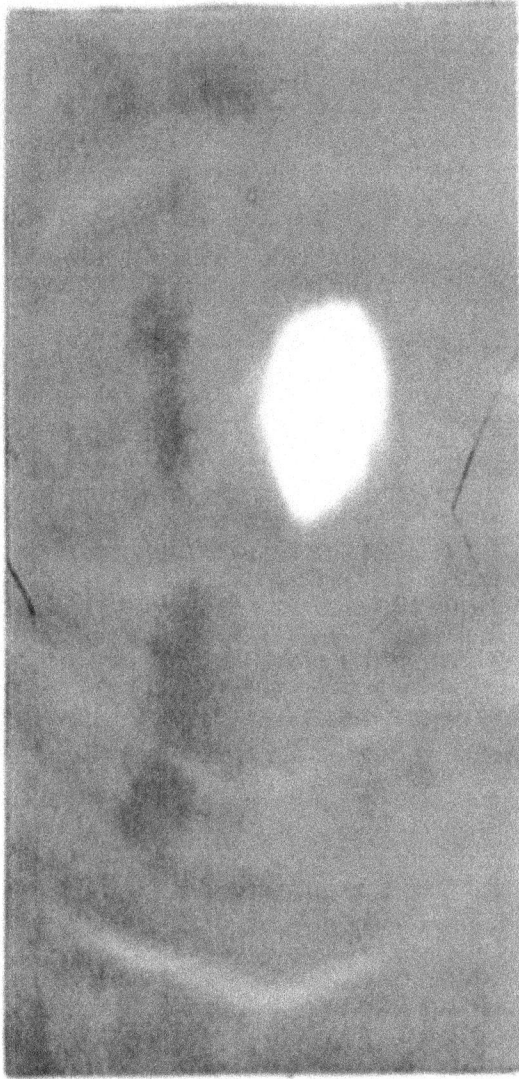

CÓPIA DE FILME S8 (Fig. 2)

Série Virgilio.

CONFIDENCIAL

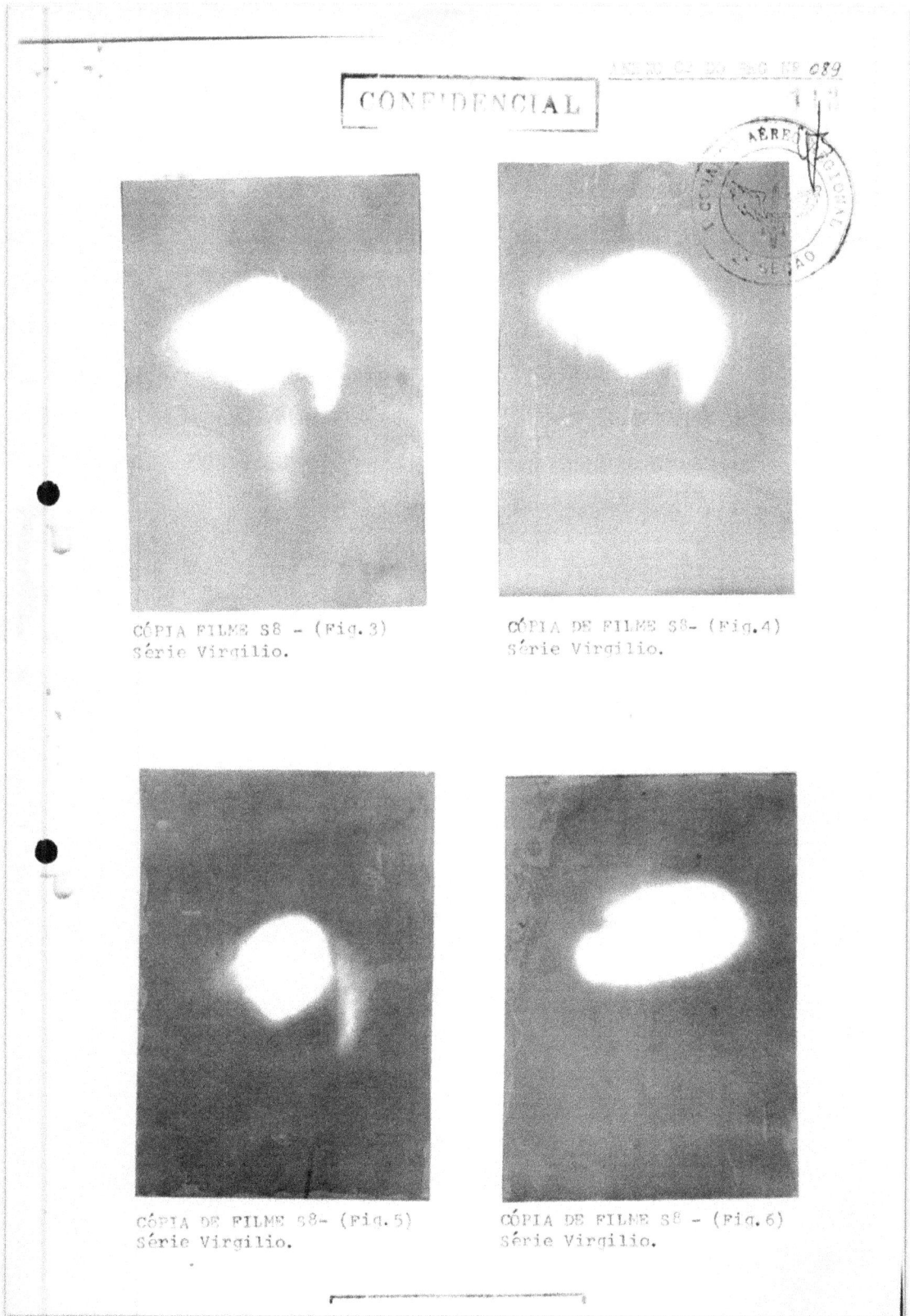

CONFIDENCIAL

CÓPIA FILME S8 - (Fig. 3)
Série Virgilio.

CÓPIA DE FILME S8- (Fig. 4)
Série Virgilio.

CÓPIA DE FILME S8- (Fig. 5)
Série Virgilio.

CÓPIA DE FILME S8 - (Fig. 6)
Série Virgilio.

The fact that Brazilian Air Force personnel filmed the strange lights that haunted riverside communities in 1977/78 during Operation Saucer, was something already accepted as absolute truth by the UFO community. The presentation of this Information Report merely provides documentary proof of its occurrence.

Unfortunately, to date, none of these videos have ever been released to the public. The Brazilian Air Force has consistently denied having such films in its archives.

Although Super 8 films are very durable, it is difficult to ensure that they have been stored correctly or that they have not been destroyed over the nearly past 50 years. Equally difficult would be to confirm that they were not sent abroad or that they are not lost in unidentified boxes in the Brazilian Air Force's archives. Another possibility is that the content of these films is so fantastic that our authorities feel that they need to remain out of public sight. There are many unanswered questions.

All we can to at present is to continue investigating and wait for these films to come to light one day and, through them, we will be able to decipher the origin of these fantastic events that have become the biggest UFO event in the entire world.

Chapter 20

The Following Years

Although the great UFO Flap related to the "*Chupa-Chupa*" phenomenon occurred in the second half of 1977 and early 1978, numerous cases took place after that period. Although Operation Saucer ended prematurely, the military continued to investigate cases in the neighborhood during the following year. In addition, numerous subsequent events attract attention.

The mysterious attacks on the population of the Amazon began in late April 1977, during the Crabs Island Incident. In May and June, new cases of attacks were repeated in the western Maranhão, near the Pinheiro River. As the weeks went by, the phenomenon moved to the state of Pará, alarmingly intensified, and making front-page headlines in the newspapers.

As the situation worsened and the risk of social chaos in the affected cities increased, the Brazilian Air Force began an operation to map the events, identify whatever was responsible for the attacks and, finally, calm the local population.

During Operation Saucer, the military personnel team led by then Captain Uyrangê Hollanda interviewed witnesses and victims, held skywatches and documented some sightings, obtaining hundreds of photographs of the phenomenon, in addition to several hours' recordings.

Although the operation was crowned with success and visibly maintained a gradual approach with the intelligence behind the phenomenon, the high command of the Brazilian Air Force (FAB) determined the end of its activities in December 1977.

Although Operation Saucer had officially ended, some members of the team continued their research independently. The team was made up by Captain Hollanda, Sergeant Flávio and a few other members. This research activity continued for some time and was successful in recording new occurrences. These records are included in the Operation Saucer reports that are now publicly available. These documents were stored at COMAR headquarters in Belém for some time.

Vitório Peret, a former flight attendant, knew the members of this informal team of military personnel and joined them for some of the group's skywatches. Peret says that one night the group was holding a skywatch in the area where the phenomenon occurred when Sergeant Flávio arrived and declared that the operation's files had been manipulated and some files had mysteriously disappeared. Sometime later, the last remaining files were sent to Brasília, Brazil's capital, where they must still be today.

Among the members of Operation Saucer, some details should be mentioned. The first is the fact that all members of the operation need to wear glasses due to eyesight problems that arose after the mission.

The second fact is that, as far as we know, members of Operation Saucer, most notably Captain Hollanda and Sergeant Flávio, had some type of material implanted in their arms. Colonel Hollanda, in an interview to the Revista UFO, manipulated this implant in front of the cameras.

Today, the number of cases has decreased significantly. Some sporadic cases still occur, although they are not as ferocious as they were in the past. Residents of Colares, children, grandchildren and relatives of those who once experienced the unusual experiences with the "Chupa-Chupa" phenomenon, at the time of Operation Saucer, aware of the gravity of the situation and the lack of knowledge by the population regarding everything they went through, entrusted some of their memories and events that occurred at the time and later to our knowledge, in the hope that their history and culture would not be lost.

Although the illiteracy rate is almost zero, there seems to be no interest in passing on the local stories and experiences of the island's inhabitants. Inappropriately, the "Colarenses" (Those who are born in Colares) of today treat Ufology as folklore and continue to use terms such as "*visagens*" (visions), "*Matinta Pereira*" (the witch of Pará), "*Sapa*" (a female frog) and "*bola de fogo*" (fireball) among others.

Although many of the reports do not reach us Ufologists with the full name of the witness, date, time and place of the sighting or contact, it is still important to record these events. I will bring you some of the most recent reports I received.

Almost 10 years ago (2014), an old fishermen called José Mario, who still lives and fishes in Colares, said… "*There, on Machadinho beach, there is a being that appears very high up, it is made of light. People here say it is the "Matinta-Pereira, the witch of Pará. There are also small beings that have even commanded and taken people to other places. He took my father off the boat in the high seas and threw him on the Machadinho beach. He said they were very small beings. He does not like to talk about this very much. These beings controlled his mind, there were many of them, they had slanted eyes and lizard skin or overalls. That is how he told us. He became dizzy and forgetful*".

Reports say that many residents and fishermen disappeared (or were abducted) at that time, and neither their bodies were found nor any information collected, and their whereabouts remain unknown to this day.

Once, a woman was returning home at 4h30 a.m. when she saw a "skinless being" that looked like a "grey" sitting on the carpet at the local market.

Another witness, Maria Helena, reported that on a certain night with a full moon in the late 1990s, in November, she was returning home alone from a party at 12:45 a.m. when she saw from afar, two blocks ahead of her, a "*giant man*" "*the size of the lamppost*" that stood in front of her house. She described the being as "*a man, because he was wearing long pants*", "*very tall, thin, who at one point turned his head and you could see his face, but he seemed to have no face*", "*wearing very bright, very shiny clothes, and he kept looking directly at the moon, almost the entire time I was there*". She stood there for almost 2 hours watching the "man" who did not move, and, worried about the time predetermined by her mother of 1h a.m. for her return home, luckily some fishermen were heading to the beach, and she accompanied them on the way to her house, which was halfway between them. The moment she "joined" the group of fishermen, the being disappeared, she went home scared, and her mother was awake, waiting for her. She told her mother what had happened, justifying her delay, which her mother believed and said that it was a rebellious spirit, that it beat people who walked alone in the streets at night and that it "came from *Machadinho*". She found the situation strange and didn't know why he hadn't done anything to her. "*I told her everything, in great detail,*" "*I spent a long time looking at that man,*" "*everyone here believed me and my mother added that he used to appear as a very luminous being, that the natives feared and that his name was "Dah".*"

There is also a being that in the 1980s was called "Sapa" (female frog), which according to reports, had been observed in several places simultaneously. It looked like a large man, with long, thin legs that were half-bent outwards, and that screamed a loud sound like the croaking of a frog. It was "*smooth, skinless and some people even saw the veins on its body. They also say that when it was seen, it was always heading towards the sea and diving in there*". These "sightings" became so well-known that the inhabitants asked the local police for help to catch or get rid of this being that was terrorizing the population. They set up "alerts" on the shore where it appeared, but they never even saw it, because it would let out its croaking cry in one part of the city and almost immediately do the same on another side of the city, leaving everyone disoriented. Just as reports of sightings of this being began suddenly, they were extinguished and no longer mentioned.

In the late 1960s and early 1970s, a surprising event was witnessed by an old and well-known fisherman, near the old "Cabanas bridge", from the *Cabanagem* movement (It was a revolt that took place in Grão-Pará, between 1835 and 1840, during the Regency Period), in the Bacuri neighborhood. He saw a "sign", "a huge number in the sky". His son tells the story: "*My father is 80 years old and he saw this when he was almost 30. At that time, Colares was a village and the only way to travel was by sea. There was a light on the sand, and it was so bright that even the grains of sand could be seen, and he looked up at the sky and saw these huge golden numbers with red edges. He tells this story a lot. He tells this story every time... I grew up hearing it.*"

Another story came to me from a 78-year-old woman named Flora Alcelinda. One day, at dawn, in 2009, when it was unusually hot, something almost unnatural for the place, she woke up feeling uneasy and went to her balcony to cool off. When she looked at the square in front of her, there was a half-man, half-horse being drinking water from the fountain in the square. She told her relatives and friends what had happened, and on the following day they tried in vain to see the creature again, but it never appeared again.

One late afternoon in 2020, a couple riding a motorcycle towards Humaitá Beach, a paradisiacal place, while passing one of the roads that lead to the beach, had a big scare: their hair was "electrified", literally standing on end (with the hair standing straight up) as an effect they felt. A little further on, they returned to normal, then they retraced their moves and the same thing happened a second time. They were terrified, thinking that it was just the old "*visagens*" and returned to the city, ending the tour without ever returning there. They passed through a "vortex or flow of energy".

Figure 101. Humaitá Beach, in Colares. (Credit: Internet)

Figure 102. George Knapp and Thiago Ticchetti at Humaitá Beach, in Colares, 2024. (Credit: Author's file).

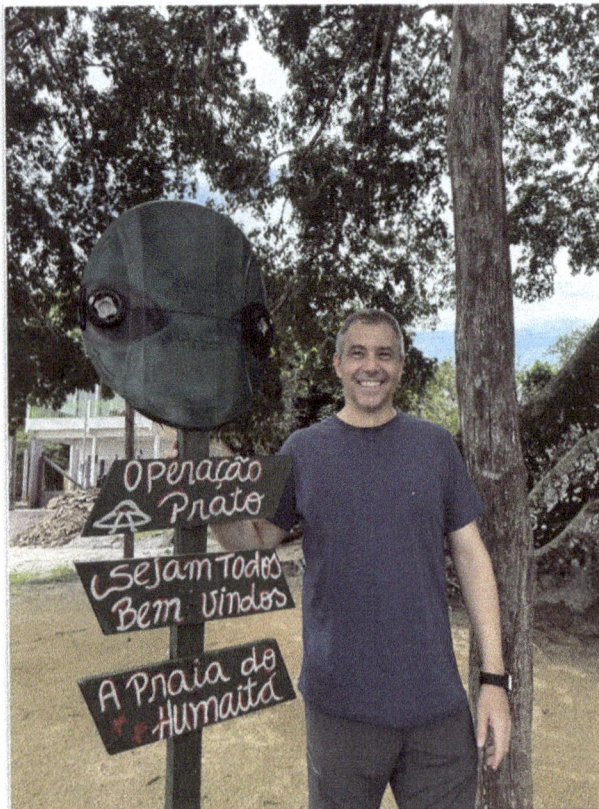

In early June 2022, family members saw a triangular-shaped craft with colored lights. Around 1:30 a.m., two adults and their 14-year-old nephew witnessed a solid, triangular-shaped object. They filmed the UFO and their nephew made a drawing of it.

Figure 103. Thiago Ticchetti at Humaitá Beach. The signs read: Operation Prato. Welcome. Humaitá Beach. This location was one of the epicenters of the "Chupa-Chupa" sightings and attacks and a skywatching place for the mission's military personnel. (Credit: Author's files).

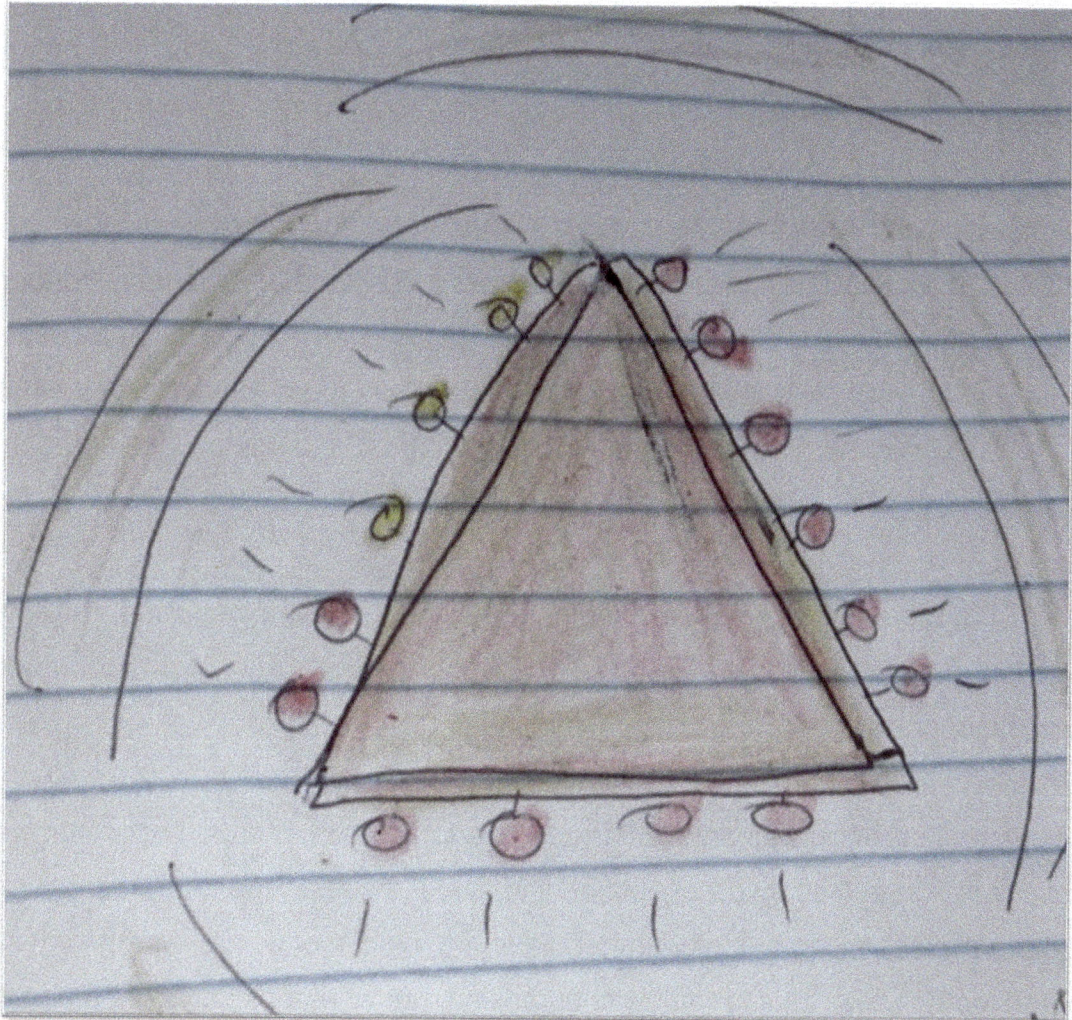

Figure 104. Triangular UFO seen and filmed by three people. (credit: www.fenomenum.com.br).

In 2024, a new mission brings unprecedented information

In January 2024, I was in Pará, where I visited the cities of Colares, Barcarena and Belém in search of new reports involving UFOs in that region. The place is rich in incidence. We found recent cases and a previously unknown witness to the "*Chupa-Chupa*" attacks in 1977.

At 11 a.m. on January 24th, I arrived at Belém International Airport. The late morning was very humid and hot, as usual at this time of the year in the state of Pará. I was there to investigate the famous incident researched by famous journalist George Knapp and invite him to participate in the recordings of one of the episodes of his new series for a streaming channel. The theme of this episode will be the "*Chupa-Chupa*" attacks and Operation Saucer, both in 1977, in addition to the investigation of new cases that have still occurred in the region.

The next morning, we left Belém early heading to Barcarena. There are approximately 252 km, between highway and ferry. The state of Pará is huge and the closest cities are so far away and it is almost always necessary to cross rivers. To Barcarena we traveled for almost 1h30 by ferry. But, why go to Barcarena? Although the city was not one of the focuses of the "*Chupa-Chupa*" attacks, the phenomenon was also recorded there. Nevertheless, what really attracted us were two more recent cases that came to us. Sightings of lights, marks on the ground, small dead animals, unknown beings, and a photo of a UFO!

We arrived in Barcarena in the early afternoon and moved to Trambioca Island. The weather was cloudy, and it was raining a little. The production team quickly set up their equipment and I took the opportunity to collect data, testimonies, and evidence.

The first person I spoke to was artisan Selma Furtado. In 2009, she and her sister went out to collect seeds and fruit on a soccer field behind some houses. When they were there, they noticed three circles, and in one of them a hole in the middle, of different sizes drawn in the grass of the football field. This in itself was surprising, but what mostly caught their attention was that around these circles there were several dead water cockroaches, which despite being native to the region, do not have the habit of staying together and much less to be found in that field. Added to this mystery is the fact that all water cockroaches had two holes on their backs.

Selma then looked towards the goal posts on the football field and had a brief glimpse of a "small man", and the two women fled the scene in fear. Selma told her cousin, and along with other people they went to the location.

Another interesting aspect of this sighting of the little man is that Selma said that the tops of the trees were dented. I asked if they were burned, but she reconfirmed, "*no, they were really crushed*".

Figure 105. Selma Furtado saw a humanoid in a football field and a column of smoke over the water, surrounded by spherical objects (Credit: Author's files).

I asked if what she had seen was a light, but she said no, that it was a small being, but the elders in the neighborhood always said that there was a light that appeared in the field.

Still regarding the marks on the football field, when asked how long they were visible, she told me that for some time, but that the next day, within the smaller circular mark, other marks had appeared with a triangle and a straight line in the middle!

Three months after the incident on the football field, Selma was fishing again with her sister when they saw a waterspout. Around the trunk there were several lights. The two women were very scared, took their *rabeta* (a small boat with a motor) and rowed to the beach. Once there, they called other people, but the formation had already disappeared. I asked how big the lights would be and she told me they were big. Selma told me that another resident of the island had also seen lights over the water and forest.

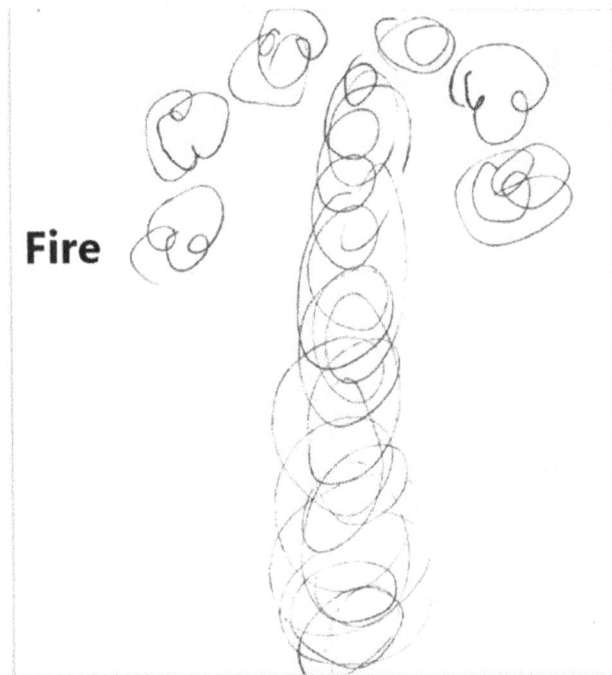

Figure 106. Drawing of the column of fire that was surrounded by small luminous spheres, made by Selma (Credit: Author's files).

Figure 107. Jhonny Monteiro photographed a UFO over the trees of the football field in Barcarena. He also saw, like Selma, three mysterious circles and dead animals on the field. (Credit: Author's files).

At the end of the interview, I asked her what she thought the luminous spheres she saw could be and what could have made the marks on the football field. "Ah, I think it was a flying saucer thing." I asked Selma to draw what she had seen in both cases.

But the cases didn't stop there. Then, I spoke with Selma's godson, Jhonny Monteiro, who had not only sightings, but photo and film evidence!

Jhonny was also an eyewitness to the marks on the ground on the football field. But it happened after he witnessed a UFO over the place.

According to Jhonny, in the early morning hours of 2009 he was taking photos of small animals in the forest, when he came across an object hovering over the Island's football field. He took a photo and ran from the scene. He was afraid that the object might come towards him and take him away. The UFO would be between five and seven meters long. Soon afterwards, when he told relatives and others, he learned that other people had also seen the object. The UFO photo is clear, despite the low resolution, and shows a dark discoid object over the treetops surrounding the football field. Jhonny also said that the tops of the trees on the following days remained crushed, not broken or burned.

Figure 108. UFO photo taken by Jhonny (credit: Author's files).

Preliminary analysis of the photo indicates that there was no manipulation of the image but given the low resolution and the lack of the original file, it is not possible to say whether the object in the image is real, or whether it is an interpretation of something in the photo or whether it has been altered somehow.

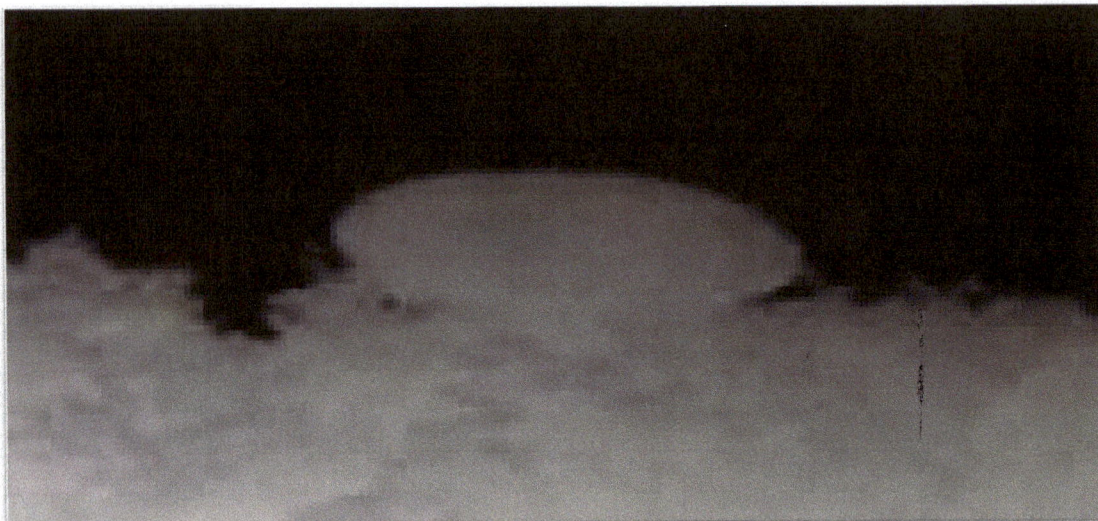

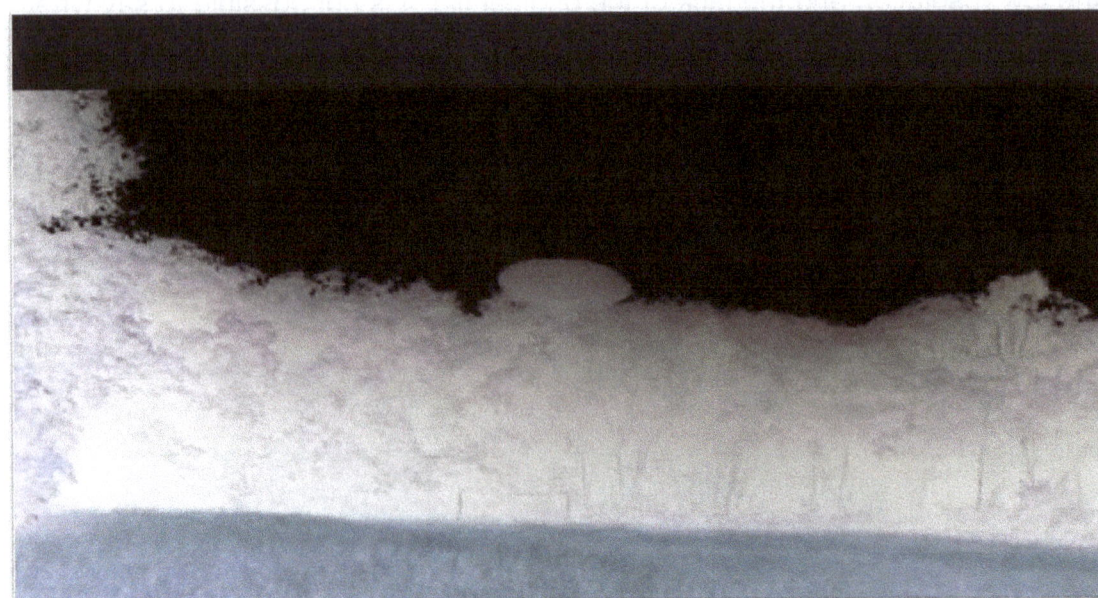

Metallic
Static
No sound emitting
No windows

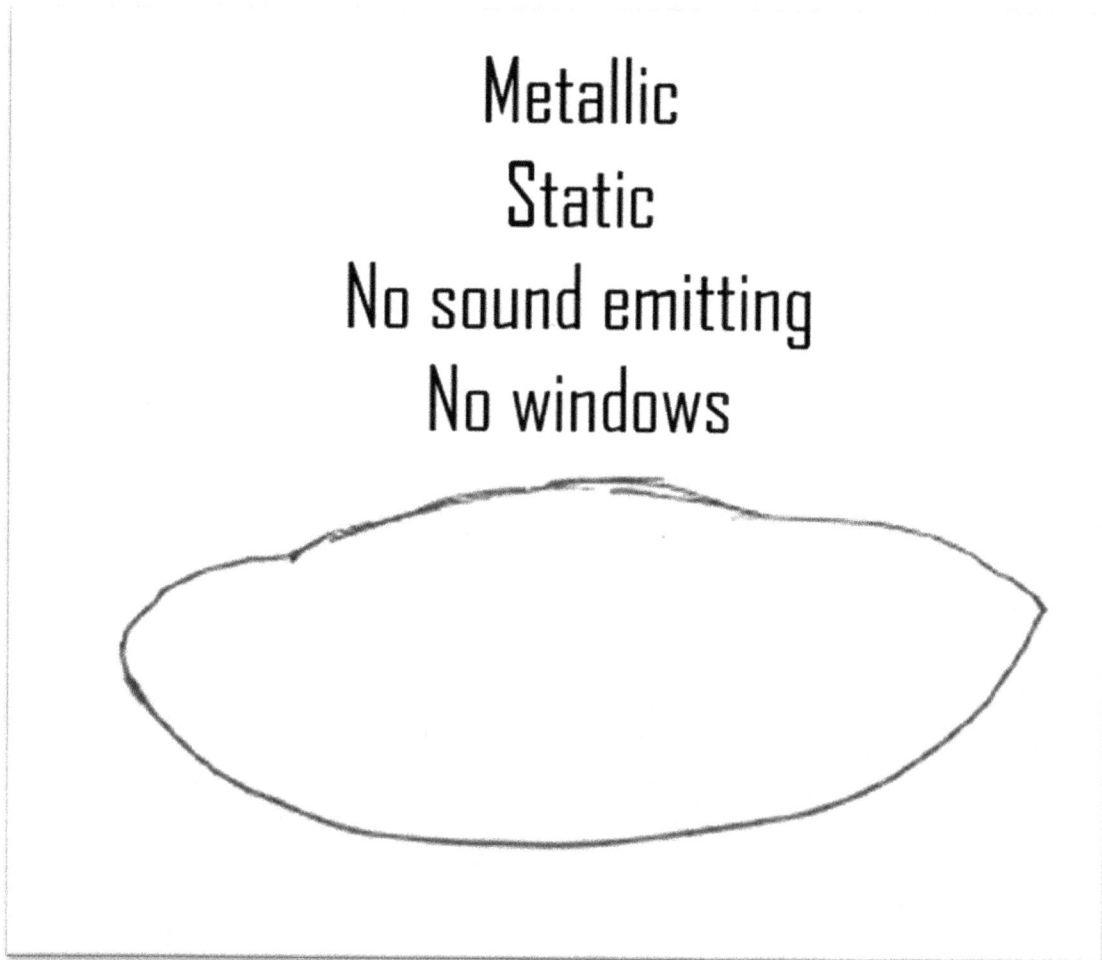

Figure 109. Analysis of the image did not indicate manipulation. UFO drawing by Jhonny (Credit: Author's files).

The next day, he returned to the place with other people because his godmother (Selma) had had bee shouting that she had seen three circular marks on the football field. Jhonny told me he could say it was that object that made the marks. This is his conviction; there is no way to prove it.

Regarding the circles, Jhonny said that the largest circle was between 10 and 11 meters, the middle circle measured between five and seven meters and the third also measured between 10 and eleven meters and had a triangle inside it and a straight line in the middle (same story as Selma).

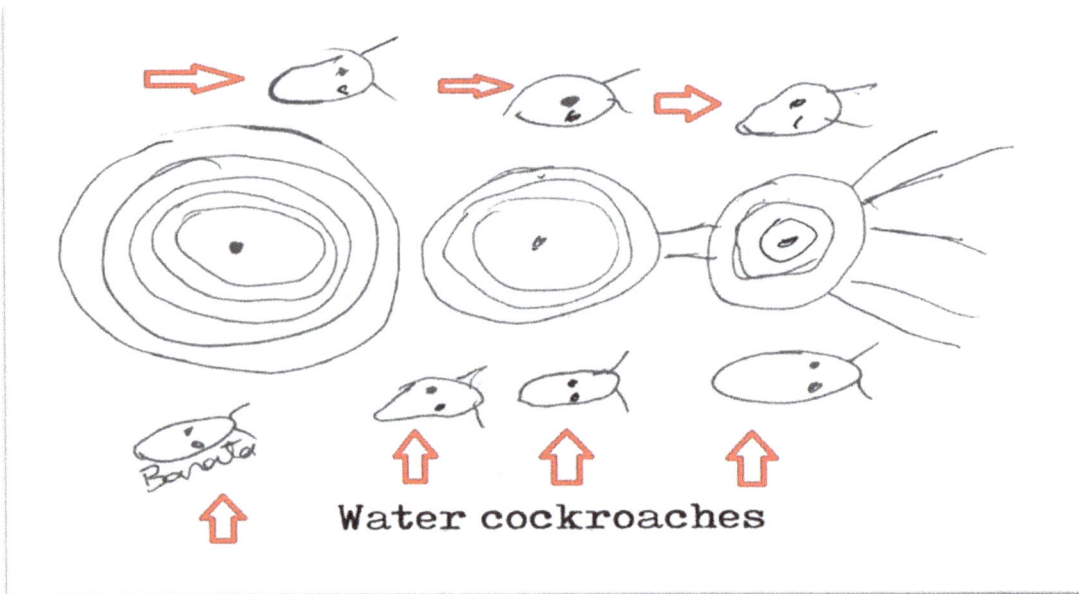

Figure 110. Drawing made by Jhonny (Credit: Author's files).

Figure 111. Photo taken by Jhonny of one of the circles found on the soccer field (Credit: Author's files).

Jhonny said that he also saw dead water cockroaches around the circles, and took pictures of them. In addition, he found several other dead animals near the circles, such as a frog, a scorpion, a beetle, a cricket and two snakes, as if they passed by and died. He reinforced

248

the detail that all cockroaches had two perforations in their shells. It's a curious detail. In the center of the larger circle a mushroom grew.

Figure 112. Water bugs with holes in their backs (Credit: Author's files).

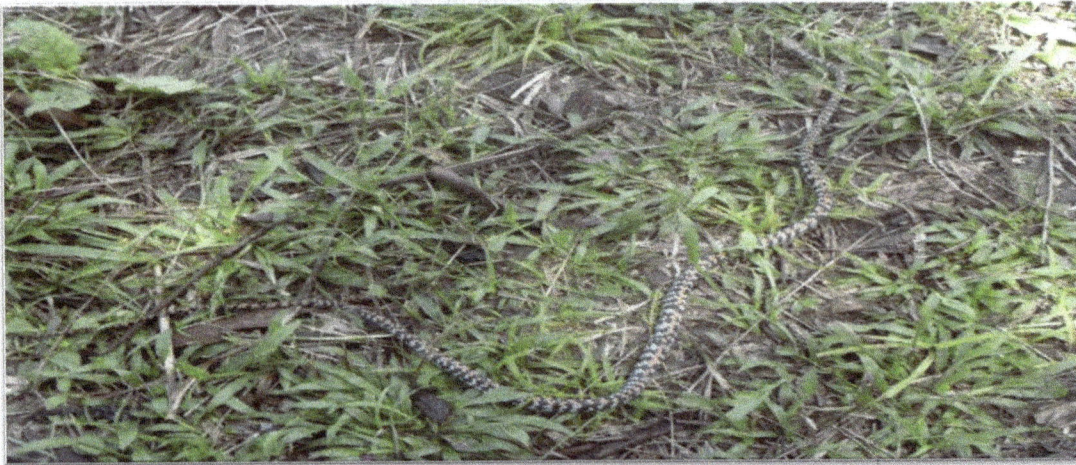

Figure 113. Dead snakes were found near the circles. It appeared that they died after crossing them.
(Credit: Author's files).

Figure 114. UFO recorded by Jhonny. (Credit: Author's files).

I asked Jhonny if there were ants or other critters feeding on these carcasses of these dead animals. He said no, but that after returning to the field, about two weeks later – as everyone was still afraid of what they had seen – there was no trace of any carcass.

It is important to emphasize that the two statements were collected independently, that is, one at a time, without showing what each one said and drew; and that during my stay on the island I asked some questions again to check if there were any changes in the reports. But the two witnesses kept what they had said.

But Jhonny had another case to report to me. He said he filmed a "fireball" on December 23, 2023. He and the people present thought it could be a plane or meteor that was coming towards them, but the object made a 90° turn. Dismiss these possibilities.

250

Shortly after making the turn, the UFO continued a few meters further into the sky and reduced its brightness until it completely went out.

I asked if it could be a flare or someone playing a joke. He replied no. "*In no way, especially as a flare wouldn't have gone up, increased the brightness and blacked out. What we saw came towards us, as if it had been released from above and then made a very large and rapid curve. This is not the first report of lights here on Trambioca Island.*"

Figure 115. The beach at Trambioca Island. (Credit: Author's files).

A skywatch was already planned at the football field, and the testimonies only reaffirmed that we were in the right place. We set up our equipment at the place where the circles were seen and there, we waited for something to happen even with a rainy night and an overcast sky. At around 2 a.m. we gathered everything back and left Barcarena behind. Next destination: Colares.

Figure 116. The skywatch was set up on the soccer field. The photo shows where the UFO was seen by Jhonny. (Credit: Author's files).

The next day George Knapp and I explored Colares Island. We went to Humaitá Beach, the scene of dozens of "*Chupa-Chupa*" phenomena. According to some reports we collected, the flying devices (of various sizes and shapes) came from the river's horizon – other times they came out of the water – to shoot their rays of light onto the beach's stones and sand. Residents lit fires and fired firecrackers and shotguns at the lights and unknown aircraft.

Humaitá Beach is beautiful. One of the most prominent features is a Samaúma tree, a century-old Amazon native tree. Legend has it that mystical beings – and extraterrestrials – come from its roots. Colares embraced the theme "Chupa-Chupa", extraterrestrials, "Vampire Lights" and so on, even after its population had suffered so much at that time.

Figure 117. Humaitá beach, one of the hotspots for UFO sightings during the "Chupa-Chupa" attacks. The military from Operation Saucer held several skywatches at the site. (Credit: Author's files).

From the beach you can see an island, known as *Farol de Colares* (Colares Lighthouse), which measures around 200 meters in length where the Operation Saucer military set up an observation base. From there they had a privileged view of what was coming from the horizon across the river, what was "coming out" of the river and what was flying towards the city.

Figure 118. View of Humaitá Beach from Colares lighthouse. (Credit: Author's files).

A lot has changed in the city since the end of the 1970s, but one building has remained the same since it was built and was part of the history of the "*Chupa-Chupa*" phenomenon: the Igreja Matriz de Colares (Colares Main Church). It was built in the 16th and 17th centuries, by architects Pedro Nunes Tinoco and João Antunes and master bricklayer André Duarte. The temple is dated 1755*, shortly after the city was shaken by an earthquake that destroyed...

Lisbon, in Portugal. That is right: the earthquake crossed the Atlantic Ocean and reached the north coast of Brazil.

This church, during the "Vampire Lights" attacks, served as protection for the population. Women and children were taken inside, while the men stayed outside banging pots, setting off firecrackers, shooting and lighting fires to scare away the UFOs.

Figure 119. Igreja (Church) Matriz de Colares. Women and children hid inside the church during the "Chupa-Chupa" attacks, while the men stayed on the beach with weapons, fireworks and pots to scare away the objects. (Credit: Author's files).

Then I saw an older man and asked if he knew about the "*Chupa-Chupa*" phenomenon. He replied, "*Not only do I know, but I have also been an eyewitness to many sightings of those lights*". Mr. José Francisco was a direct witness of that time!

"*We were on the island. And that thing landed right there*", he pointed to a construction site close to the two of us. "*The airship took some construction material and left, rotating the lights around it. The guy who was with me took a photo of the UFO. People say it is a lie, but many people who are new in this neighborhood do not know what happened here.*

Figure 120. Mr. José Francisco was an eyewitness to a UFO attack. (Credit: Author's files).

I then asked if two people had died because of the attack (before I started recording, he had told me that). *"Yes, two people who were 'sucked'. When the device took away the person's light*, it left a trail of blood along the way. At that time all houses were made of wood. 'Mrs Mirossa' and another lady died."*

The attacks, according to José Francisco, lasted about a month, until people from the Brazilian Air Force arrived. *"Around each corner of the city there were two of them. They had those small beds and radio transmitters and when the device passed by it would point upwards."*

I asked him if there were any Americans involved in the research after Operation Saucer – something that many of us, researchers, believe to be true. Mr. Francisco said yes. And how did he know? *"They spoke 'foreign', and no one understood anything."* Nowadays, people continue to see strange lights in the skies and over the waters of Humaitá Beach, but nothing compared to what happened between 1977 and 1978.

I collected another testimony from the son of the house owner, Arthur, who came to me to report his experience. It wasn't something related to the *"Chupa-Chupa"* phenomenon because he wasn't even born in 1977, but his sighting was very similar to what people saw at that time.

According to Arthur Farias Barros, one night he and some friends were at Humaitá Beach, when he noticed a light moving in the sky while people were playing the guitar. He thought it was a person playing with a cell phone or flashlight, but if that were the case, he would have been able to see the projection where the light was coming from.

Suddenly, according to Arthur, the light threw a beam in his face. It was a huge scare and his body started to heat up. He began to wonder how that was possible. The light was distant. How could it have hit him? Still stunned by the flash, he saw the light go out.

About 40 minutes passed, and he was talking to a friend about what had happened. They were facing the beach, when suddenly, from behind his friend, between 50 and 100 meters away, a luminous object appeared with two cylindrical lights; it was long and a "V" shape, and from its middle came another light that seemed to track the soil. The object moved lightly across the beach, from one side to the other, randomly. The sighting lasted about 15

Figure 121. Arthur saw a triangular object appear on Humaitá beach and light up the sands there. (Credit: Author's files).

minutes. The two friends then decided to leave, as they were getting a little "impressed" (scared) by it.

I asked Arthur what color the light was. He told me that it was white, and the beam had a bluish-red color. The object made no sound. When they returned with more friends, the UFO had already disappeared.

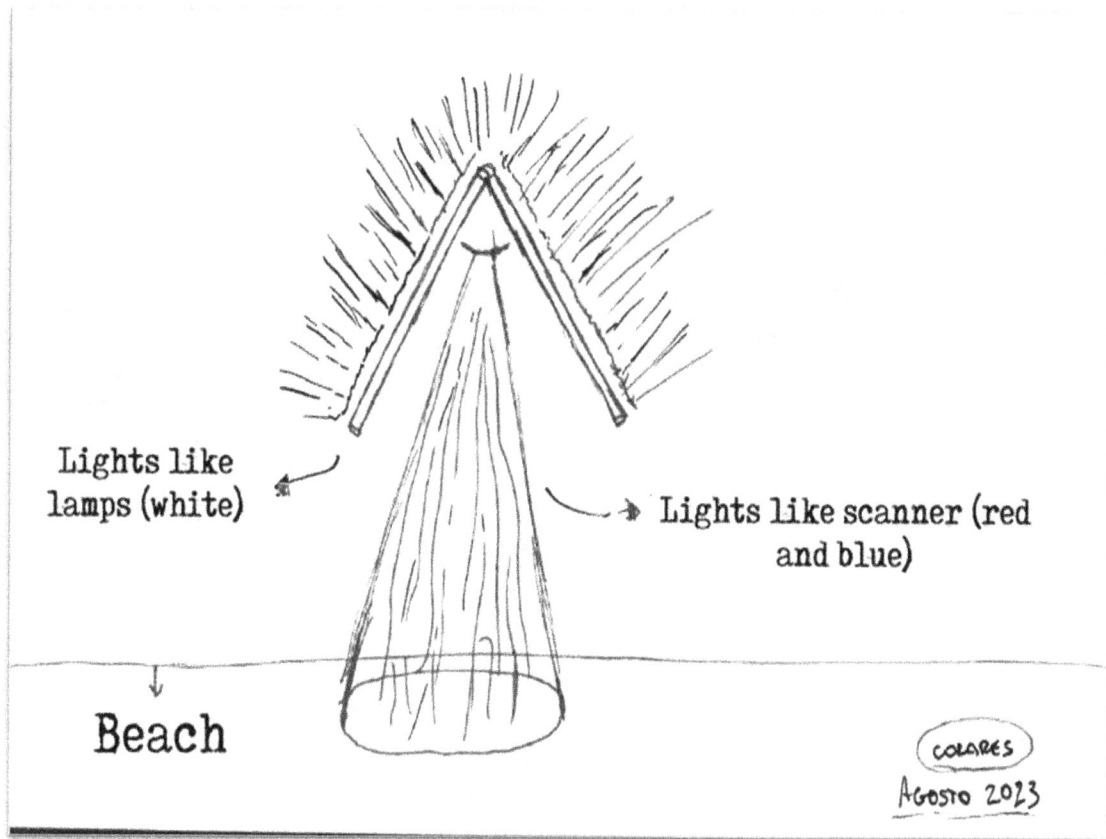

Figure 122. UFO drawing by Arthur. (Credit: Author's files).

Towards the end of Saturday, we headed for Belém. We arrived at night. The following morning, we would have the last three recordings with some of the most important characters in the history of the "*Chupa-Chupa*" phenomenon and *Operação Prato*.

On Sunday morning, we went to meet D. Aurora Fernandes. Hers is the most iconic photo, the one that best represents the attacks that occurred in 1977.

George and I arrived and were most welcome by her and her family. We sat down on two sofas in the living room and started asking what had happened that night.

Aurora Fernandes was 18 years old at the time. She and her sister wanted to go to an amusement park in the city of Belém. Her mother, Mrs. Eunice, who passed away in 2023, would not have allowed it, as the "*Chupa-Chupa*" was already in full swing. Many people had already been attacked. Furthermore, the two sisters had not washed the dinner dishes.

However, the two young women decided to leave anyway and went to the amusement park. Suddenly a turmoil began. People were shouting that Chupa-Chupa was attacking. *"It was around 9p.m. Many people started running, screaming, fleeing the amusement park. I grabbed my sister, and we ran home"*, Aurora told us.

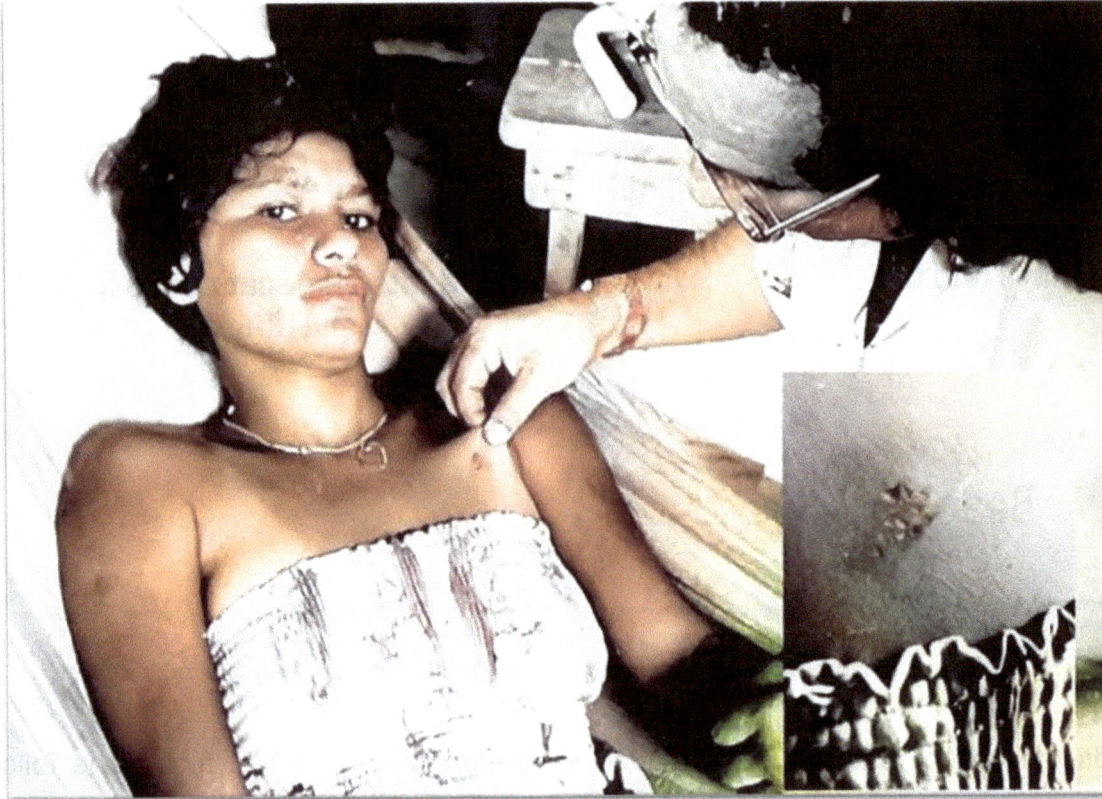

Figure 123. Aurora Fernandes being examined by Doctor Orlando Zoghbi. Detail of the three perforations in her skin. (Credit: Author's files).

When they arrived home, they told their mother what had happened and as punishment for disobeying, they were forced to wash the dinner dishes and then go straight to bed.

"At that time, our house didn't have a sink in the kitchen. The dishes were washed in a water basin outside the house", said Aurora. *"So, we were outside. Suddenly I saw a very strong red light appear behind a chestnut tree. It felt like it was enveloping me. It was big and I knew it was Chupa-Chupa, but I only got scared when it quickly approached me"*.

The light then shot a solid, bluish-white ray, which hit Aurora just above her left breast. She felt several "stings" as if a needle was piercing her. The moment she was hit she felt burning and hot. She started screaming for her mother and sister, who had their backs to her. D. Aurora then fainted.

"*When I woke up, I was in the hospital. I had a terrible headache and dizziness*", said Aurora. When leaving the hospital, she could not walk, as her legs were weakened. She was examined by doctors Wilton Reis and Orlando Zoghbi. "*For a long time after the attack, a doctor came to see me at my house to check on me.*" The brand was visible for about a month. For a long time after the attack, she was afraid to leave the house, like many of the people who were victims of Chupa-Chupa.

When asked how she feels about what happened to her in retrospect, she said she didn't want that to happen to anyone. "*It was really bad. I became very sick, scared, at home. Why did it happen to me and not my sister who was next to me? I do not know the answer*".

We said goodbye to D. Aurora and set off to find another victim of the "Vampire Lights. This time I would go alone with part of the production team to talk to Newton de Oliveira Cardoso, the "Lieutenant".

Newton lives in Colares but is awaiting surgery; he relocated temporarily to Belém so he can be closer to the hospital and have better medical care.

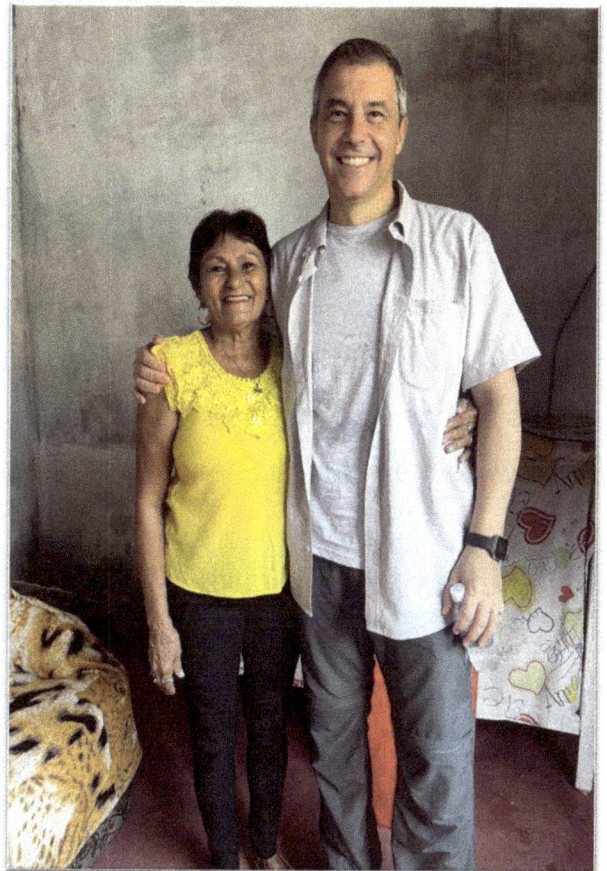

Figure 124. Aurora and Thiago Ticchetti, in her house. (Credit: Author's files).

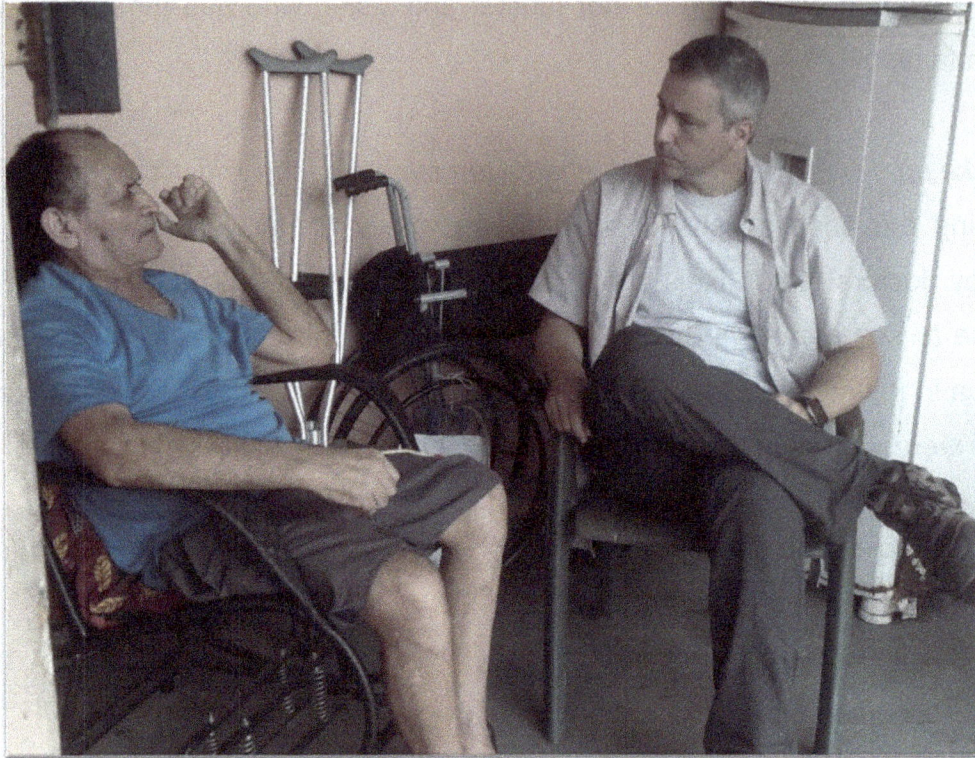

Figure 125. Newton Cardoso and Thiago Ticchetti, in Belém, 2024. (Credit: Author's files)

I was welcomed by him and his daughter Gleice on the porch of their house. I sat next to him and while the production team was arranging the equipment, we started making small talk.

When everything was ready, we started. The first thing I asked was about that night in 1977 in Colares.

Newton told me that he had gone fishing, that was his profession, and that when he returned, he found out about several attacks that had happened in Colares. Worried that he might become the next victim, he decided to go to a nearby village called Mocajuba. When his girlfriend, Normalina (who would later become his wife) arrived, she said that many more attacks were happening there than in Colares.

The terror was so great that as soon as it got dark there was no one left on the streets. Residents closed the doors and windows of their homes. Some men stayed outside, defying their fear, ready to "face" those lights and flying objects that were "sucking" people's blood.

They lit fires, lamps, and lanterns; they banged pans; They fired into the air, but nothing seemed to stop the attack.

In such a scenario, Newton went to sleep. He was tired from a whole morning fishing in the river. He was at his mother-in-law's house and at that time (and because of the customs of the time) he slept in a bed, his future mother-in-law in the middle and his future wife on the other side, all in the same room.

"After falling asleep, what I remember was waking up to screaming inside the house. My wife was screaming, there were people inside the room. And I was feeling something in my neck; a stinging, a burn", said Newton.

He felt there was a sting on his neck. Newton was weak, dizzy, and had a headache. He asked what had happened. His wife said that *Chupa-Chupa* had caught him. Neighbors had seen a light, a very bright flying object, over the house. This object shot a light that crossed the screen and, in front of it, hit Mr. Newton. His mother-in-law and his wife saw the light. His mother-in-law even had part of her body hit by the light, causing her to feel hot.

Even though he was staggering and had a headache, Newton wanted to leave. There, he would no longer stay. His wife tried to convince him to stay, and they finally reached an agreement: he would sleep at his sister-in-law's house.

I asked Newton how long the mark was visible and how long the after-effects of that attack lasted. He replied that the mark was visible for about a week and then disappeared, but the consequences remained for a long time.

"For a long time, I slept under a piece of furniture in my house. I was very afraid that I might be attacked again. I also had a headache for a long time. What happened to me affected me deeply," said Newton.

But the following statement was something I had never heard before. *"I was healed by the aliens,"* said Newton.

"A few years ago, I was approached by a person from Acre (a state in the north of Brazil). He told me that the aliens wanted to talk to me, they wanted to cure me. He then took me to a shed I had at the back of my house in Colares. There he asked me to close my eyes and seconds later I felt sucked into another place. When I opened my eyes, I was in a place with rivers, with wooden houses and I saw small, blue beings with long fingers. Soon after that I came back. I couldn't speak in public; I couldn't communicate with anyone. I wouldn't be here talking to you so calmly. They cured me of that. Today I speak without any shyness."

Well, according to the dictionary, the meaning of "heal" is a way to combat a disease; restoration of health; process of recovering or improving something; improvement, regeneration. Could extraterrestrial beings have taken away Newton's "block" from speaking in public? For many this may seem silly, but for him it was something that hindered his life, something like an illness. And, according to Newton, these beings cured him of this "disease", that is, they did something to improve something. It is not for me to dispute that conviction. Whether or not he had this contact with blue beings from another planet, I cannot confirm or deny.

Shortly afterwards, we finished recording. We thanked Newton, his wife and daughter Gleice and left for our last task: interview journalist Carlos Mendes, the person who investigated the entire *"Chupa-Chupa"* phenomenon in 1977 and 1978.

The meeting with Carlos took place in the historic center of Belém. Then a young journalist during the dictatorship years, Carlos Mendes came under strong pressure from the military to stop publishing his articles about *"Chupa-Chupa"*.

He was constantly followed when he went to the attacked areas. *"They knew who I was and accompanied me all the time, and what made them very angry was that several times I arrived before them at the sites of attacks or sightings. In fact, I had started investigating the phenomenon four months earlier."* He came to live with Captain Uyrangê Hollanda, then captain, and describes him – unlike many others of his contemporaries – as a strong and oppressive man, determined and a dictator. *"The captain was a very difficult man, practically inaccessible and unapproachable."*

Figure 126. Thiago Ticchetti, George Knapp and Carlos Mendes. (Credit: Author's files).

Mendes was accompanied by his colleagues Biamir Siqueira, also a journalist, and Photographer José Ribamar dos Prazeres. Both came into close contact with "*Chupa-Chupa*", and Ribamar took hundreds of photos of the phenomena. Carlos stated that many of the photos that the air force claims were his were in fact taken by Ribamar, who, on one occasion, had his photos and even negatives confiscated by the military. "*It was a very strong and intimidating threat.*"

When we asked Carlos where the photos and negatives were, he sadly replied, "*my newspaper editor at the time couldn't stand the pressure from the military and handed over everything. Luckily, Ribamar and I kept some photos and others that been published in the newspaper and a few more.*"

We then addressed cases involving people who had been attacked by lights, ships, and unidentified flying objects. Was there an estimate of how many people would have seen the lights and how many would have been attacked? Without thinking twice Carlos said, "*in two* states, *Pará and Maranhão, I estimated that more than 250.000 people were witnesses to the "Chupa-Chupa" phenomenon, and around a 2.000 people were attacked*". George and I were silent. This number had never been raised. Nothing like this was ever estimated.

With so many people attacked, finding their medical records shouldn't be a problem, so where would the medical records of these thousand people be? Carlos replied, "*They're gone. All medical records disappeared from the forensic medical institute hospitals. They were handed over to the Americans.*"

It doesn't surprise me that North Americans were involved in the investigation of the "*Chupa-Chupa*" phenomenon, also because Captain Hollanda himself said so. Other witnesses, such as José Francisco, said that people speaking "foreign" were in Colares.

And Carlos went further: he said that he watched a video by Hall Puthoff, an American researcher and laser specialist, where he stated that he studied the reports of the "*Chupa-Chupa*" attacks, classifying them as very good. "*And folks, Dr. Puthoff had ties with the CIA.*" As he said this, George's eyes lit up. "*I know Puthoff. He is friends with Robert Bigelow, who is a very good friend of mine. I'm going to look for these documents.*"
"*I tell you that the CIA was involved in the investigation of the Chupa-Chupa phenomenon and wanted at all costs to take all the evidence we had,*" said Carlos. "*I discovered a CIA agent, who claimed to be Peruvian, infiltrated in our newspaper to steal the photos and everything we had there. His name, if true, was Juan. He showed up there asking for a job, as he had just moved to Belém*".
This suspicion corroborates what we ufologists have always thought. Operation Saucer ended unexpectedly and abruptly, but the "*Chupa-Chupa*" attacks continued long after, and American involvement in the investigation and/or theft of documents and photos becomes an increasingly true possibility. Another Brazilian Air Force mission was sent to the state in 1978, but without the same "success" as the previous one.

After a few more hours of interviews, with new revelations that can be seen when the episode of George Knapp's series airs, we finished our interview.

Chapter 21

What Could the "Chupa_Chupa‹ be?

So far, this question has not been answered, and since there are no convincing answers, official or unofficial, to explain what really happened, several hypotheses have emerged. The most common ones have to do with secret tests, secret aircraft, and an experiment by the US or even the former Soviet Union. Why did they do that in the states of Pará and Maranhão? What was the aim of it? For what purpose?

If we think along these lines, at least three crimes were committed against the Brazilian State.

Figure 127. Could the "Chupa-Chupa" phenomenon be an extraterrestrial activity? A secret project from another country? A communist invasion? (illustrative image).

The first crime was the violation of Brazilian airspace and its transformation into a testing ground for a bizarre experiment, using human guinea pigs. There was an attack on the sovereignty of one country by another that entered it without asking for permission. On the other hand, did the invader ask for permission, obtain it from the Brazilian government, but then everything was hushed up? Secondly, the Brazilian people, more precisely from the North of the country, were cowardly attacked, by force, subjecting themselves to an unusual blood extraction test, suffering physical, moral and psychological harm.

If this happened, it is clear that the Brazilian people should file two lawsuits with the international courts of justice: the first, of a criminal nature, would include as victims, according to estimates, more than 2,000 people, in Pará and Maranhão, attacked by the murderous lights in their homes, on roads, fishing boats, plantations in the countryside or during a simple walk somewhere when they were surprised and attacked by unknown machines and rays. The second lawsuit would be of a civil nature, for compensation for material, physical, emotional and moral damages. The victims of the lights and weapons coming from space, in several cases reported by the press and reports to the police, had to leave their homes in a hurry to run away from something they did not understand and that clashed even with their religious beliefs. They left their possessions and family behind. These people were also publicly ridiculed by some federal, state and municipal authorities, scholars, and even by more skeptical sectors of the press. They were called crazy, and became the butt of jokes and ridicule by people who did not believe their stories and doubted what had happened. They were nicknamed sons or daughters of aliens, lovers of beings from other worlds, idiots and fanatics.

Communists over Brazil?

Another hypothesis for the attack of the lights, this one started by the military, was that there was a new communist infiltration, this time no longer in the south of the state, where in 1974 the Araguaia Guerrilla, ideologically inspired by the Communist Party of Brazil (PCdoB), had been officially extinguished, but in the northeast of Pará, a vast region, 300 km in a straight line, that stretches from the municipality of Viseu, on the border with Maranhão, to Colares, just 63 km from the capital, Belém.

The communists wanted to overthrow the Brazilian military government established in 1964 and settled in Pará in 1969 to raise awareness among farmers and poor people that the seizure of power, by force of arms, would have to start from the countryside and move to the

Brazilian cities. The Armed Forces infiltrated agents and violently repressed the guerrillas, killing 69 of them in the forests of southern Pará.

However, dismantled and annihilated, would there be any new pockets of resistance, this time in Colares, Vigia, Santo Antônio do Tauá and even in Mosqueiro, areas which are closer to an invasion of Belém by supposed guerrillas? As incredible as it may seem, this was a hypothesis considered by Captain Hollanda, when he took charge of the second phase of Operation Saucer.

Hollanda was from the 2nd Section of the 1stRegional Air Command (COMAR), the so-called A2 of the Air Force in Belém, led by Colonel Camilo Barros. In Belém, one of the captain's activities was to monitor demonstrations by students, social movements and intellectuals from Pará who frequently took to the streets to protest against the military regime. Hollanda was known for his anti-communism and for being a military man who studied communist literature.

The communist hypothesis for the apparitions and attacks of the lights, however, was soon discarded by Brigadier Protásio Lopes de Oliveira, commander of the Brazilian Air Force in Belém, by Colonel Camilo Barros and, even more so, by Captain Hollanda himself. The rare boats that arrived at the coast of Vigia and Colares, with some weapons, were simply following the old smuggling route, which was repressed by the Federal Police. The weapons were destined for some stores in Belém. The most widely accepted theory, even by the victims of the lights themselves — and, what is even more incredible, by the military personnel responsible for investigating the facts — based on the testimony of Uyrangê Hollanda, from the episode on the Guajará-Mirim River — is the most frightening of all. Moreover, it is this theory that, almost 40 years later, revives the flame of what happened as a disturbing light that never leaves the conscious minds of the people attacked: the extraterrestrial origin. It is impossible to say that this would be impossible to happen, but there is also no way to rule it out with coherent explanations capable of disproving the victims themselves.

Secret Experimental Aircraft?

Yellow, blue, red, green, white, orange, light gray light. Color variations from yellow to red, from green to red, from red to orange, from blue to yellow. The craft that emitted these lights had different shapes: cigar, stingray, police van, gas cylinder, disc. In addition, they attacked mainly at night, in areas with little or no electricity. When they descended from the sky, according to the dozens of reports, they attacked in streets, roads, rivers, forests, sometimes first hovering over the homes of riverside dwellers and farmers. Then, they fired laser rays at people.

From a larger beam of light, after the victim was paralyzed, although still standing, but conscious and with their eyes open, smaller and very thin beams of light would come out, as if they were a kind of laser pen, which were directed, in the case of men, to the neck, thighs, arms and abdomen, and in the case of women to the breasts, arms, hands or legs. The marks of these attacks on the skin resembled small punctures, one next to the other. Soon after the apparitions and attacks, the victims most directly hit by the rays would faint and when they regained consciousness, the lights had already disappeared into the sky, at high speed, according to reports from neighbors and friends who ran to help the victims.

What was the objective, in the most specific cases, of the rays on people's bodies? To suck their blood for some genetic reproduction experiment in an environment totally foreign to that known to Earth science? And why just scare some people, without removing the blood from their bodies, sparing them from the invasion of their individuality? Were they not the right people for the kind of experience that was being conducted? These are questions gone with the wind and the tide, which however never bring the answers we so desire.

And if they were secret experimental aircraft, where would they be today? The technology that these objects presented allowed them to fly easily in the sky and navigate underwater in rivers. Do we have anything that can do this today?

In addition, if we consider the shapes of the objects seen, what technology are they using, which relies on no turbines, propellers, wings, tails and rudders? How can a cylindrical object fly and maneuver? And what's more, objects shaped like the "current" Tic-Tac UFOs, but which were seen almost 50 years ago. Can we then conclude that the Tic-Tac UFOs seen, photographed and filmed in 1977 and 1978 are the object seen by US Navy pilots in 2004?

After all, were they ships from other worlds or from our own, from the interior of the Earth, as old theories insist about the hollow Earth at the poles, through which advanced ships of supposed peoples, much more intelligent than us, would enter and leave, but who, officially, never gave any sign of their presence? Do the reports of some people and the testimony of military personnel and journalists, as occurred in Baía do Sol, in June 1978 and September 1979, about the diving of UFOs in the Pará River and then the vision of these same ships suddenly leaving the bottom of the river towards the atmosphere, confirm the existence in that region of a portal that would lead to the interior of the Earth?

Geologists vehemently deny that the rivers in that area are deeper than 40 m or that there are cracks in the surface of the rivers that could lead to depths still unknown. For them, it is all nothing more than delusions, speculations and crazy theories about things that, scientifically, cannot be proven. In fact, there is no shortage of theories about what happened in the region during the appearance of the lights.

Some say that Operation Saucer was in fact a fraud to cover up secret military tests, mainly by the US government, a partner of Brazil and supporter of the military regime that ruled our country for 21 years. However, if it was a fraud, how can we explain the fact that fishermen and farmers gave strong accounts of the attacks they suffered? Scorned for being illiterate, ignorant, mystical, malnourished and uninformed, were they used as pawns to manipulate even journalists and the media that widely reported the events, in many cases suffering all kinds of censorship and intimidation so that they would not do their job and leave the places where the lights appeared? You can say whatever you want, speculate as much as you want, but underestimate people's intelligence.

Evidence of an Enigma

There are hundreds of photographs, thousands of pages of documents from the Brazilian Air Force, hundreds of witnesses, dozens of newspaper covers, and more than 20 hours of footage (still "missing") of UFOs that prove that something happened in the states of

Maranhão and Pará between 1977 and 1978. But beyond all of this, there are peculiar and unmistakable characteristics of the phenomenon that became known worldwide as "Chupa-Chupa":

1. The UFO wave of the "Chupa-Chupa" phenomenon constituted a broad activity, lasting approximately six months (July-December 1977) basically affecting the main basins of northern Brazil and its surroundings.

2. The sightings and Close Encounters of the Second and Third Kinds (CE II & CEIII) are, in their entirety, nocturnal.

3. Spherical, cylindrical and fish-shaped objects are common.

4. Most UFOs move from airspace to land or from the ocean to the continent.

5. During their nocturnal evolutions, UFOs preferentially fly over small coastal and rural communities, often attacking human beings through powerful light projections with paralyzing effects.

6. The victims of the "Vampire Lights" are adults of both sexes and the attacks do not occur randomly. The injuries are first-degree burns, no longer than 15 centimeters in length, located, in most cases, over the thoracic region. During the incidence of the "Vampire Lights", general paralysis of the limbs, loss of speech, a sensation of shock, progressive numbness from the feet to the head, heat and the sensation of stinging in the region affected by the light are observed.

7. After the attacks, victims complain of dizziness, body pains, tremors, lack of energy, drowsiness, weakness, hoarseness, hair loss, peeling of the damaged skin and frequent headaches.

8. Sightings of the crew of these objects are rare, and the few reports describe creatures very similar to humans and of average height. A few cases report hairy, humanoid beings with large heads and black eyes.

Chapter 22

UFO/UAP Official Photos

MILITARY OPERATION: Operation Saucer – I COMAR - FAB MILITARY **REGISTRATION**: 024

LOCATION: Colares – Pará **DATE**: Saturday, November 5th, 1977 **TIME**: 18:26

EQUIPMENT: Minolta SRT 101 **LENS**: Rokkor zoom 100/200 mm

FILM: RE-2475 – 1000 ASA **APERTURE**: 5.6 **SHUTTER SPEED**: 1/30 sec

HEIGHT: 1,500 m **DISTANCE**: 1,200 m **APPARENT SIZE**: 2.5 m

SPEED: Medium (over 300 km/h) **TRAJECTORY**: Straight **COLOR**: Reddish yellow

SHAPE: Undefined **DESCRIPTION**: Luminous bodies

PHOTOGRAPHER AND/OR OBSERVERS: Military team from A2 – I COMAR

OBSERVATIONS: Composed of 5 photographs in the military report. Electrostatic and magnetic effects were observed for a period of 20 minutes. They followed a southwest/northeast direction.

Military Registry - FAB nº 024. Colares, Pará state. November 5th, 1977. Time: 06h26 P.M. Photographer: A2 Military Team (credit: I COMAR)

MILITARY OPERATION: Operation Saucer – I COMAR - FAB MILITARY **REGISTRATION**: 041

LOCATION: Baía do Sol, Mosqueiro - Pará

UFO Attacks in Brazil

DATE: Tuesday, November 22nd, 1977 TIME: 05:18

EQUIPMENT: Yashica Eletro TLS 35 **LENS**: 400 mm

FILM: 1000 ASA **APERTURE**: 6.3 **SHUTTER SPEED**: 1/30 sec

HEIGHT: 1,500 m **DISTANCE**: 1,500 m **APPARENT SIZE**: 2 and 3 cm respectively

SPEED: Low (200 km/h) **TRAJECTORY**: One UFO was rectilinear and the other irregular (curved)

COLOR: Dull yellow **SHAPE**: Undefined **DESCRIPTION**: Luminous bodies (2)

PHOTOGRAPHER AND/OR OBSERVERS: 1S C AT MT FLÁVIO COSTA and civilians UBIRATAN PINON FRIAS, Mrs. F.M. FREITAS COSTA and Miss CÉLIA MONTENEGRO

OBSERVATIONS: The official report includes: 6 photographs. One of the UFOs flew in a south/north direction, while the other object flew in a south/northeast direction. At a certain point, one UFO crossed the path of the other flying object. At some points, one of the UFOs accelerated at a constant rate.

Military Registry - FAB nº 041. Baía do Sol, Mosqueiro, Pará state. November 22nd, 1977. Time: 05h18 A.M. Photographer: 1st Flávio Costa (credit: I COMAR)

UFO Attacks in Brazil

MILITARY OPERATION: Operation Saucer – I COMAR - FAB MILITARY **REGISTRATION**: 062

LOCATION: Baía do Sol, Mosqueiro - Pará

DATE: Friday, December 9th, 1977 **TIME**: 23:50

EQUIPMENT: Yashica Eletro TLS 35 **LENS**: 400 mm **FILM**: 1000 ASA

APERTURE: 6.3 **SHUTTER SPEED**: 1/30 sec **HEIGHT**: 2,000 m

DISTANCE: 3,000 m **APPARENT SIZE**: 5 cm **SPEED**: Average (over 300 km/h)

TRAJECTORY: Slightly curved to the right **COLOR**: Reddish yellow

SHAPE: Undefined **DESCRIPTION**: Luminous body

PHOTOGRAPHER AND/OR OBSERVERS: 1S C AT MT FLÁVIO COSTA and civilian UBIRATAN PINON FRIAS

OBSERVATIONS: The official report includes: 1 photograph. The UFO followed in a southwest/east direction. The UFO moved as if driven by impulses, and the light intensity varied greatly. No noise was heard. The change in route occurred after the observers attempted to get closer.

Military Registry - FAB nº 062. Baía do Sol, Mosqueiro, Pará state. December 9th, 1977. Time: 11h50 P.M. Photographer: 1st soldier Flávio Costa (credit: I COMAR)

MILITARY OPERATION: Operation Saucer – I COMAR - FAB MILITARY **REGISTRATION**: 065

LOCATION: Rio Guajará, Ananindeua – Pará **DATE**: Saturday, December 10th, 1977 **TIME**: 00:50

EQUIPMENT: Minolta SRT 101 **LENS**: Rokkor Zoom 100/200 mm

FILM: RE-2475 - 1000 ASA **APERTURE**: 5.6 **SHUTTER SPEED**: 1/30 sec

HEIGHT: 50 m **DISTANCE**: 2,000 m **APPARENT SIZE**: 25 cm

SPEED: Variable (acceleration impulses)

TRAJECTORY: Irregular (varying in zigzag) **COLOR**: Reddish yellow

SHAPE: Undefined **DESCRIPTION**: Body luminous

PHOTOGRAPHER AND/OR OBSERVERS: Military team from A2 - I COMAR

OBSERVATIONS: Composed of the military report: 9 photographs. The UFO made several evolutions, constantly changing direction. Total absence of noise or air movement. The UFO followed in a Southeast/Northwest direction.

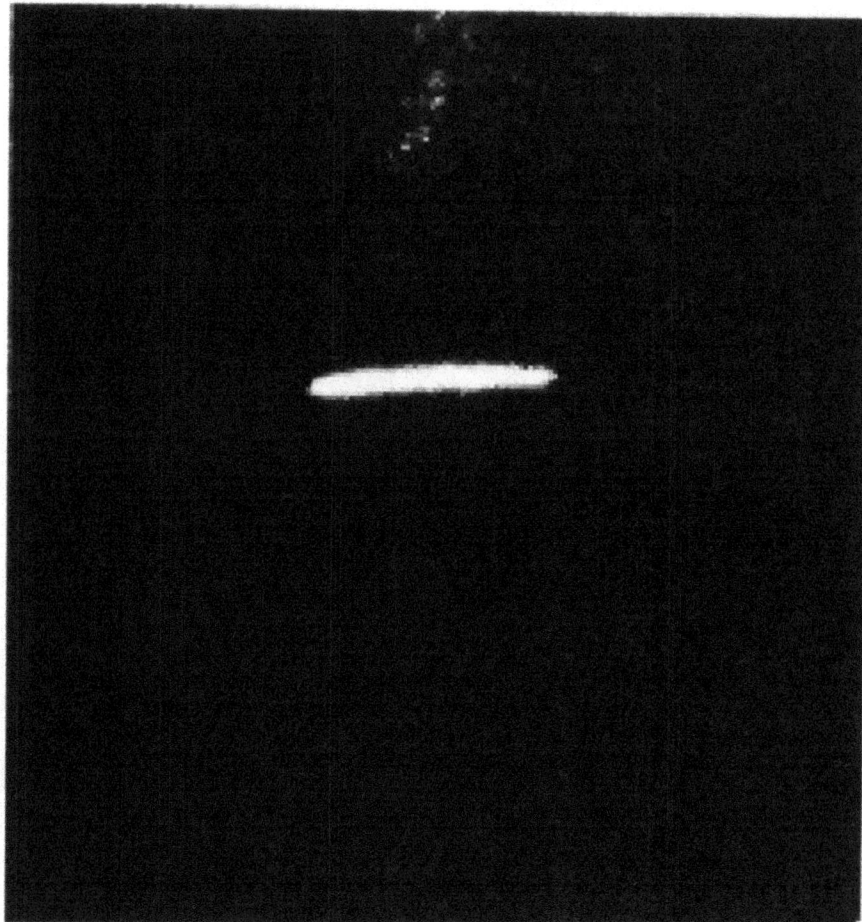

Military Registry - FAB nº 065. Guajará River, Ananindeua, Pará state. December 10th, 1977. Time: 00h50. Photographer: A2 military team (credit: I COMAR)

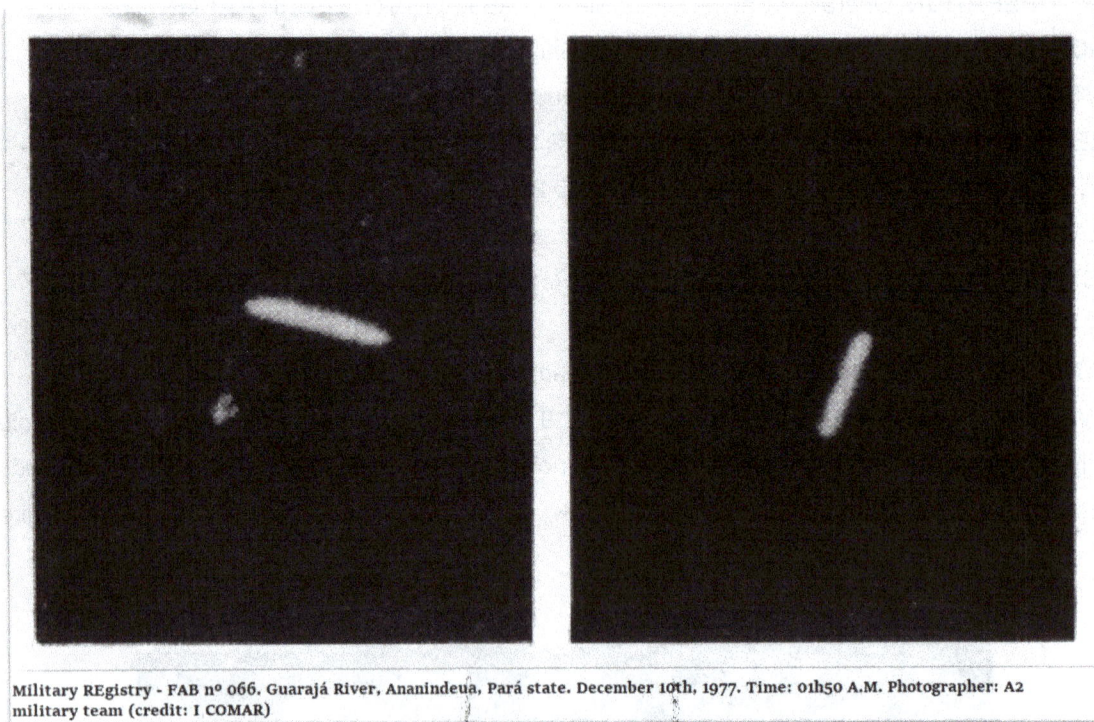

Military REgistry - FAB nº 066. Guarajá River, Ananindeua, Pará state. December 10th, 1977. Time: 01h50 A.M. Photographer: A2 military team (credit: I COMAR)

Military Registry - FAB nº 066. Guarajá River, Ananindeua, Pará state. December 10th, 1977. Time: 01h50 A.M. Photographer: A2 military team (credit: I COMAR)

278

MILITARY OPERATION: Operation Saucer – I COMAR - FAB MILITARY **REGISTRATION**: 067

LOCATION: Baía do Sol, Mosqueiro – Pará (over Ponta do Machadinho – Colares Island)

DATE: Saturday, December 10th, 1977 **TIME**: 20:30 **EQUIPMENT**: Yashica Eletro TLS 35 **LENS**: 400 mm **FILM**: RE-2475 - 1000 ASA **APERTURE**: 6.3 **SHUTTER SPEED**: 1/30 sec

HEIGHT: 300 m **DISTANCE**: 3,000 to 3,500 m **APPARENT SIZE**: Undefined (the flash covered an area of 50 cm) **SPEED**: Stationary **TRAJECTORY**: Stationary in the North sector

SHAPE: Undefined **DESCRIPTION**: Body luminous **COLOR**: Undefined (like fireworks)

PHOTOGRAPHER AND/OR OBSERVERS: 1S Q AT MT FLÁVIO COSTA and civilians A. N. Dias, F. M. Costa Dias and Mrs. F.M. Freitas Costa

OBSERVATIONS: Composed of the military report: 1 photograph. Short period of visibility (explosions). Possible change of position. At 9:30 p.m., a phenomenon with the same characteristics was observed in the Southeast sector. The UFO remained stationary in the North sector. It remained visible from 8:30 p.m. to 8:35 p.m.

Military Registry - FAB nº 067. Baía do Sol, Mosqueiro, Pará. December 10th, 1977. Time: 08h30 P.M. Photographer: 1st soldier Flávio Costa (credit: I COMAR)

MILITARY OPERATION: Operation Saucer – I COMAR - FAB MILITARY **REGISTRATION**: 068

LOCATION: Baía do Sol, Mosqueiro – Pará (over Colares Island – Jejutaua River)

DATE: Sunday, December 11th, 1977 **TIME**: 03:25 **EQUIPMENT**: Yashica Eletro TLS 35 **LENS**: 400 mm **FILM**: RE-2475 - 1000 ASA **APERTURE**: 6.3 **SHUTTER SPEED**: 1/30 sec

HEIGHT: 3,000 m **DISTANCE**: 1,500 m **APPARENT SIZE**: 10 cm **SPEED**: Low (-200 km/h)

TRAJECTORY: Straight trajectory **COLOR**: Light yellow (pale) **SHAPE**: Undefined **DESCRIPTION**: Body luminous **PHOTOGRAPHER AND/OR OBSERVERS**: 1S Q AT MT FLÁVIO COSTA and civilians A. N. Dias, F. M. Costa Dias and Mrs. F.M. Freitas Costa

OBSERVATIONS: Composed of the military report: 11 photographs. Variation of luminous intensity in a measured (pulsating) manner. The larger UFO was accompanied by a smaller one.

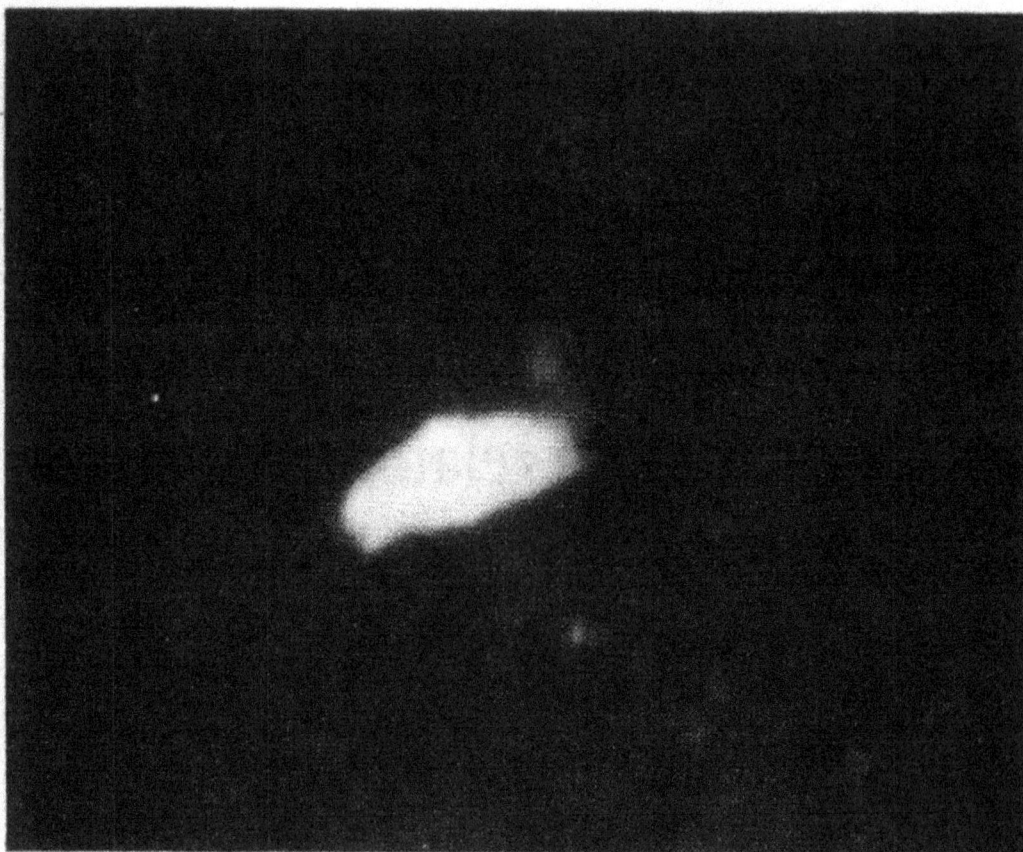

Military Registry - FAB nº 068. Baía do Sol, Mosqueiro, Pará state. 11th of December 1977. Time: 03h25 A.M. Photographer: 1º soldier Flávio Costa. (credit: I COMAR)

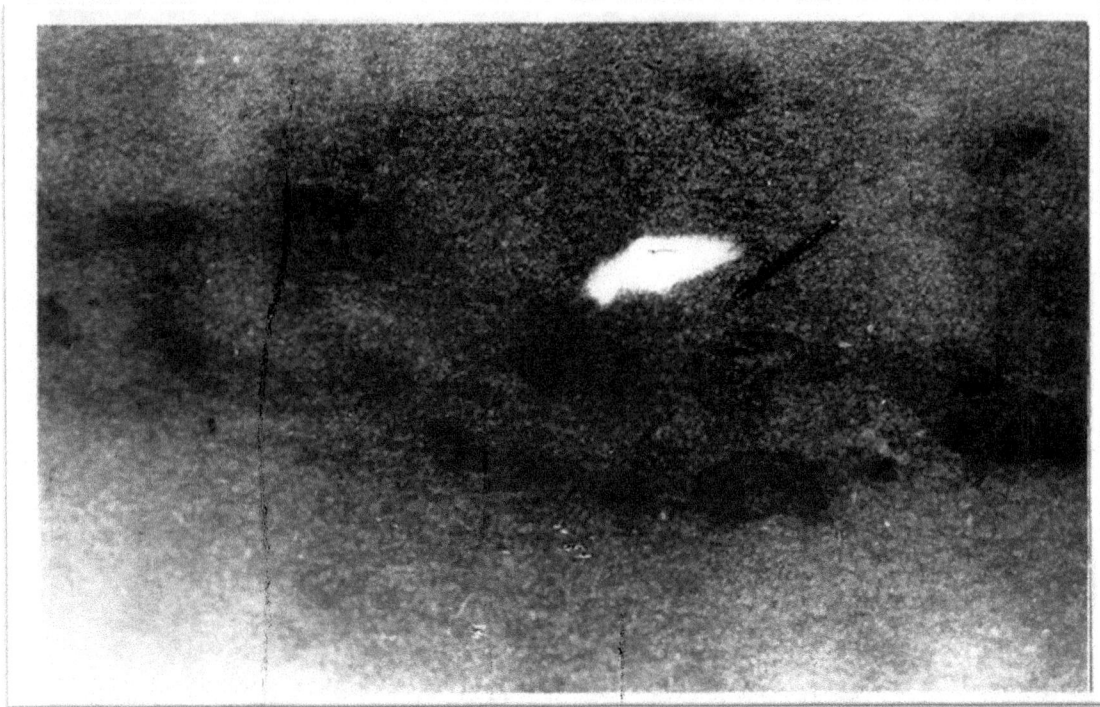

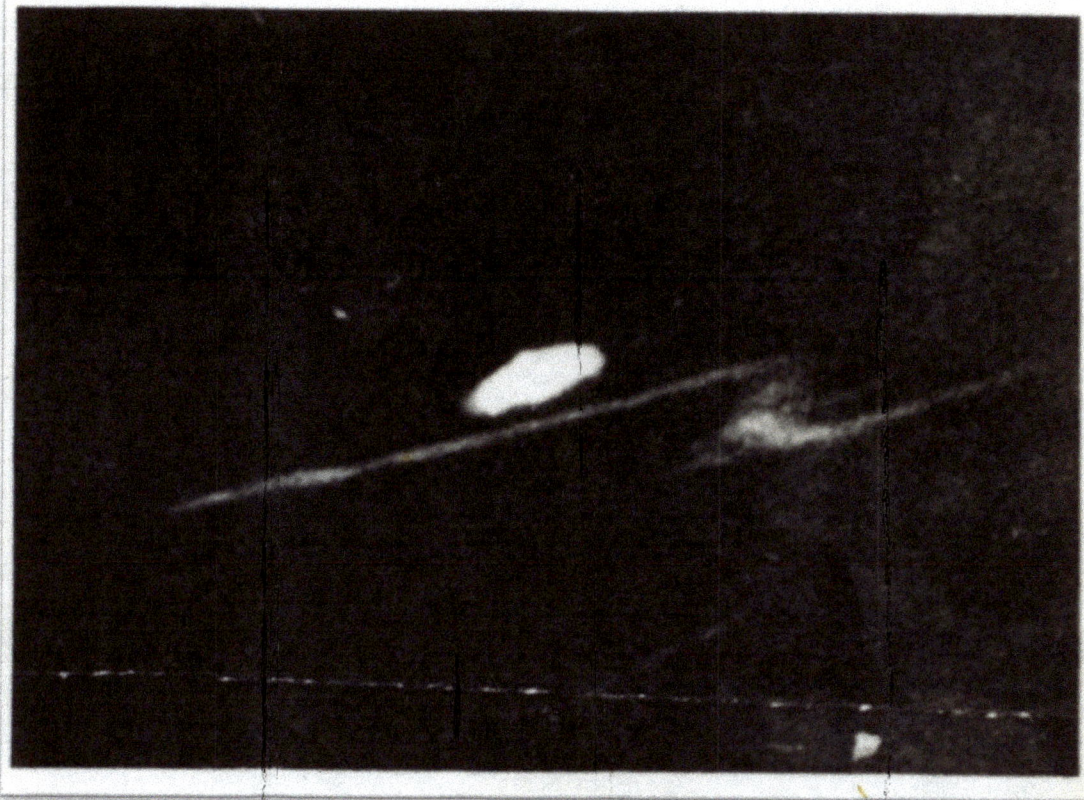

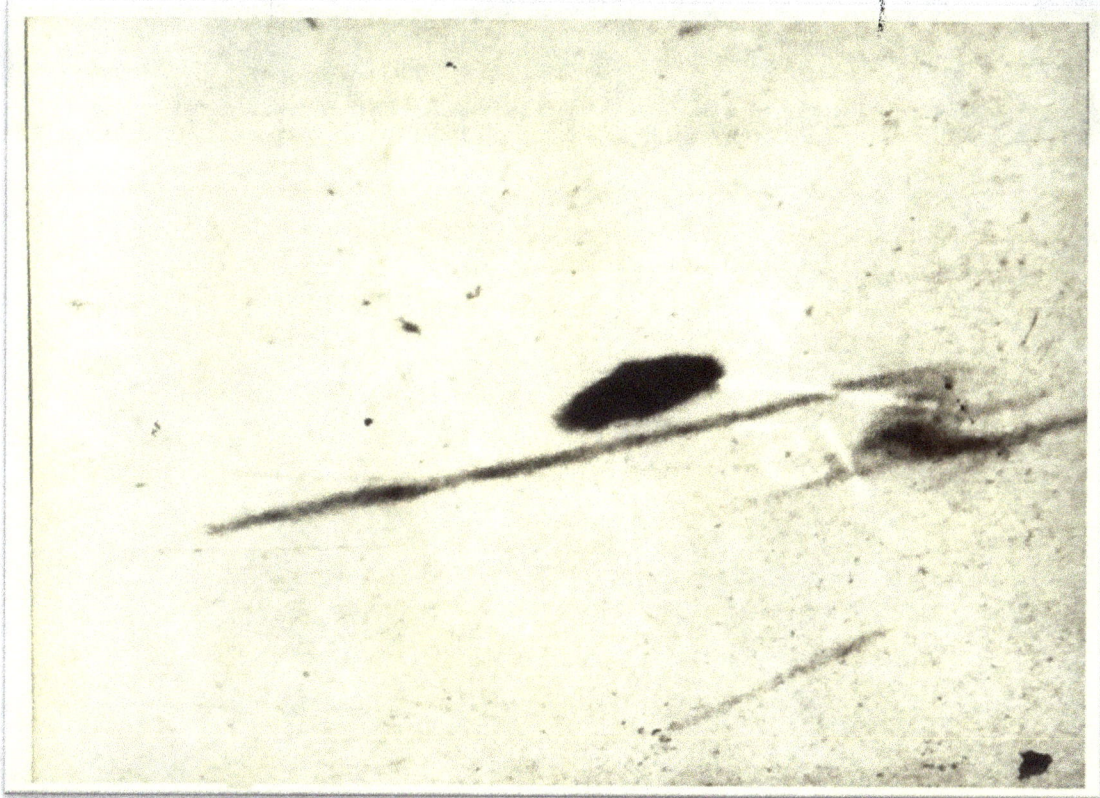

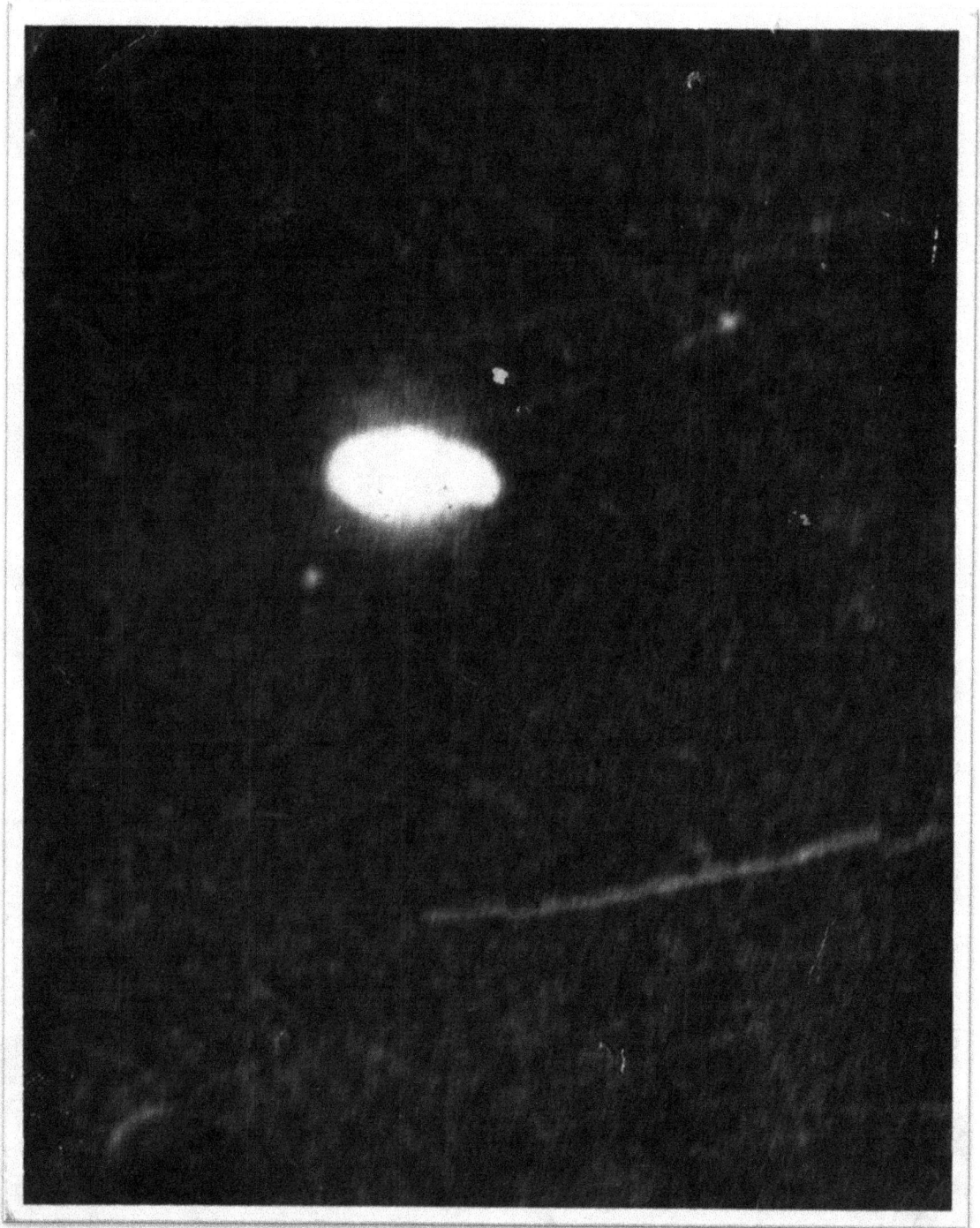

MILITARY OPERATION: Operation Saucer – I COMAR - FAB MILITARY **REGISTRATION:** 070

LOCATION: Baía do Sol, Mosqueiro – Pará **DATE**: Tuesday, December 13th, 1977 **TIME**: 21:05

EQUIPMENT: Yashica Eletro TLS 35 **LENS**: 400 mm **FILM**: RE-2475 - 1000 ASA

APERTURE: 6.3 **SHUTTER SPEED**: 1/30 sec **HEIGHT:** 1,000 m **DISTANCE**: 500 m **APPARENT SIZE**: 30 cm

SPEED: Stationary **TRAJECTORY**: Apparently stationary **SHAPE**: Undefined **DESCRIPTION**: Luminous body

COLOR: Reddish yellow, intense bluish reflections and variations in light intensity, in short periods

PHOTOGRAPHER AND/OR OBSERVERS: 1S Q AT MT FLÁVIO COSTA and civilians UBIRATAN PINON FRIAS, Mrs. F. M. Freitas Costa and Miss CÉLIA MONTENEGRO **OBSERVATIONS**: The military report includes: 3 photographs. No noise was heard, nor was any marked movement observed. It disappeared in the same position. UV filter – ROKKOR (adapted to the camera lens) was used.

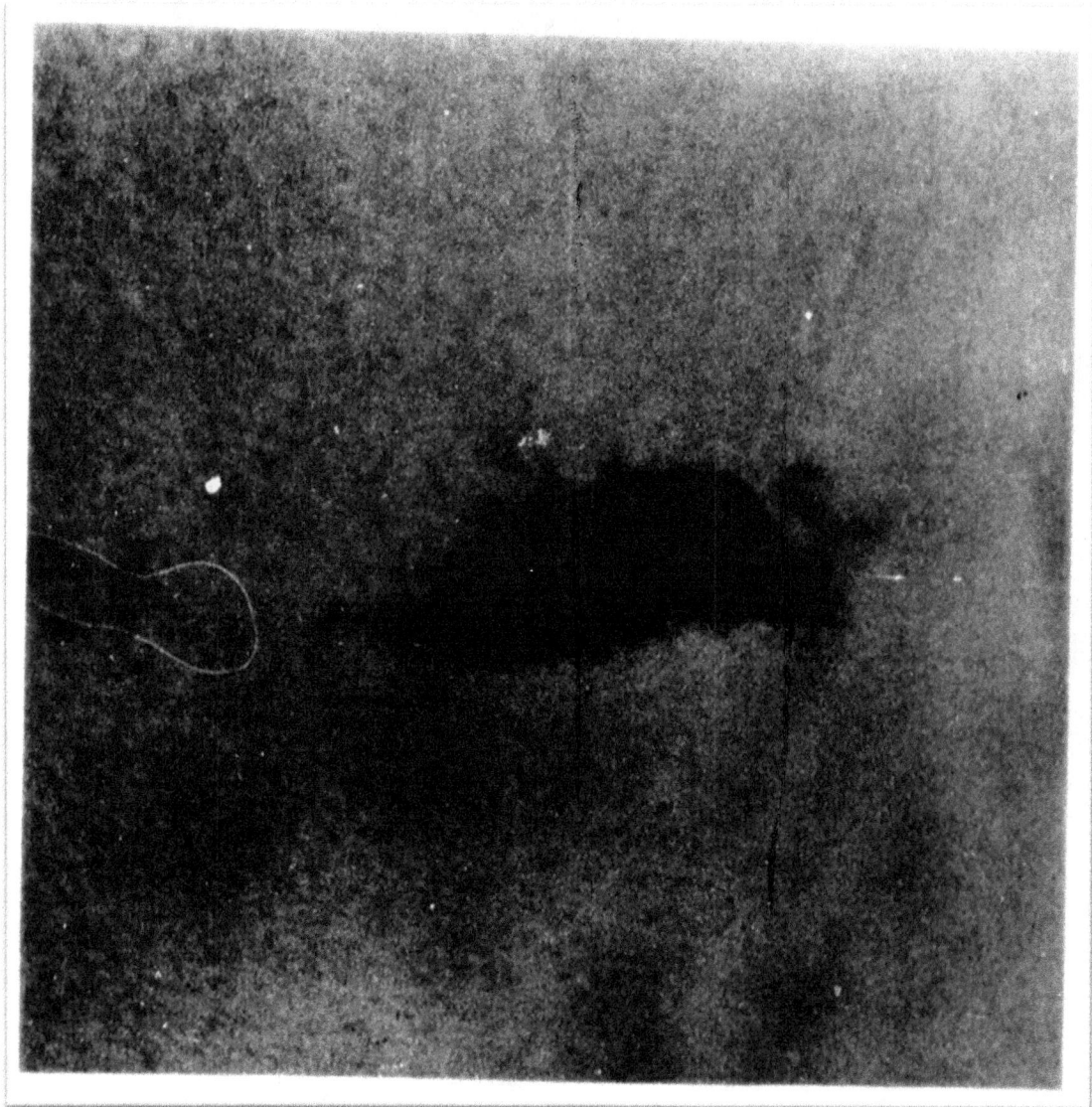

UFO Attacks in Brazil

MILITARY OPERATION: Operation Saucer – I COMAR - FAB MILITARY **REGISTRATION**: 071

LOCATION: Baía do Sol, Mosqueiro – Pará (over Colares Island – Jejutaua River) **DATE:** Tuesday, December 13th, 1977 **TIME**: 23:55 **EQUIPMENT:** Yashica Eletro TLS 35 LENS: 400 mm **FILM:** RE-2475 - 1000 ASA **APERTURE**: 6.3 **SHUTTER SPEED**: 1/30 sec **HEIGHT**: 2,000 m **DISTANCE**: 3,000/4,000 m **APPARENT SIZE**: 5 cm **SPEED**: Medium (over 300 km/h)

TRAJECTORY: Straight COLOR: Reddish yellow SHAPE: Undefined

DESCRIPTION: Luminous body **PHOTOGRAPHER AND/OR OBSERVERS**: 1S Q AT MT FLÁVIO COSTA and civilians UBIRATAN PINON FRIAS and Mrs. F. M. Freitas Costa **OBSERVATIONS**: Composed of the military report: 1 photograph. UFO followed in a Northwest/Southeast direction.

Military Registry - FAB nº 071. Baía do Sol, Mosqueiro, Pará state. December 13th, 1977. Time: 11:55 P.M. Photographer: 1st soldier Flávio Costa. (credit: I COMAR)

MILITARY OPERATION: Operation Saucer – I COMAR - FAB MILITARY **REGISTRATION**: 074

LOCATION: Guajará River, Ananindeua – Pará **DATE**: Wednesday, December 14th, 1977
TIME: 22:45 **EQUIPMENT**: Minolta SRT 101 **LENS**: ROKKOR ZOOM 100/200 mm
FILM: RE-2475 - 1000 ASA **APERTURE**: 5.6 **SHUTTER SPEED**: 1/30 sec **HEIGHT**: 500 m **DISTANCE**: 1,000 m **APPARENT SIZE**: 10 cm**SPEED**: Variable (high-speed bursts) **TRAJECTORY**: Straight**COLOR**: Reddish yellow **SHAPE**: Undefined **DESCRIPTION**: Luminous body**PHOTOGRAPHER AND/OR OBSERVERS**: Military team from A2 – I COMAR

OBSERVATIONS: The military report includes: 1 photograph. The UFO followed a northeast/southwest direction. It made several evolutions, occasionally emitting flashes of high-intensity blue light (electric welding). It made no noise or air displacement.

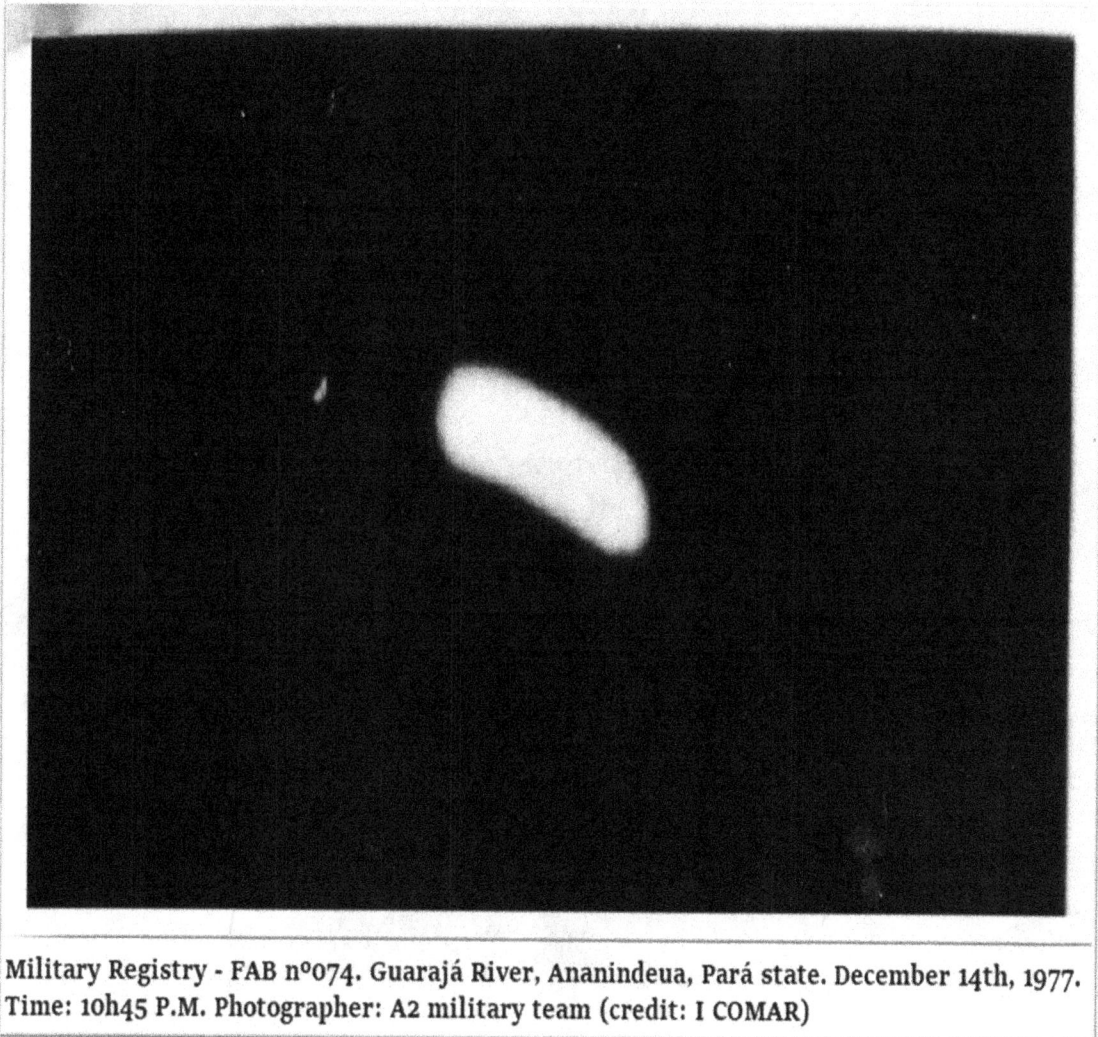

Military Registry - FAB nº074. Guarajá River, Ananindeua, Pará state. December 14th, 1977. Time: 10h45 P.M. Photographer: A2 military team (credit: I COMAR)

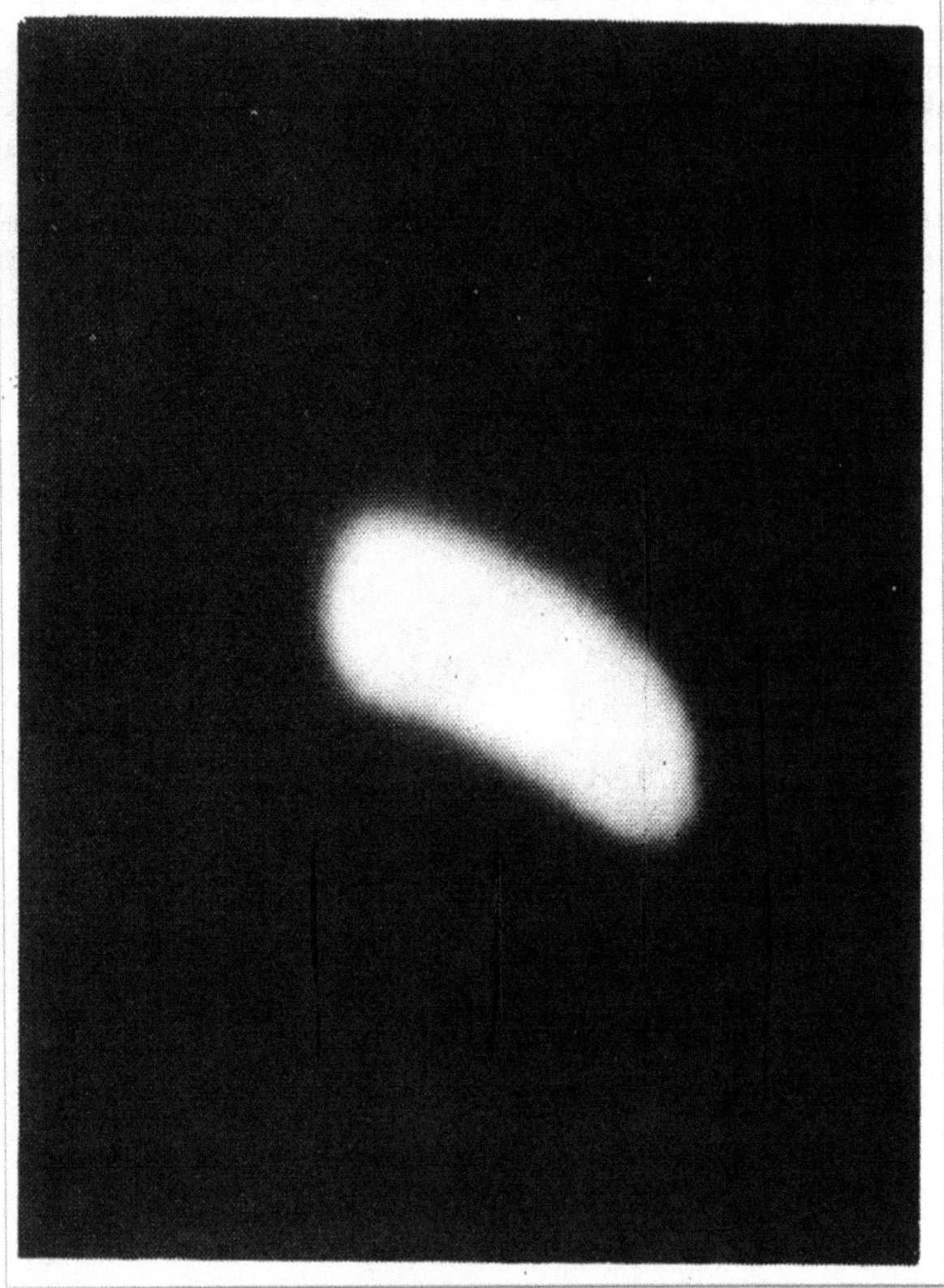

LOCATION: Highway PA-47 – Jejú/Piçarreira Farm – São Domingos do Capim – Pará

DATE: Friday, December 16th, 1977 **TIME**: 23:50 **EQUIPMENT**: Minolta SRT 101 **LENS**: ROKKOR ZOOM 100/200 mm **FILM**: RE-2475 - 1000 ASA **APERTURE**: 5.6

SHUTTER SPEED: 1/30 sec **HEIGHT**: 4,000 m **DISTANCE**: 2,500 m

APPARENT SIZE: 2 to 3 cm **SPEED**: Average (over 300 km/h) **TRAJECTORY**: Straight and descending

COLOR: Light yellow, bright **SHAPE**: Undefined **DESCRIPTION**: Luminous body

PHOTOGRAPHER AND/OR OBSERVERS: Military team from A2 – I COMAR

OBSERVATIONS: The military report includes: 6 photographs. The UFO followed in a Southeast/Northwest direction. From a distance, the UFO looked like a very large and bright star. When it leveled off (300 m), it was covered by tall grass. The team moved on foot to get a better look.

Military Registry - FAB nº 075. Highway PA-47, Fazenda Jejú / Piçarreira, São Domingos do Capim, Pará state. December 16th, 1977. Time: 11h50 P.M. Photographer: A2 Military Team (credit: I COMAR)

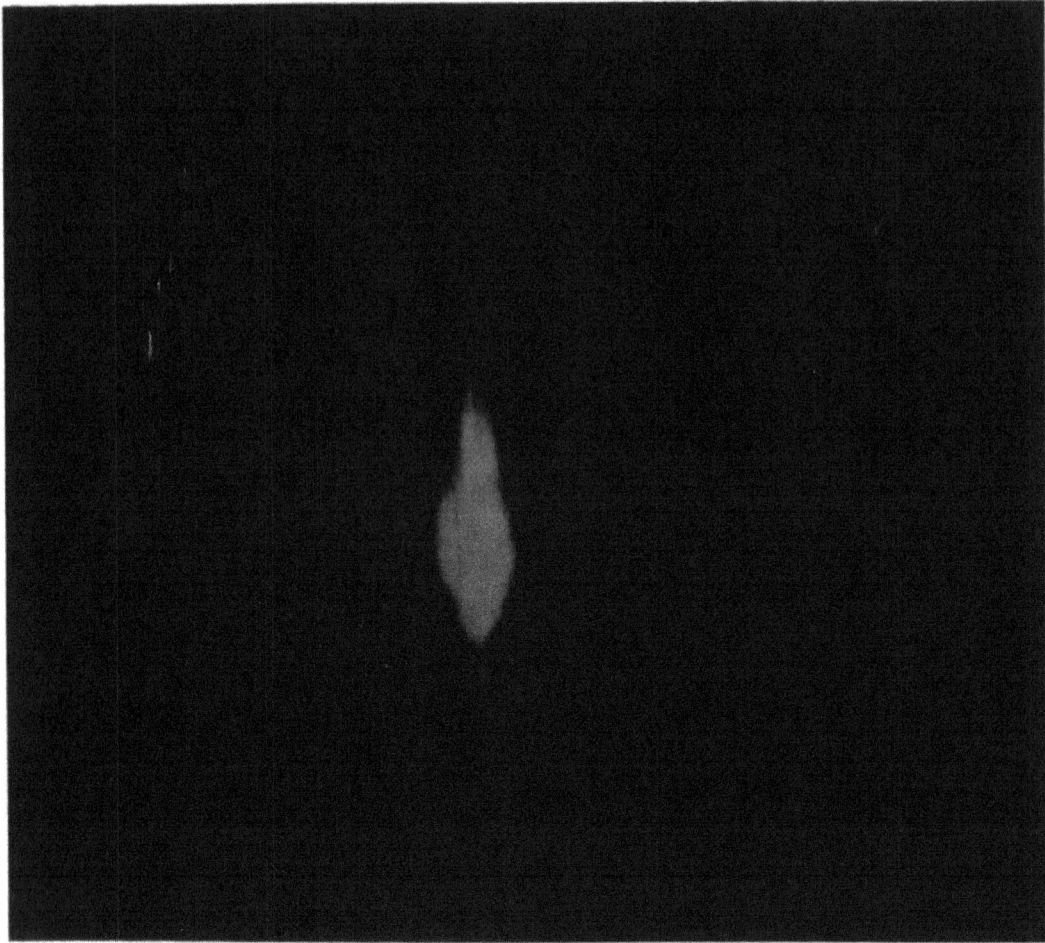

Military Registry - FAB nº 075. Highway PA-47, Fazenda Jejú / Piçarreira, São Domingos do Capim, Pará state. December 16th, 1977. Time: 11h50 P.M. Photographer: A2 military team (credit: I COMAR)

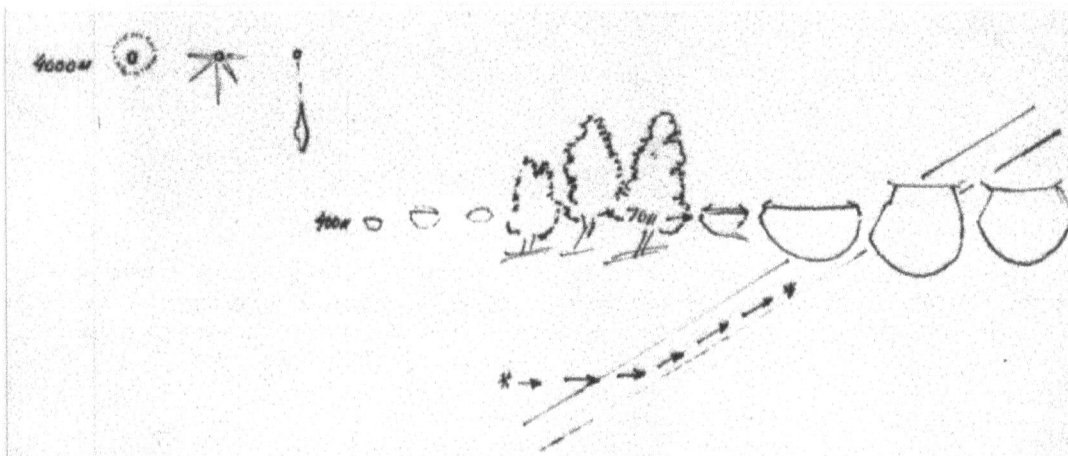

Figure 128. Trajectory and "transformation" of the UFO, according to the military report. (Credit: 1st COMAR).

MILITARY OPERATION: Operation Saucer – I COMAR - FAB MILITARY **REGISTRATION**: 076

LOCATION: Highway PA-47 – Jejú/Piçarreira Farm – São Domingos do Capim – Pará

DATE: Saturday, December 17th, 1977 **TIME**: 00:30 **EQUIPMENT**: Minolta SRT 101

LENS: ROKKOR ZOOM 100/200 mm **FILM**: RE-2475 - 1000 ASA **APERTURE**: 5.6 **SHUTTER SPEED**: 1/30

sec **HEIGHT**: 200 m **DISTANCE**: 300 m **APPARENT SIZE**: 50 cm

SPEED: Variable. Slow climb. Accelerated, increasing in ascent

TRAJECTORY: Spinning (top) and undulating **COLOR**: Reddish yellow, with bluish reflections

SHAPE: Circular (tapered) **DESCRIPTION**: Luminous body

PHOTOGRAPHER AND/OR OBSERVERS: Military team from A2 – I COMAR

OBSERVATIONS: Composed of 26 photographs in the military report. The UFO followed in a northwest/southeast direction. The UFO changed color, going from a very intense bluish white to a reddish red, varying its brightness and speed. The team members did not observe its ascent. It disappeared through the interposition of trees. The UFO descended from a height of 400 meters to 70 meters.

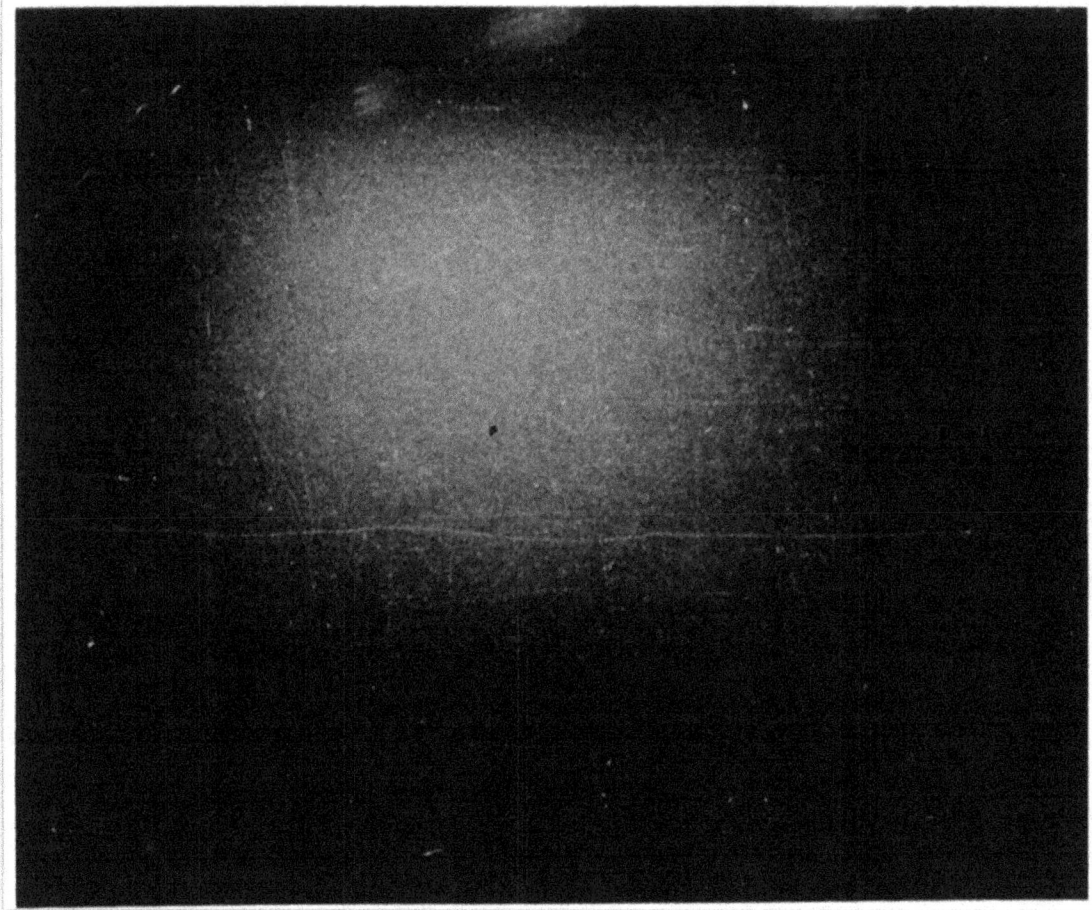

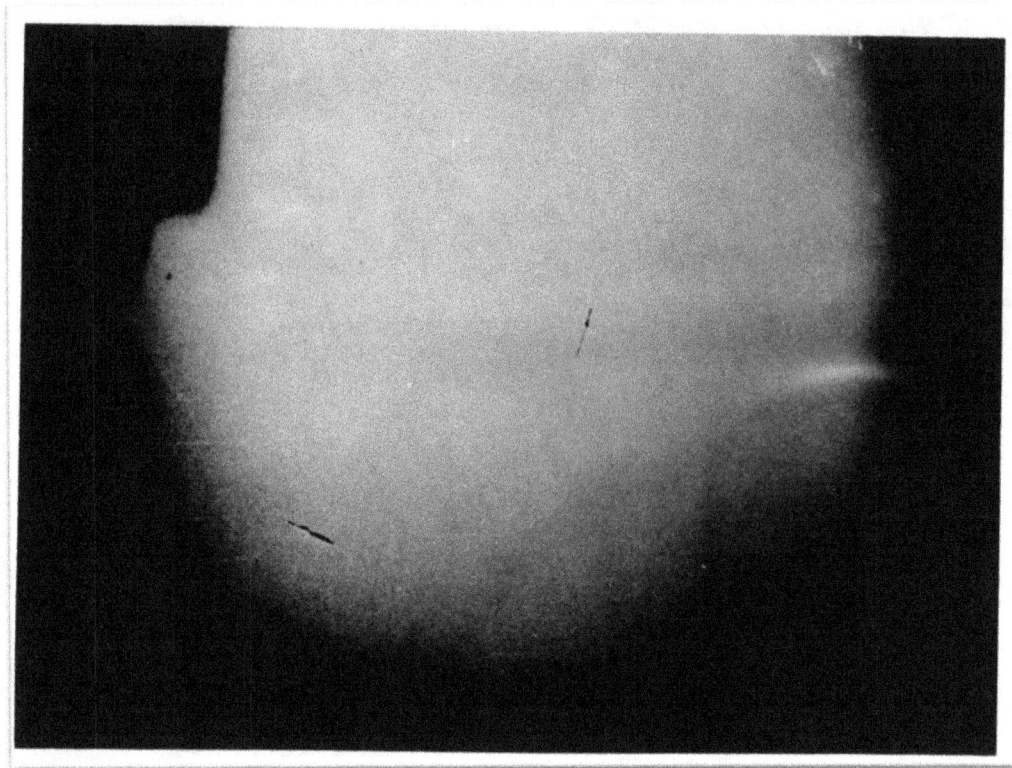

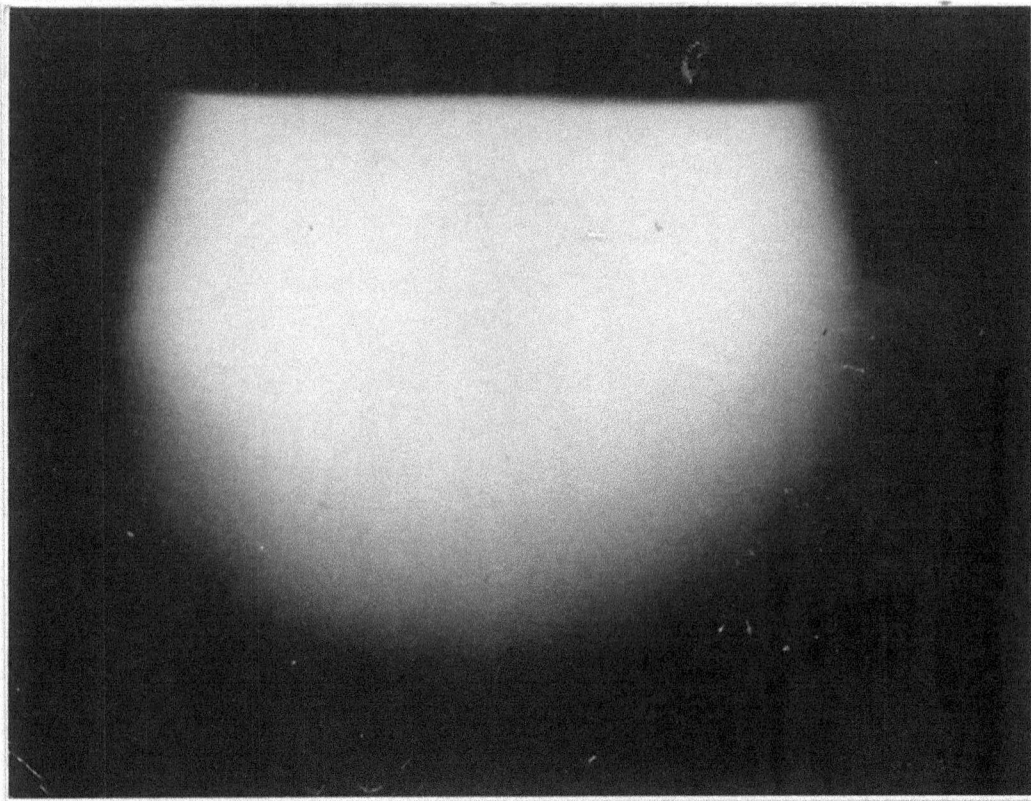

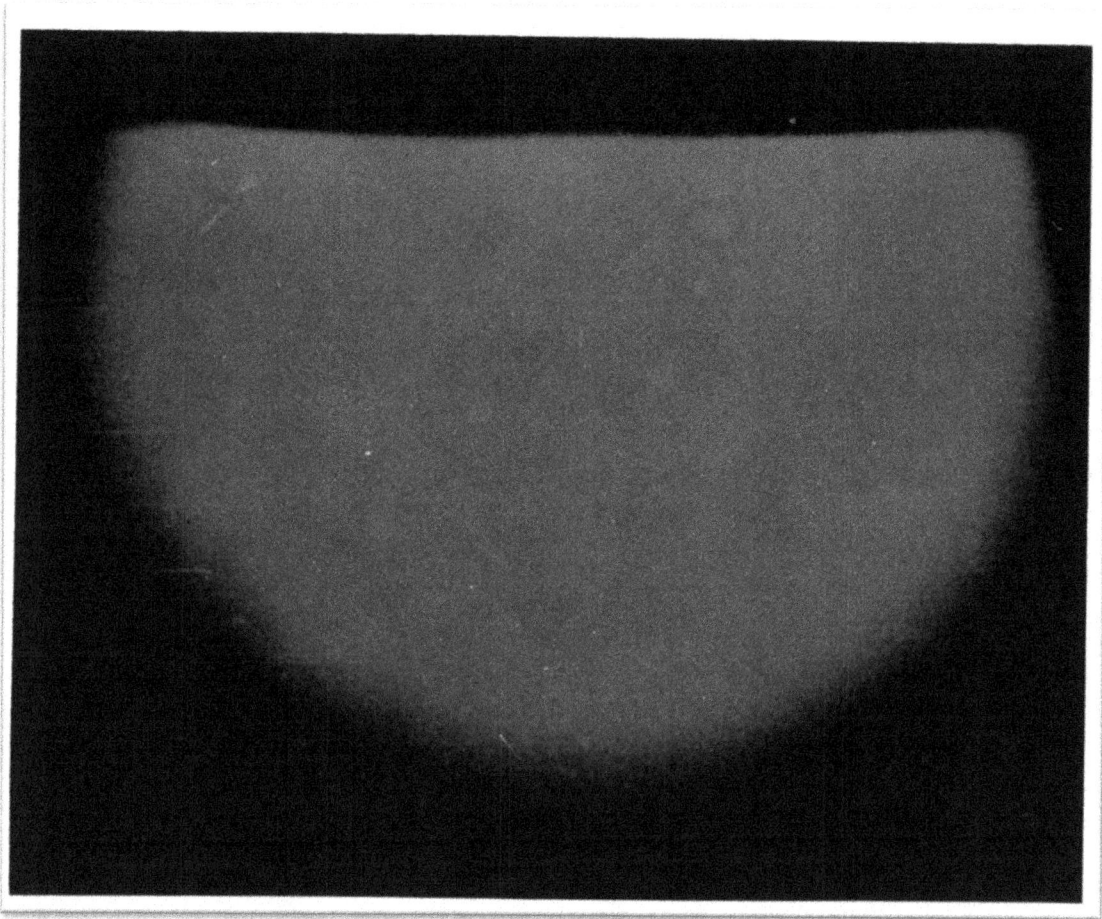

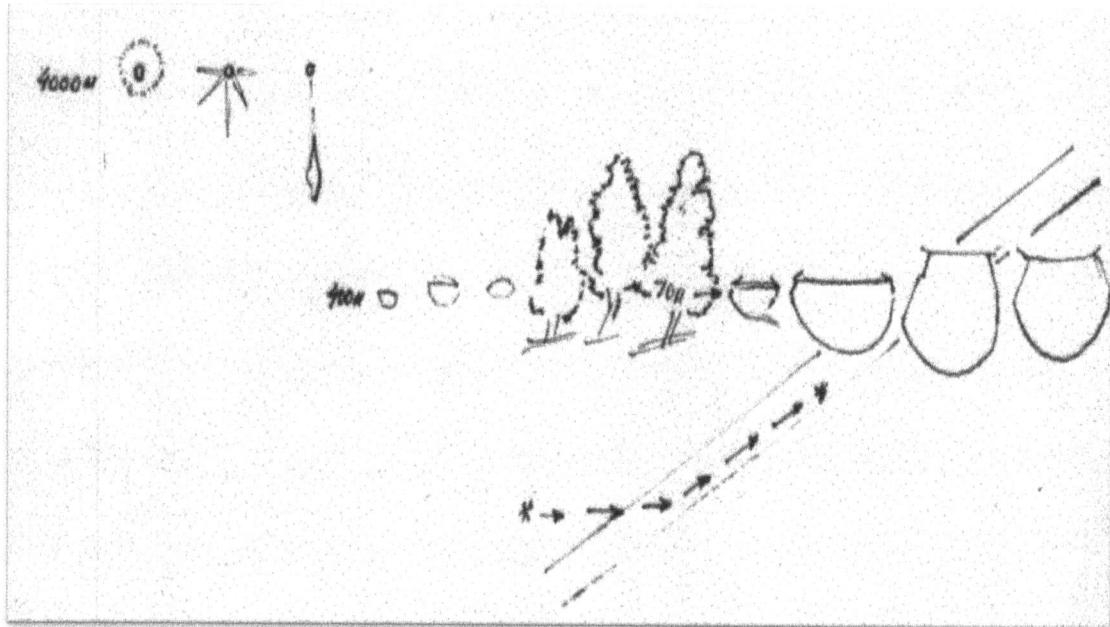

Figure 129. Trajectory and "transformation" of the UFO, according to the military report. (Credit: 1st COMAR).

MILITARY OPERATION: Operation Saucer – I COMAR - FAB MILITARY **REGISTRATION**: 076

LOCATION: Highway PA-47 – Jejú/Piçarreira Farm – São Domingos do Capim – Pará

DATE: Saturday, December 17th, 1977 **TIME**: ? **EQUIPMENT**: Olimpus Trip 35

LENS: ? FILM: 100 ASA **APERTURE**: ? **SHUTTER SPEED**: ?**HEIGHT**: 0 m **DISTANCE**: ? **APPARENT SIZE**:

?**SPEED**: ? **TRAJECTORY**: Landed **COLOR**: Reddish yellow, with bluish reflections **SHAPE**: Circular (tapered)

DESCRIPTION: Luminous body

PHOTOGRAPHER AND/OR OBSERVERS: Military team from A2 – I COMAR

OBSERVATIONS: Composed of 26 photographs in the military report. The UFO landed on the Jejú Farm, owned by Mr. Expedito, leaving marks on the ground. Two 40 x 40 cm tides were found, with a depth of 50 cm, bordered by an external circle approximately 2.5 meters in diameter.

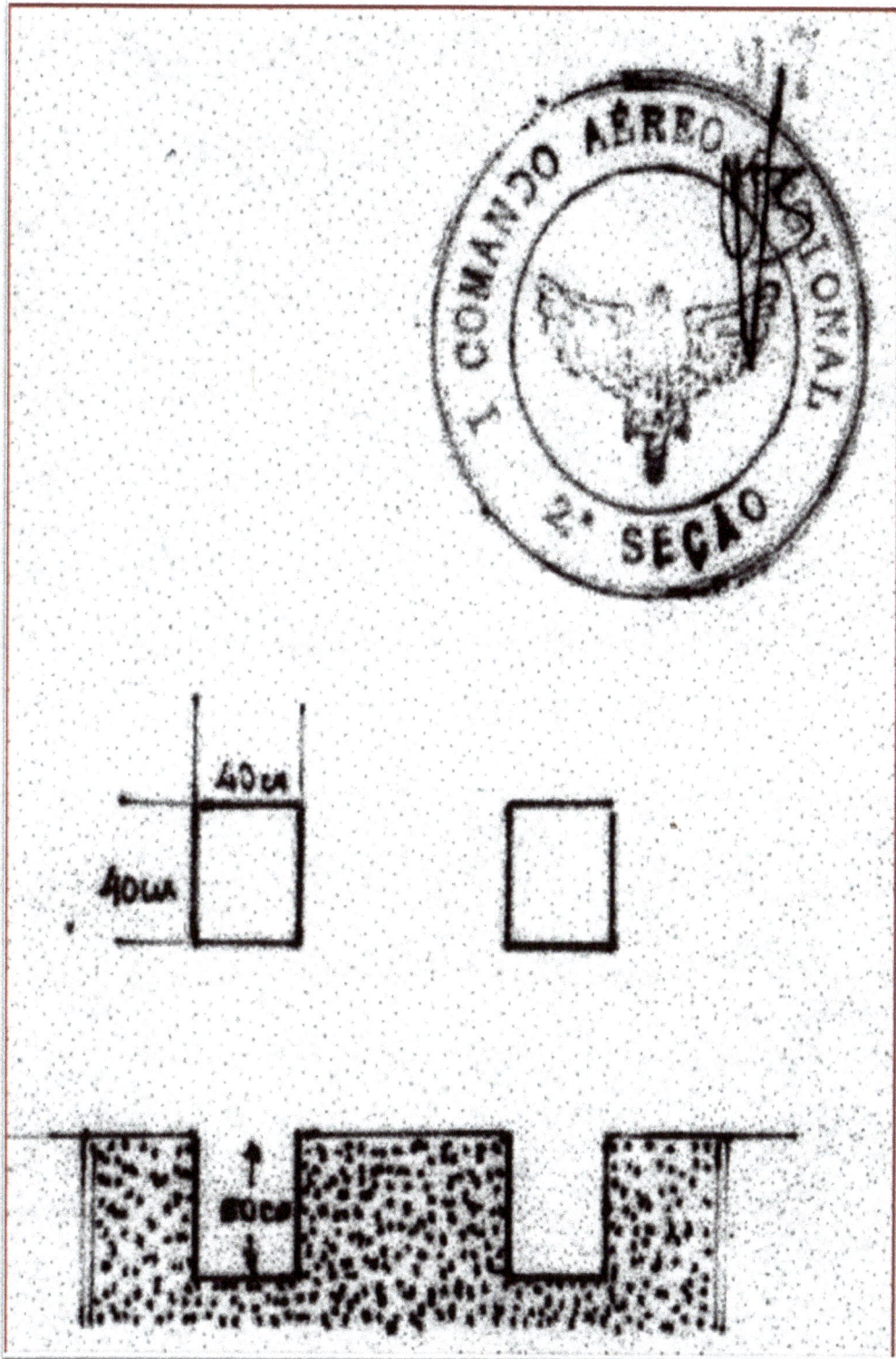

MILITARY OPERATION: Operation Saucer – I COMAR - FAB MILITARY **REGISTRATION**: 080 **LOCATION**: Baía do Sol, Mosqueiro – Pará **DATE**: Thursday, December 22nd, 1977 **TIME**: 01:30

EQUIPMENT: Minolta SRT 101 LENS: ROKKOR ZOOM 100/200 mm **FILM**: RE-2475 - 1000 ASA **APERTURE**: 5.6 **SHUTTER SPEED**: 1/30 sec **HEIGHT**: 1,100 m **DISTANCE**: 1,000 m **APPARENT SIZE**: 15 cm **SPEED**: Average (500 km/h) **TRAJECTORY**: Straight and then curved **COLOR**: Reddish yellow **SHAPE**: Undefined DESCRIPTION: Luminous body

PHOTOGRAPHER AND/OR OBSERVERS: 1S Q AT MT FLÁVIO COSTA and civilians UBIRATAM PINON FRIAS, Mrs. F. M. FREITAS COSTA and Miss. CÉLIA MONTENEGRO

OBSERVATIONS: Composed of the military report: 4 photographs. UFO followed East/Southwest direction, making a turn to Southeast. Accelerated to supersonic speed. The UFO quickly disappeared.

Military Registry - FAB nº 080. Baía do Sol, Mosqueiro, Pará state. December 22nd, 1977. Time: 01h30 A.M. Photographer: 1st soldier Flávio Cavalcante (credit: I COMAR)

MILITARY OPERATION: Operation Saucer – I COMAR - FAB MILITARY **REGISTRATION**: 089

LOCATION: Laranjeira River, CIMATRO – VOLKSWAGEN - Belém – Pará

DATE: Saturday, January 28th, 1978 **TIME**: 01:08 **EQUIPMENT**: ELMO 360 (X4) Camcorder

LENS: ? **FILM**: G160 Ektacrome (160 ASA) **APERTURE**: ?

SHUTTER SPEED: ? **HEIGHT**: 300 m **DISTANCE**: 1,500 m **APPARENT SIZE**: 10 cm

SPEED: Medium (-300 km/h) **TRAJECTORY**: Gentle curve to the left

COLOR: Reddish yellow **SHAPE**: Undefined **DESCRIPTION**: Luminous body

PHOTOGRAPHER AND/OR OBSERVERS: Military team from A2 – I COMAR and PLA VIRGÍLIO ERNESTO ARANTES MELLO **OBSERVATIONS**: Composed of the military report: 6 photographs of the video frames. The UFO followed a North/South direction. The UFO moved in a serpentine fashion, moved away from the city lights (Belém) to the left and disappeared at a distance of 3,000 m in front of the observers. Private film belonging to Mr. Adalberto Kovacs Nogueira.

Military Registry - FAB nº 066. Guarajá River, Ananindeua, Pará state. December 10th, 1977. Time: 01h50 A.M. Photographer: A2 military team (credit: I COMAR)

MILITARY OPERATION: Operation Saucer – I COMAR - FAB MILITARY **REGISTRATION**: 107

UFO Attacks in Brazil

LOCATION: Baía do Sol, Mosqueiro – Pará **DATE**: Tuesday, June 20th, 1978 **TIME**: 04:25 **EQUIPMENT**: Minolta SRT 101 **LENS**: Rokkor zoom 100/200 mm **FILM**: RE-2475 – 1000 ASA **APERTURE**: 5.6 **SHUTTER SPEED**: 1/30 sec **HEIGHT**: 400 m **DISTANCE**: 1,500 m

APPARENT SIZE: 20 cm **SPEED**: Average (-300 km/h) **TRAJECTORY**: Slightly curved to the left **COLOR**: Reddish yellow, matte. Variable in intensity **SHAPE**: Oval

DESCRIPTION: Luminous body **PHOTOGRAPHER AND/OR OBSERVERS**: 1S Q AT MT FLÁVIO COSTA and civilian UBIRATAM PINON FRIAS

OBSERVATIONS: Composed of the military report: 1 photograph. UFO headed southeast. The UFO moved in a zigzag pattern, as if under the action of short-term impulses, with sudden increases in luminous intensity.

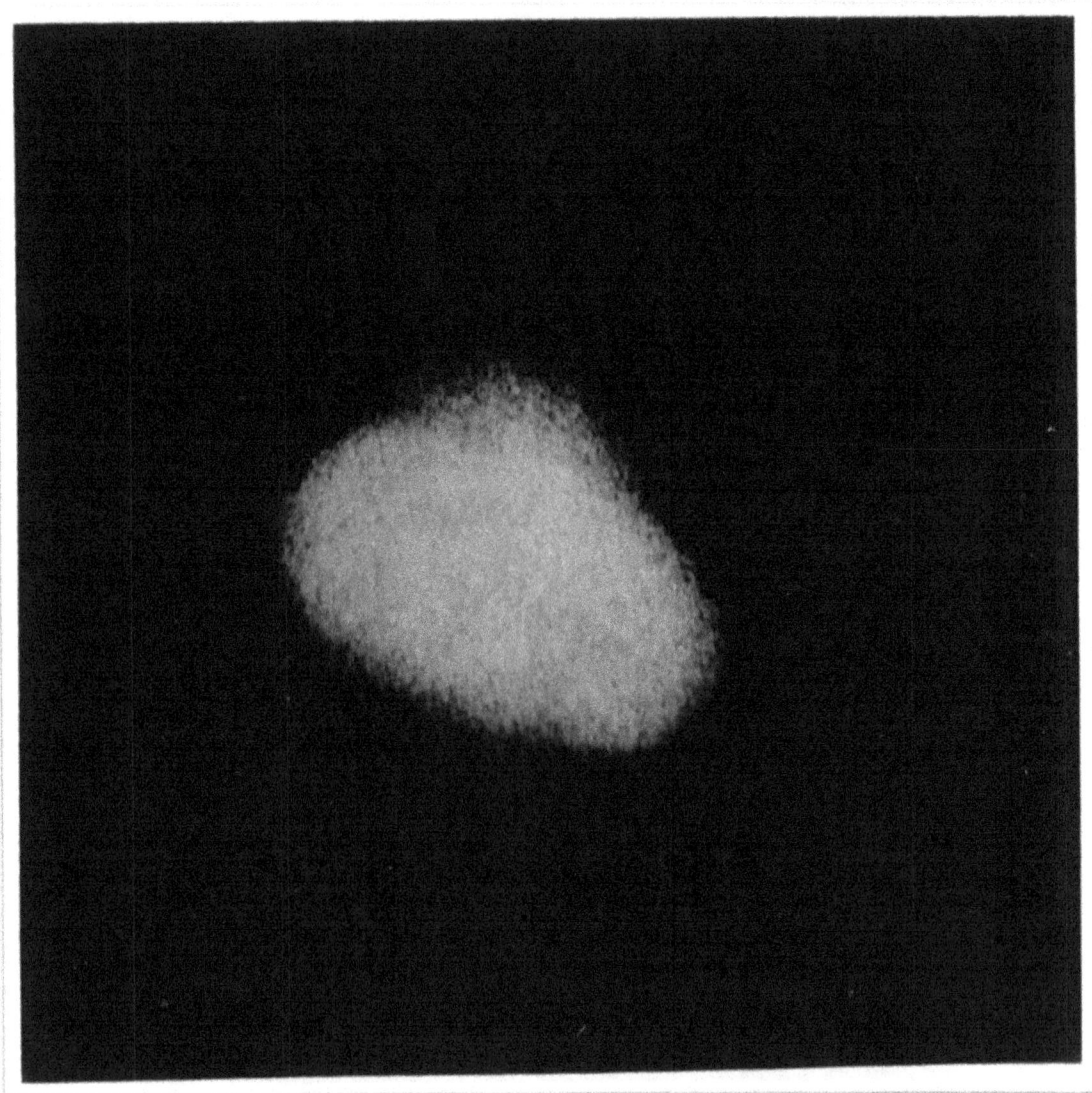

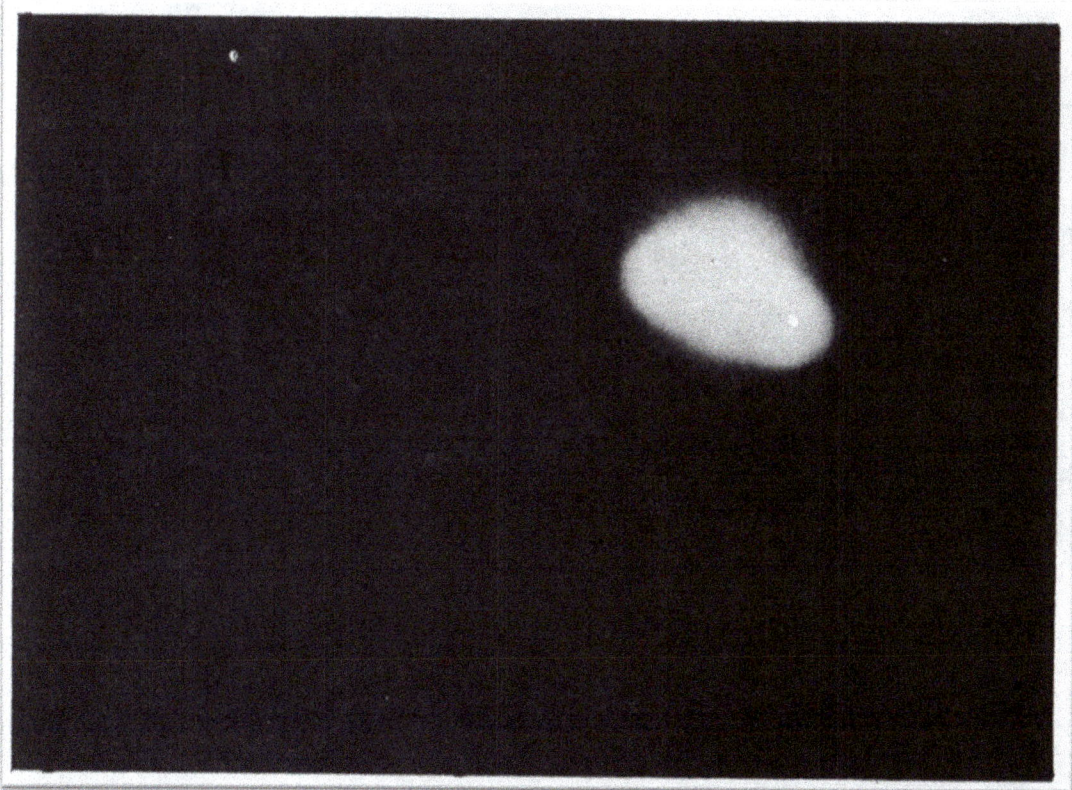

MILITARY OPERATION: Operation Saucer – I COMAR - FAB MILITARY **REGISTRATION**: ?

LOCATION: Colares – Pará **DATE**: February 1978 **TIME**: ? **EQUIPMENT**: CANON 814 (X4) Super 8 Camcorder **LENS**: ? **FILM:** G160 Ektacrome (160 ASA) **APERTURE**: ? **SHUTTER SPEED**: ? **HEIGHT**: ?

DISTANCE: ? **APPARENT SIZE**: ? **SPEED**: ? **TRAJECTORY**: ?

COLOR: ? **SHAPE**: Undefined **DESCRIPTION**: Luminous body **PHOTOGRAPHER AND/OR OBSERVERS**: Military team from A2 – I COMAR **OBSERVATIONS**: Enlarged copies of frames from the video. Projection on ORTHO Type 3 film. Equipment used is LUCKY ENLARGER 6DM with lens: 1:3.5 1:50 mm 2.6X

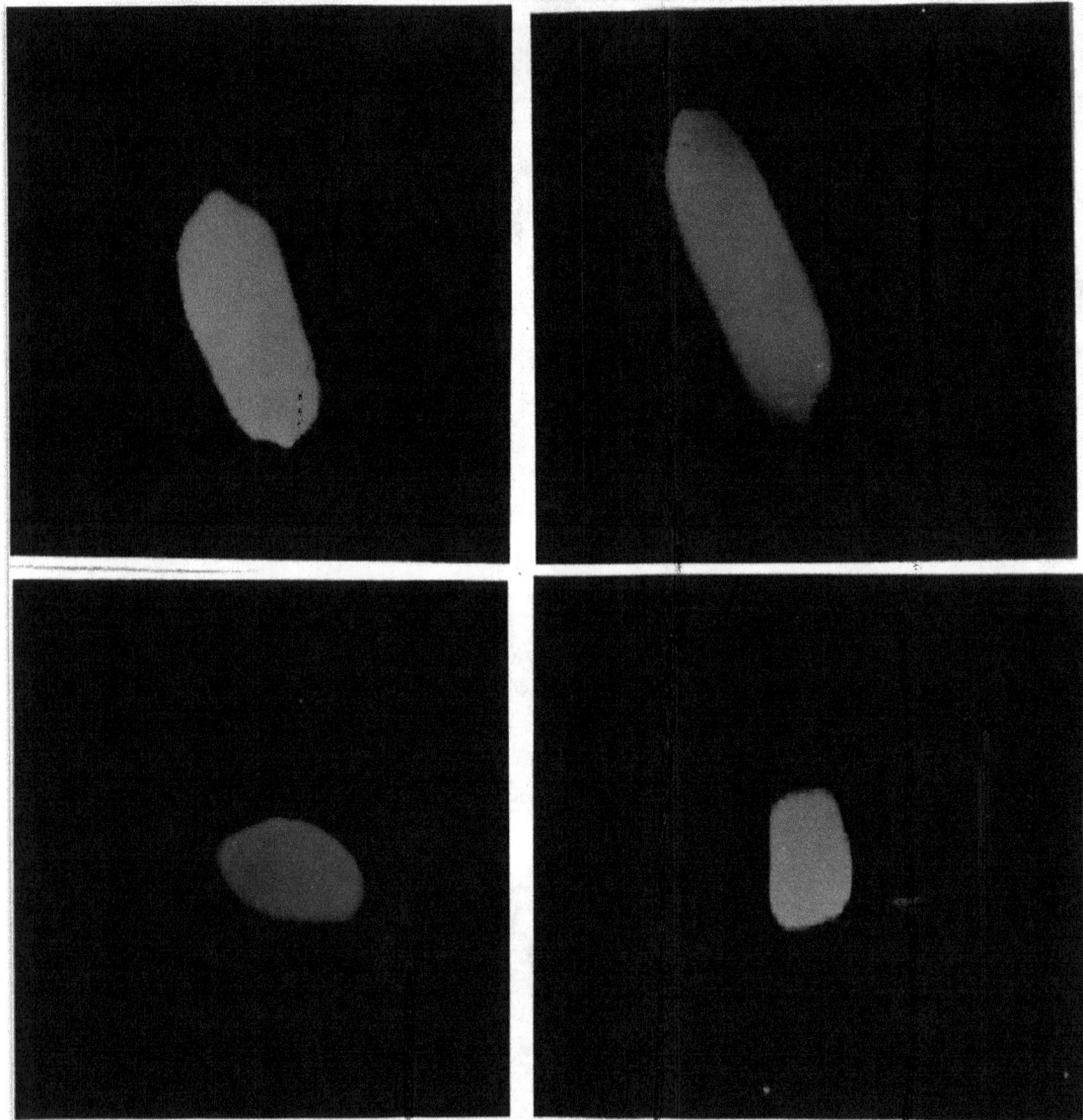

Extended video frame shots. Military Registry - FAB (unknown). Colares, Pará state. February, 1978. Time: (unknown). Photographer: A2 military team (credit: I COMAR)

MILITARY OPERATION: Operation Saucer – I COMAR - FAB MILITARY **REGISTRATION**: ?

LOCATION: Maguari River, Ananindeua – Pará **DATE**: Thursday, February 23rd, 1978 **TIME**: ?

EQUIPMENT: CANON 814 (X4) Super 8 Camcorder LENS: ?

FILM: G160 Ektacrome (160 ASA) **APERTURE**: ? **SHUTTERSPEED**: ? **HEIGHT**: ? **DISTANCE**: ?

APPARENTSIZE: ?**SPEED**: ? **TRAJECTORY**: ? **COLOR**: ?**SHAPE**: Undefined **DESCRIPTION**: Luminous body **PHOTOGRAPER AND/OR OBSERVERS**: Military team from A2 – I COMAR **OBSERVATIONS**: Enlarged copies of frames from the video. Projection on ORTHO Type 3 film. Equipment used is LUCKY ENLARGER 6DM with lens: 1:3.5 1:50 mm 2.6X

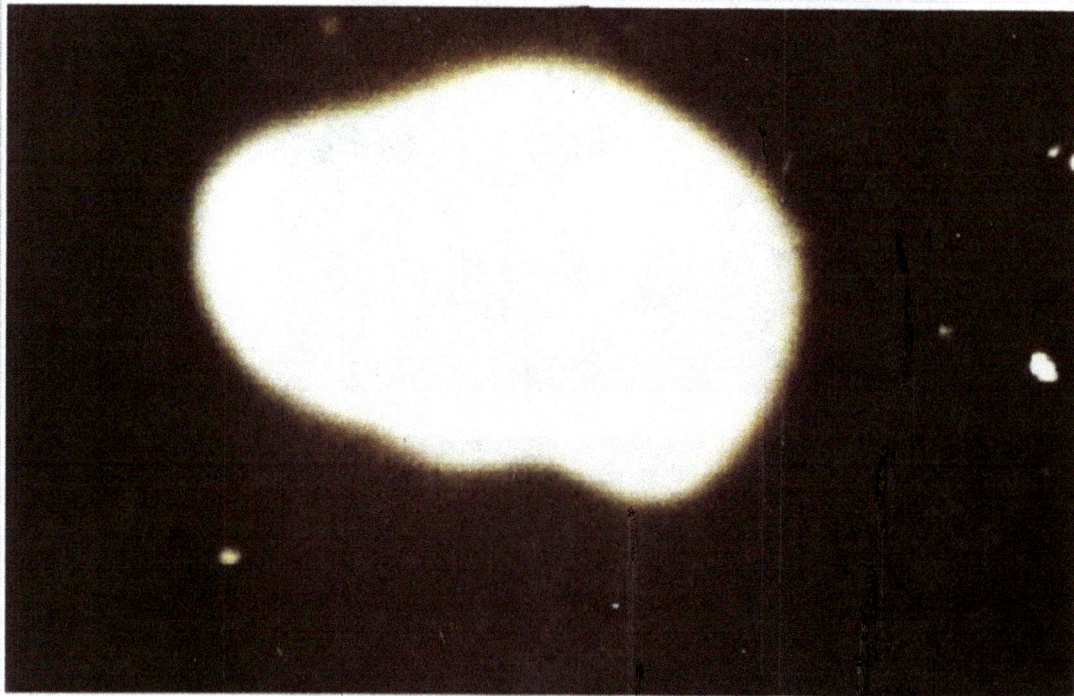

Extended video frame shots. Military Registry - FAB (unknown). Maguari River, Ananindeua, Pará state. February 23rd, 1978. Time: (unknown). Photographer: A2 military team (credit: I COMAR)

UFO Attacks in Brazil

LOCATION: Guajará River, Ananindeua – Pará **DATE**: Wednesday, December 14th, 1977 **TIME**: ?

EQUIPMENT: CANON 814 (X4) Super 8 Camcorder **LENS**: ?

FILM: G160 Ektacrome (160 ASA) **APERTURE**: ? **SHUTTERSPEED**: ? **HEIGHT**: ?

DISTANCE: ? **APPARENTSIZE**: ? **SPEED**: ? **TRAJECTORY**: ? **COLOR**: ?

SHAPE: Undefined **DESCRIPTION**: Luminous body

PHOTOGRAPER AND/OR OBSERVERS: Military team from A2 – I COMAR

OBSERVATIONS: Enlarged copies of frames from the video.

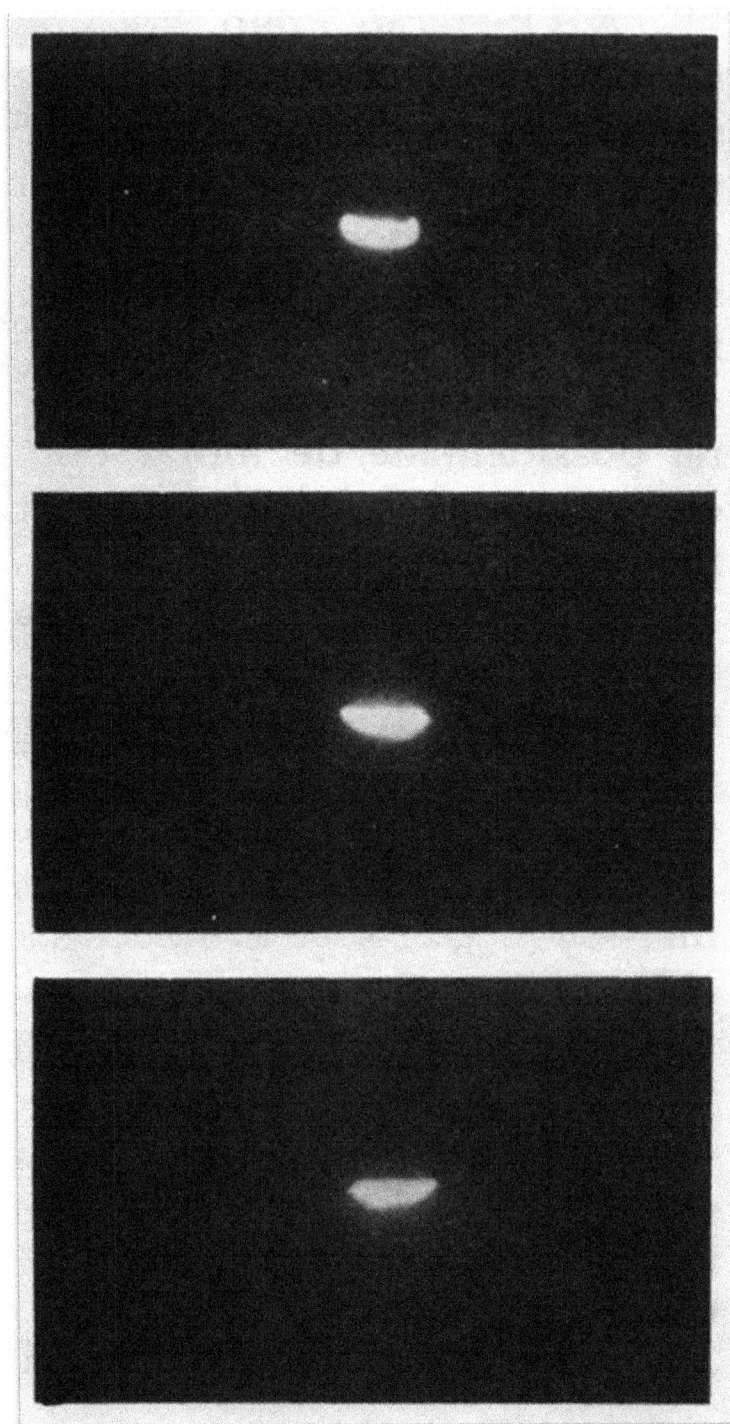

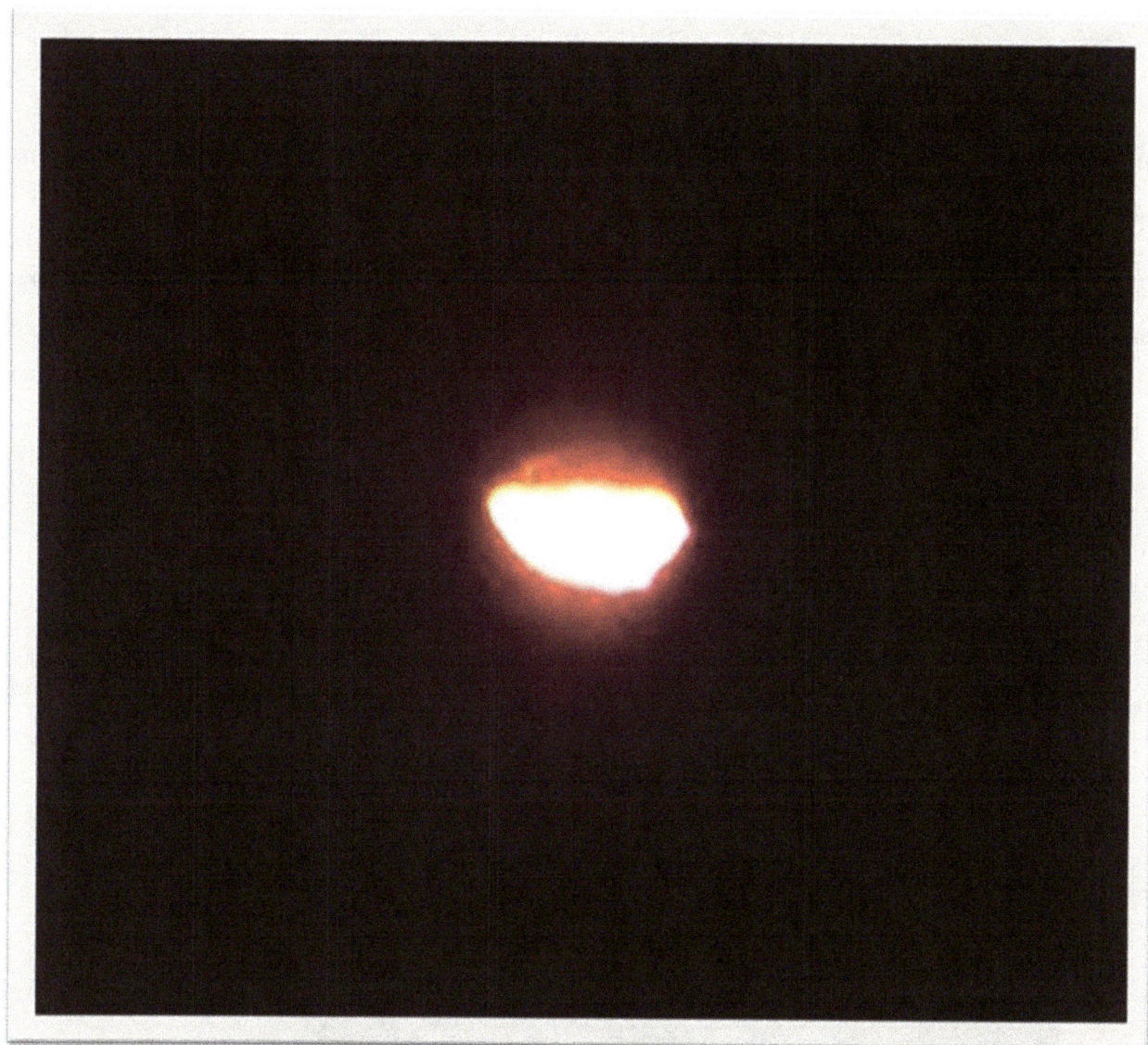

MILITARY OPERATION: Operation Saucer – I COMAR - FAB MILITARY **REGISTRATION**: ? **LOCATION**: Guajará River, Ananindeua – Pará

DATE: December 1977 **TIME**: ? **EQUIPMENT**: CANON 814 (X4) Super 8 Camcorder **LENS**: ? **FILM**: G160 Ektacrome (160 ASA) **APERTURE**: ? **SHUTTERSPEED**: ? **HEIGHT**: ? **DISTANCE**: ? **APPARENTSIZE**: ?**SPEED**: ? **TRAJECTORY**: ? **COLOR**: ? **SHAPE**: Undefined **DESCRIPTION**: Luminous body **PHOTOGRAPER AND**/OR **OBSERVERS**: Military team from A2 – I COMAR **OBSERVATIONS**: Enlarged copy of a frame from the video.

MILITARY OPERATION: Operation Saucer – I COMAR - FAB MILITARY **REGISTRATION**: ?

LOCATION: Colares – Pará **DATE**: October/November 1977 **TIME**: 01:30

EQUIPMENT: CANON Super 8 Camcorder **LENS**: ? **FILM**: Color 160 ASA **APERTURE**: ? **SHUTTERSPEED**: ? **HEIGHT**: ? **DISTANCE**: ? **APPARENTSIZE**: ? **SPEED**: ? **TRAJECTORY**: Ascending rectilinear **SHAPE**: Undefined **DESCRIPTION**: Luminous body

COLOR: Reddish yellow **PHOTOGRAPER AND/OR OBSERVERS**: Military team from A2 – I COMAR

OBSERVATIONS: Enlarged copy of a frame from the video.

MILITARY OPERATION: Operation Saucer – I COMAR - FAB MILITARY **REGISTRATION**: ?

LOCATION: Baía do Sol, Mosqueiro – Pará **DATE**: October/November 1977 **TIME**: ?

EQUIPMENT: Yachica Eletro TLS 35 **LENS**: 400 mm **FILM**: 1000 ASA APERTURE: ? **SHUTTERSPEED**: ?

HEIGHT: ? **DISTANCE**: ? **APPARENTSIZE**: ? **SPEED**: ? **TRAJECTORY**: ?

COLOR: ? **SHAPE**: Undefined **DESCRIPTION**: Luminous body **PHOTOGRAPER AND/OR OBSERVERS**: Military team from A2 – I COMAR **OBSERVATIONS**: Luminous body over the village of Baía do Sol. Probably a small UFO probe.

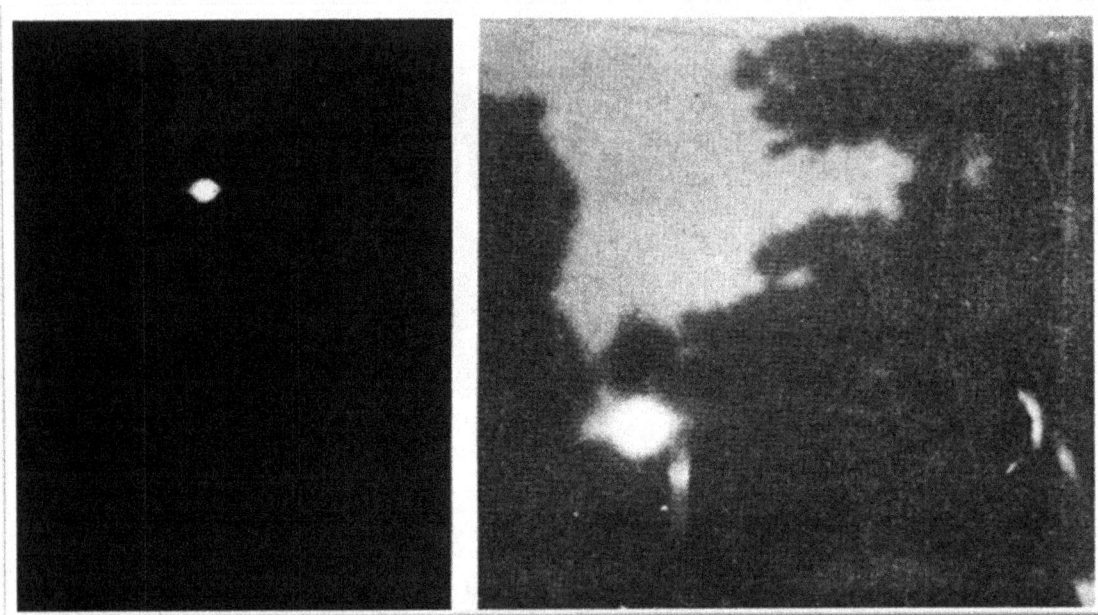

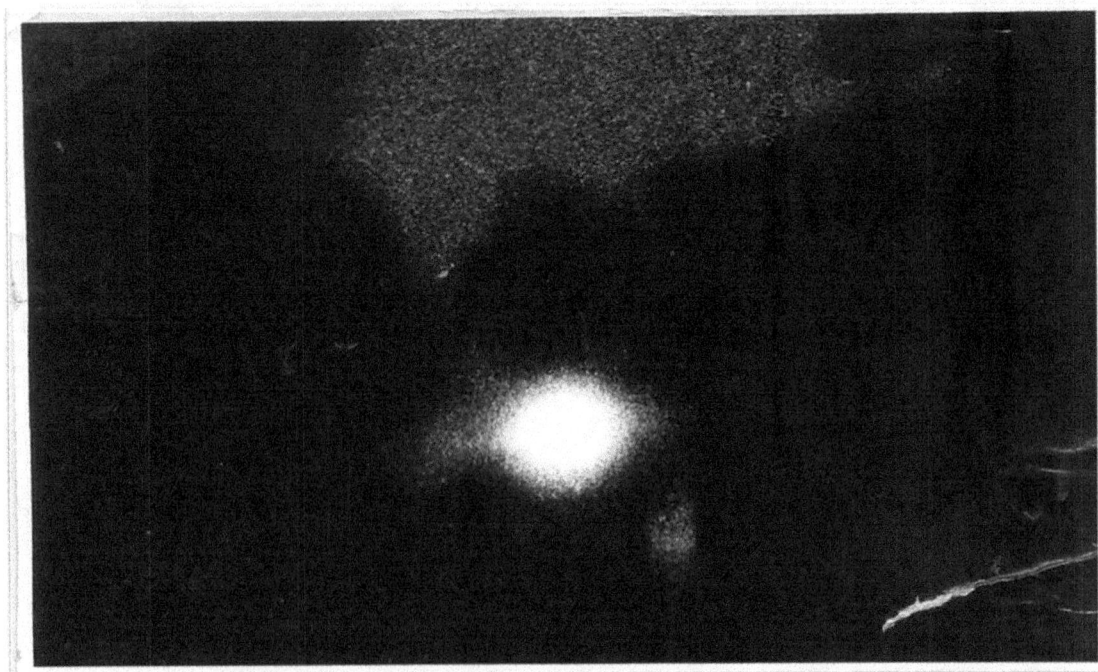

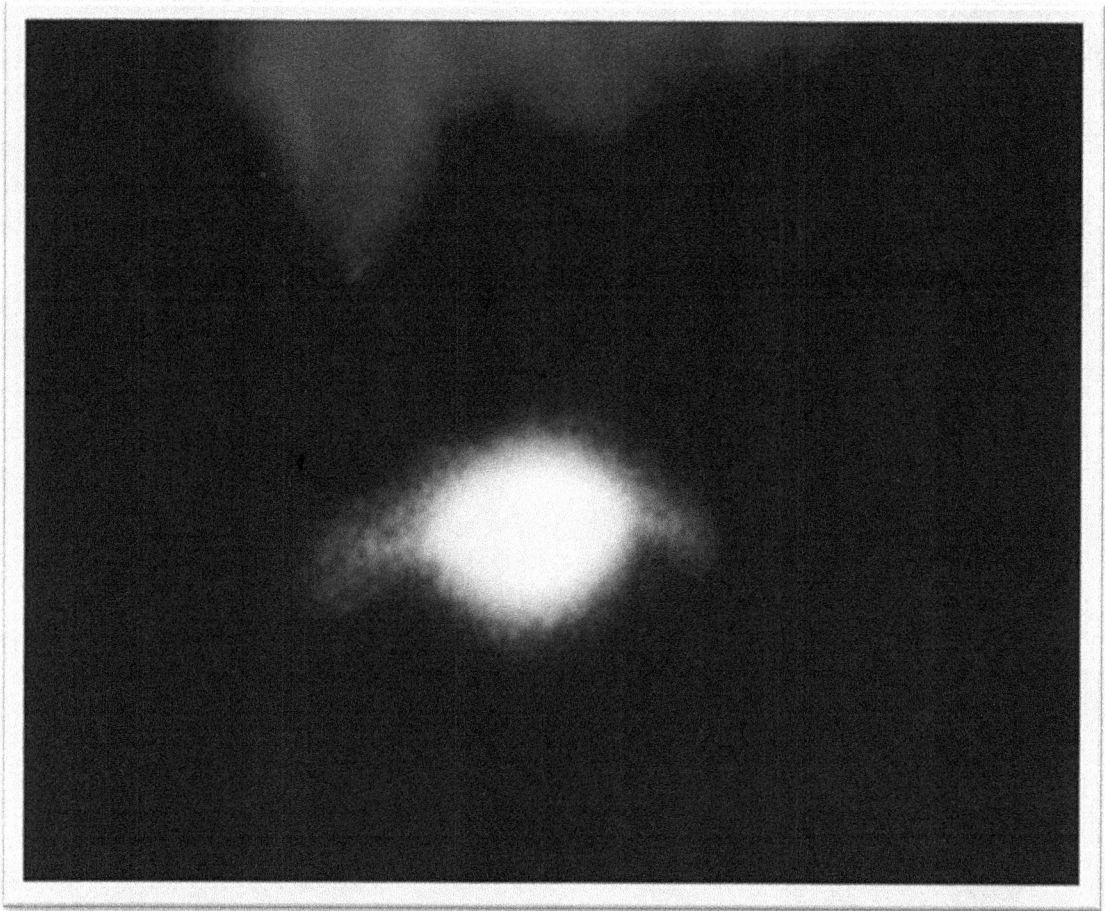

Military Registry - FAB (unknown). Baía do Sol, Mosqueiro, Pará state. October/November 1978. Time: (unknown)Photographer: A2 military team (credit: I COMAR)

MILITARY OPERATION: Operation Saucer – I COMAR - FAB MILITARY **REGISTRATION**: ?

LOCATION: Baía do Sol, Mosqueiro - Pará **DATE**: October/November 1977 **TIME**: ?

EQUIPMENT: Yachica Eletro TLS 35 **LENS**: 400 mm **FILM**: 1000 WING **APERTURE**: ? **SHUTTERSPEED**: ?

HEIGHT: ? **DISTANCE**: ? **APPARENTSIZE**: ? **SPEED**: ?

TRAJECTORY: ? **COLOR**: ? **SHAPE**: Undefined **DESCRIPTION**: Luminous body **PHOTOGRAPER AND/OR**

OBSERVERS: Military team from A2 – I COMAR **OBSERVATIONS**: Luminous body over the village of Baía do Sol. Probably a small UFO probe.

MILITARY OPERATION: Operation Saucer – I COMAR - FAB MILITARY **REGISTRATION**: ?

LOCATION: Baía do Sol, Mosqueiro – Pará **DATE**: December 1977 **TIME**: ?

EQUIPMENT: Yachica Eletro TLS 35 **LENS**: 400 mm **FILM**: 1000 ASA **APERTURE**: ?

SHUTTERSPEED: ? **HEIGHT**: ? **DISTANCE**: ? **APPARENTSIZE**: ? **SPEED**: ?

TRAJECTORY: ? **COLOR**: ? **SHAPE**: Undefined **DESCRIPTION**: Luminous body

PHOTOGRAPER AND/OR OBSERVERS: Military team from A2 – I COMAR

OBSERVATIONS: Luminous body parked over the village of Baía do Sol.

MILITARY OPERATION: Operation Saucer – I COMAR - FAB MILITARY **REGISTRATION**: ?

LOCATION: Benevides – Pará **DATE**: June 1978 **TIME**: ? **EQUIPMENT**: Minolta SRT 101

LENS: ? **FILM**: RE-2475 - 1000 ASA **APERTURE**: ? **SHUTTERSPEED**: ?

HEIGHT: ? **DISTANCE**: ? **APPARENTSIZE**: ? **SPEED**: ? **TRAJECTORY**: ? **COLOR**: ?

SHAPE: Undefined **DESCRIPTION**: Luminous body

PHOTOGRAPER AND/OR OBSERVERS: Military team from A2 – I COMAR

OBSERVATIONS: Luminous body with no defined shape over the city of Benevides - PA.

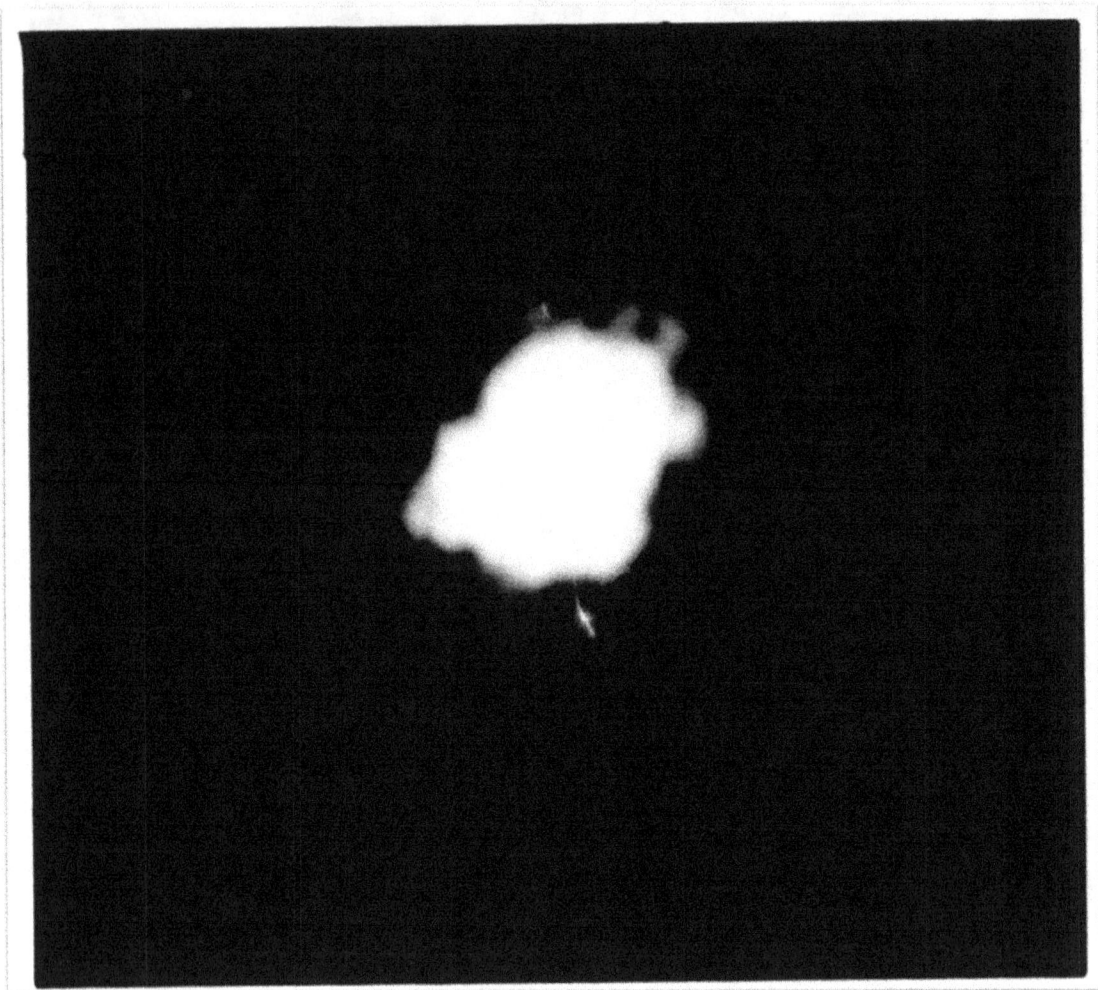

MILITARY OPERATION: Operation Saucer – I COMAR - FAB MILITARY **REGISTRATION**: ?

LOCATION: Baía do Sol – Pará **DATE**: Thursday, June 22nd, 1978 **TIME**: ?

EQUIPMENT: Nikon **LENS**: ? **FILM**: 1000 ASA **APERTURE**: ? **SHUTTERSPEED**: ?

HEIGHT: ? **DISTANCE**: ? **APPARENTSIZE**: ? **SPEED**: ? **TRAJECTORY**: ? **COLOR**: ?

SHAPE: Spinning top **DESCRIPTION**: Luminous body

PHOTOGRAPER AND/OR OBSERVERS: Military team from A2 – I COMAR and civilian JOSÉ RIBAMAR

PRAZERES **OBSERVATIONS**: Luminous body in the shape of a spinning top over Baía do Sol. Photograph published in the newspaper *O ESTADO DO PARÁ*, issue 16,542, on June 28th, 1978, on page 12.

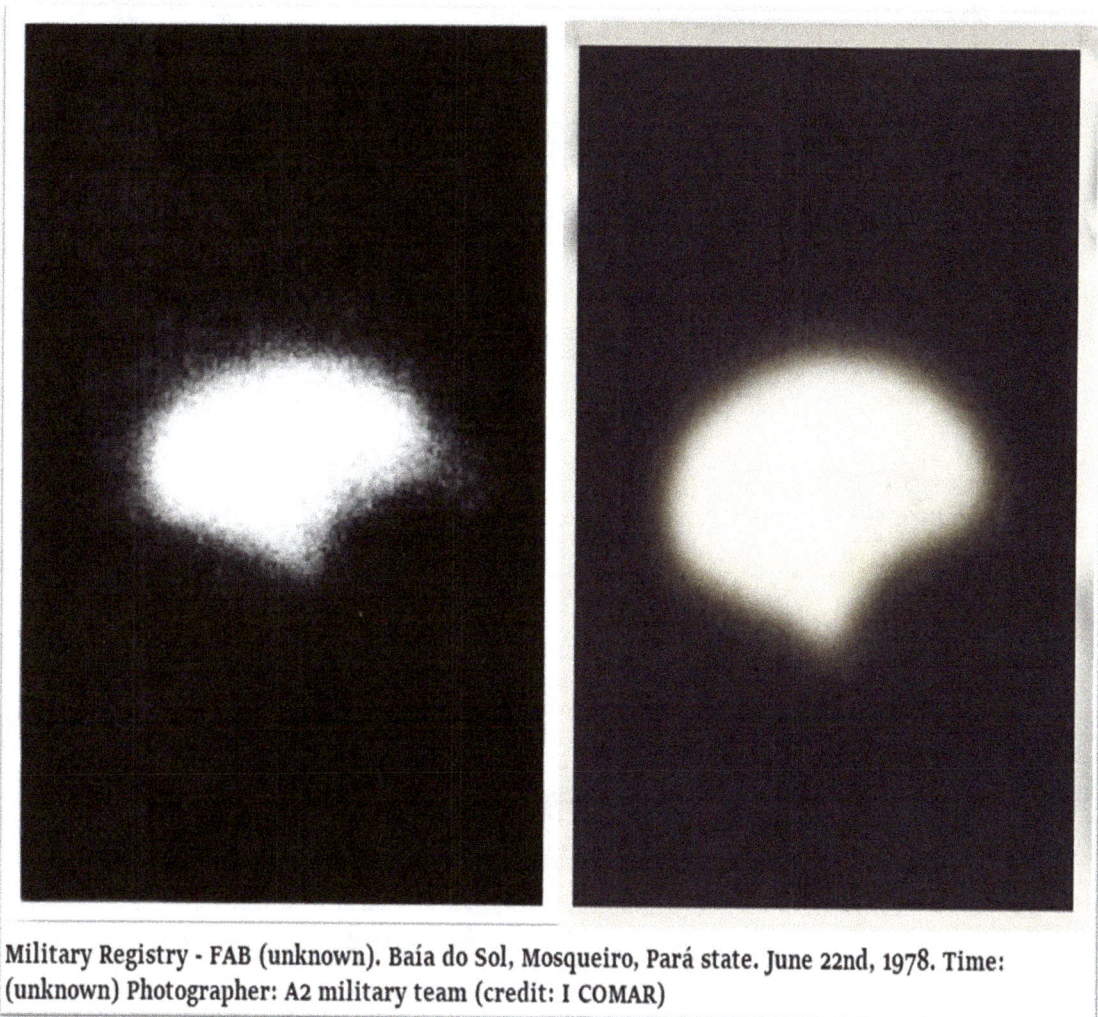

Military Registry - FAB (unknown). Baía do Sol, Mosqueiro, Pará state. June 22nd, 1978. Time: (unknown) Photographer: A2 military team (credit: I COMAR)

MILITARY OPERATION: Operation Saucer – I COMAR - FAB MILITARY **REGISTRATION**: ?

UFO Attacks in Brazil

LOCATION: Baía do Sol – Pará **DATE**: June 1978 **TIME**: ? **EQUIPMENT**: Nikon **LENS**: ?
FILM: 1000 ASA **APERTURE**: ? **SHUTTERSPEED**: ? **HEIGHT**: ? **DISTANCE**: ?
APPARENTSIZE: ? **SPEED**: ? **TRAJECTORY**: ? **COLOR**: ? **FORM**: Spinning top **DESCRIPTION**: Luminous body **PHOTOGRAPER AND/OR OBSERVERS**: Team of military personnel from A2 – I COMAR and civilian JOSÉ RIBAMAR PRAZERES **OBSERVATIONS**: Large luminous body over Baía do Sol. Photograph published in the newspaper *O ESTADO DO PARÁ*, on June 25th, 1978. At a certain point, the larger UFO released a smaller flying object.

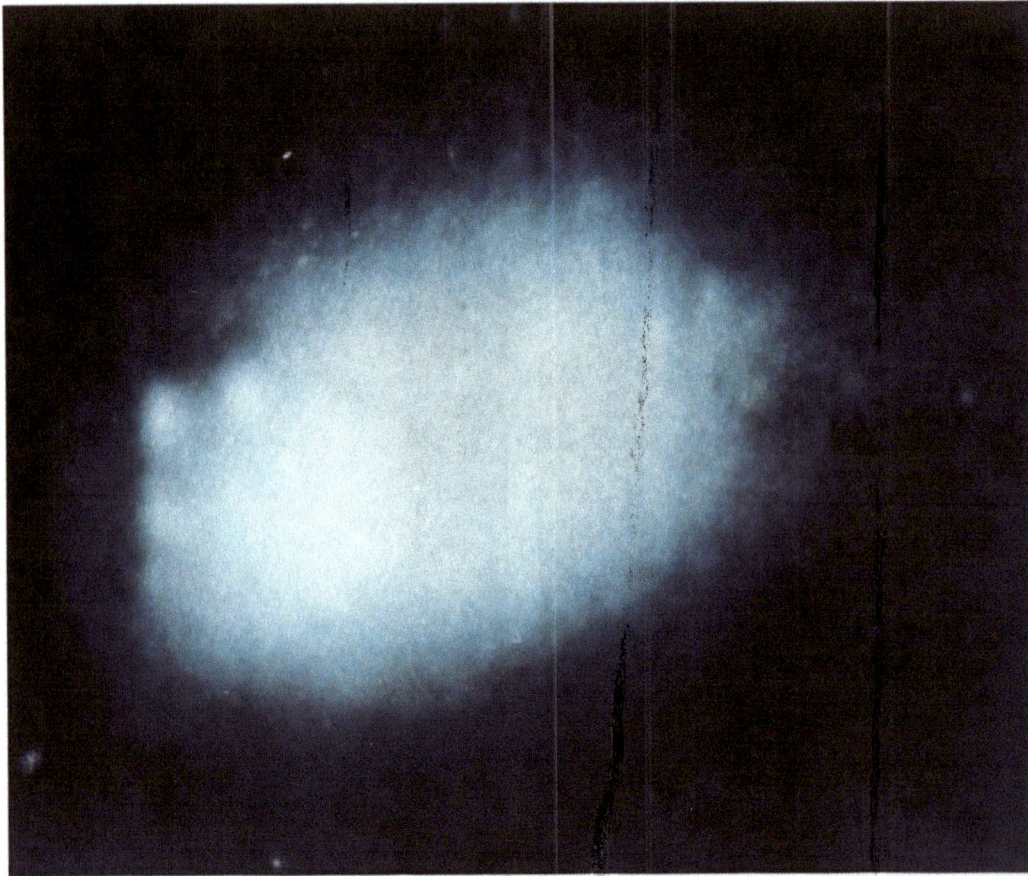

At one point the UFO released a smaller object. Military Registry - FAB (unknown). Baía do Sol, Mosqueiro, Pará state. June, 1978. Time: (unknown) Photographer: A2 military team (credit: I COMAR)

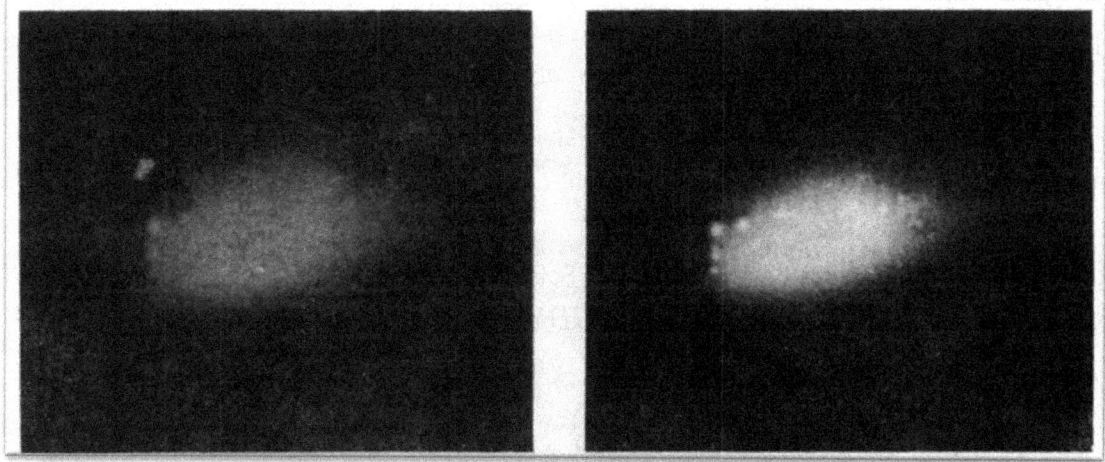

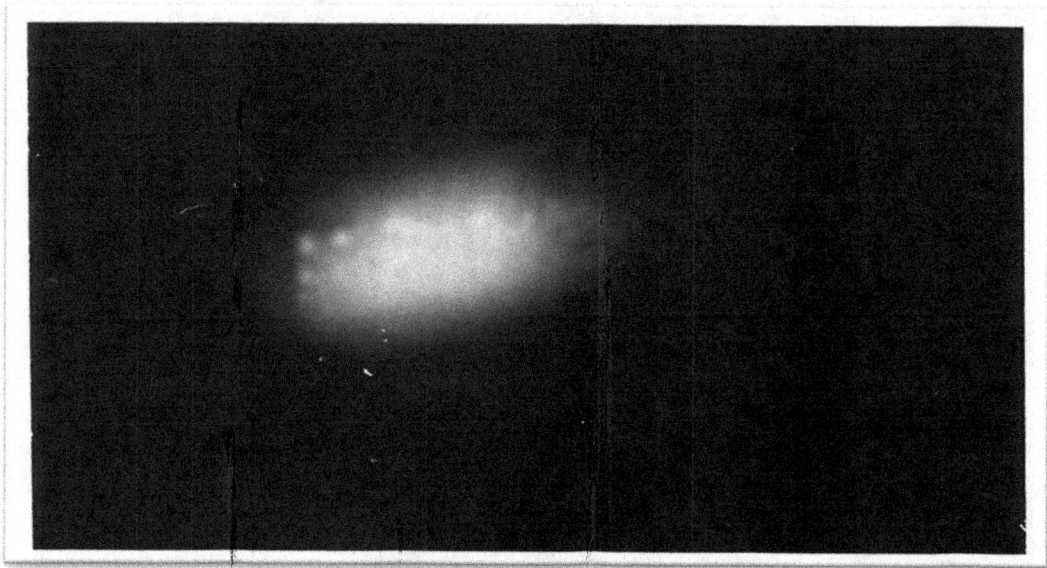

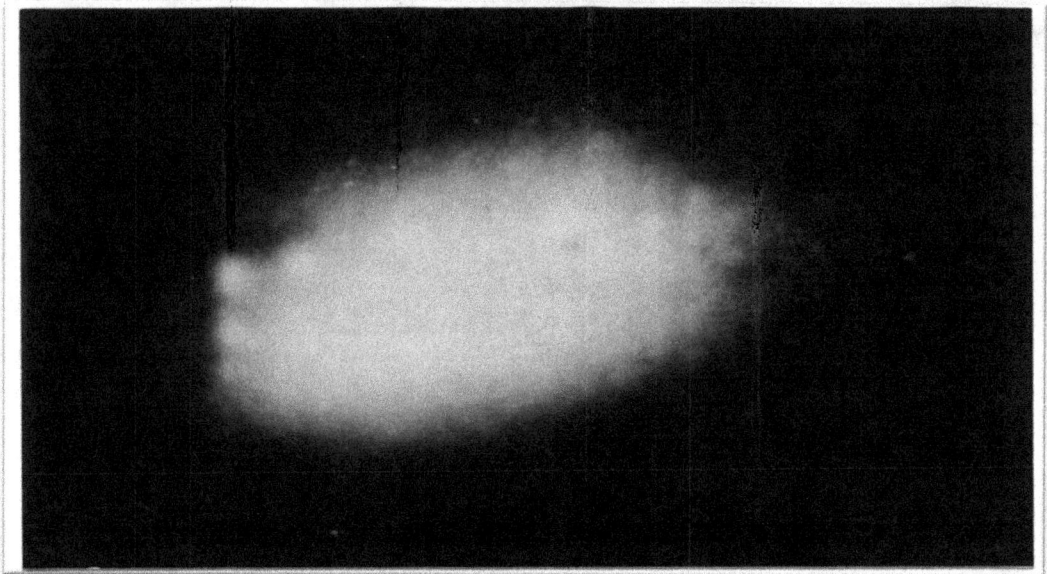

MILITARY OPERATION: Operation Saucer – I COMAR - FAB MILITARY **REGISTRATION**: ?

LOCATION: Baía do Sol – Pará **DATE**: Friday, June 16th, 1978 **TIME**: 04:35 **EQUIPMENT**: Nikon **LENS**: 400 mm **FILM**: 1000 ASA **APERTURE**: ? **SHUTTERSPEED**: ? **HEIGHT**: ? **DISTANCE**: ? **APPARENTSIZE**: ?

SPEED: ? **TRAJECTORY**: ? **COLOR**: ? **SHAPE**: Undefined

DESCRIPTION: Luminous body **PHOTOGRAPER AND/OR OBSERVERS**: Team of military personnel from A2 – I COMAR and civilian JOSÉ RIBAMAR PRAZERES

OBSERVATIONS: Large luminous body over Baía do Sol. Photograph published in the newspaper *O ESTADO DO PARÁ*, on June 25th, 1978. At a certain point, the larger UFO released small flying objects.

UFO Attacks in Brazil

MILITARY OPERATION: Operation Saucer – I COMAR - FAB MILITARY **REGISTRATION**: ?

LOCATION: Baía do Sol – Pará**DATE**: June 1978 **TIME**: 03:00**EQUIPMENT**: Nikon

LENS: 400 mm**FILM**: 1000 ASA **APERTURE**: ? **SHUTTERSPEED**: ?**HEIGHT**: ? **DISTANCE**: ?

APPARENTSIZE: ?**SPEED**: ? **TRAJECTORY**: ? **COLOR**: Reddish yellow **SHAPE**: Circular **DESCRIPTION**: Luminous body **PHOTOGRAPHER AND/OR OBSERVERS**: Military team from A2 – I COMAR**OBSERVATIONS**: Large luminous body over Baía do Sol, dropping smaller objects.

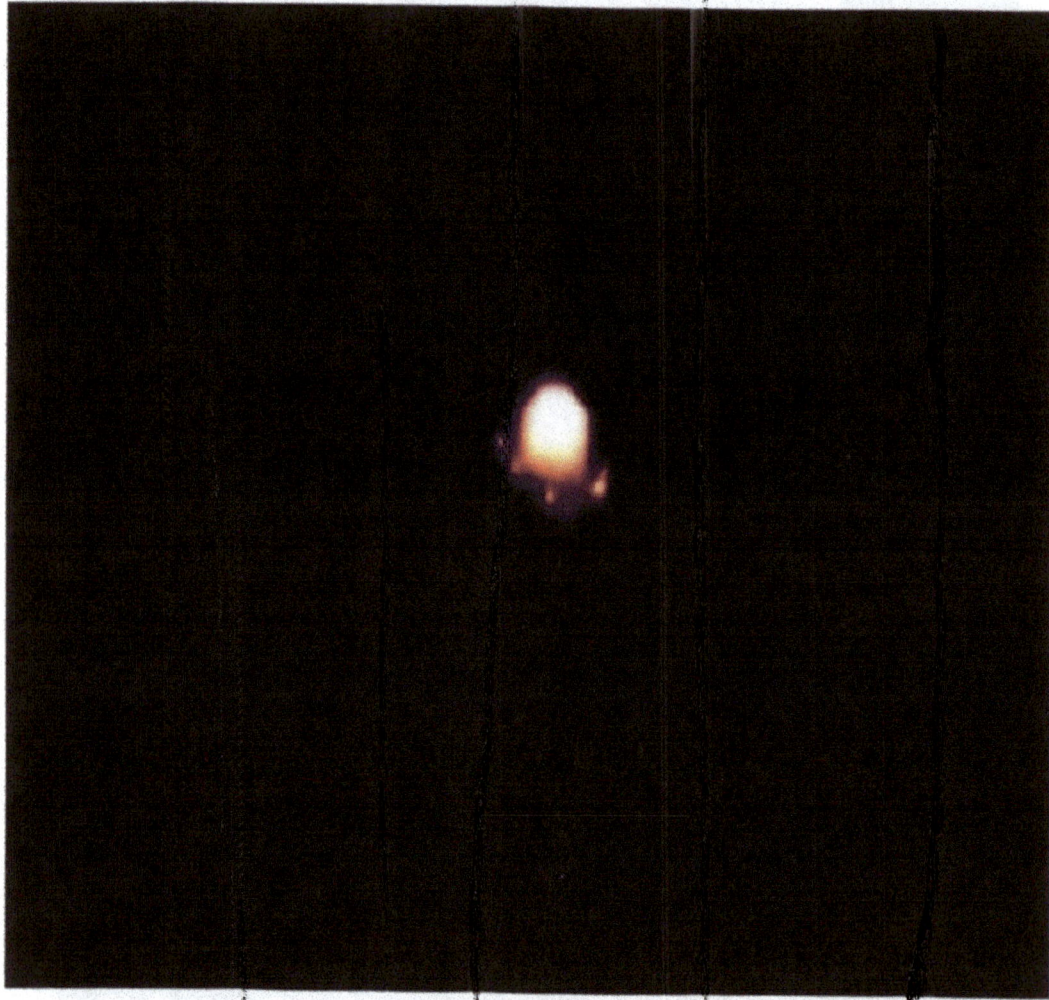

The UFO released several smaller objects. Military Registry - FAB (unknown), Baiá do Sol, Mosqueiro, Pará state. June, 1978. Time: (unknown). Photographer: A2 military team (credit: I COMAR)

MILITARY OPERATION: Operation Saucer – I COMAR - FAB MILITARY **REGISTRATION**: ?

LOCATION: Baía do Sol, Mosqueiro – Pará **DATE**: November 1977 **TIME**: ?

EQUIPMENT: Yashica **LENS**: 400 mm **FILM**: 1000 ASA **APERTURE**: ? **SHUTTERSPEED**: ?

HEIGHT: ? **DISTANCE**: ? **APPARENTSIZE**: ? **SPEED**: ? **TRAJECTORY**: ?

COLOR: Reddish yellow **SHAPE**: Undefined **DESCRIPTION**: Luminous body

PHOTOGRAPER AND/OR OBSERVERS: Military team from A2 – I COMAR

OBSERVATIONS: Luminous body moving over Baía do Sol.

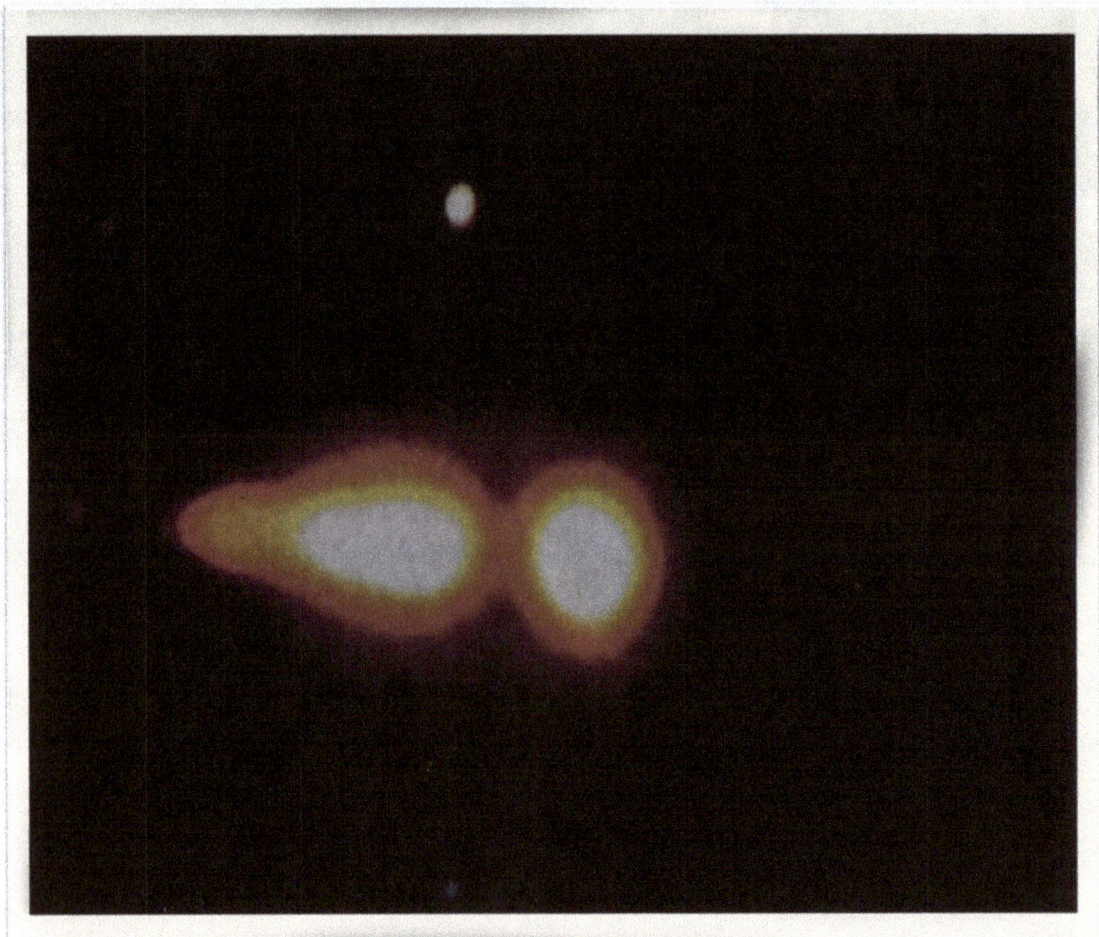

Military Registry - FAB (unknown). Baía do Sol, Mosqueiro, Pará state. November, 1977. Time: (unknown) Photographer: A2 military team (credit: I COMAR)

MILITARY OPERATION: Operation Saucer – I COMAR - FAB MILITARY **REGISTRATION**: ?

LOCATION: Baía do Sol – Pará **DATE**: Tuesday, June 20, 1978 **TIME**: ?

EQUIPMENT: Nikon **LENS**: 400 mm **FILM**: 1000 ASA **APERTURE**: ? **SHUTTERSPEED**: ?

HEIGHT: ? **DISTANCE**: ? **APPARENTSIZE**: ? **SPEED**: ? **TRAJECTORY**: ? **COLOR**: ?

FORM: Undefined **DESCRIPTION**: Luminous body **PHOTOGRAPER AND/OR OBSERVERS**: Team of military personnel from A2 – I COMAR and civilian JOSÉ RIBAMAR PRAZERES

OBSERVATIONS: Large luminous body over Baía do Sol. Photograph published in the newspaper *O ESTADO DO PARÁ*, issue 16,542, June 26th, 1978.

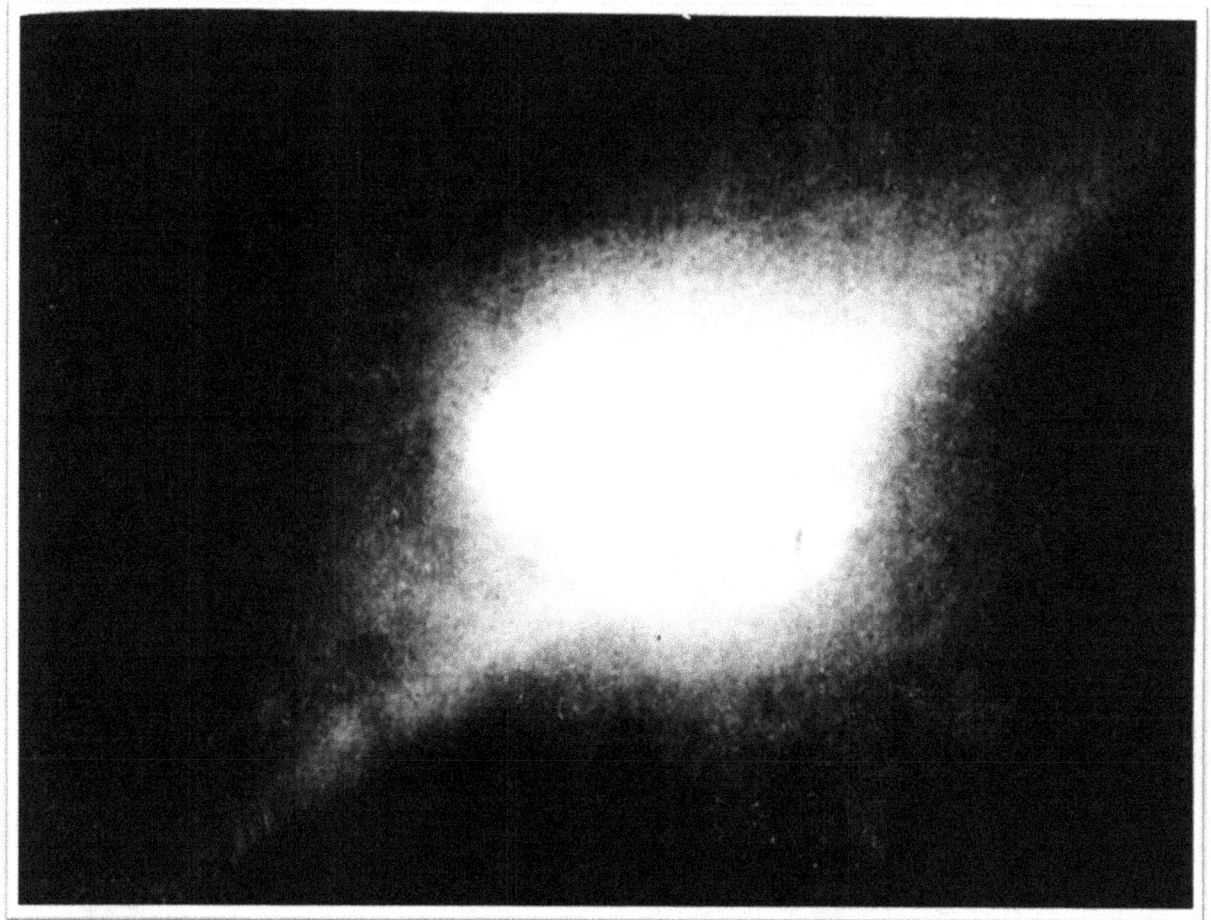

UFO Attacks in Brazil

MILITARY OPERATION: Operation Saucer – I COMAR - FAB MILITARY **REGISTRATION**: ?

LOCATION: Baía do Sol, Mosqueiro – Pará **DATE**: December 1977 **TIME**: ?

EQUIPMENT: Yashica Eletro TLS 35 **LENS**: 400 mm **FILM**: RE-2475 - 1000 ASA **APERTURE**: ?

SHUTTERSPEED: ? **HEIGHT**: ? **DISTANCE**: ? **APPARENTSIZE**: ? **SPEED**: ?

TRAJECTORY: ? **COLOR**: ? **SHAPE**: Undefined **DESCRIPTION**: Luminous body

PHOTOGRAPER AND/OR OBSERVERS: Military team from A2 – I COMAR

OBSERVATIONS: Luminous UFO in Baía do Sol with no defined shape.

MILITARY OPERATION: Operation Saucer – I COMAR - FAB MILITARY **REGISTRATION**: ?

LOCATION: Baía do Sol, Mosqueiro – Pará **DATE**: December 1977 **TIME**: ?

EQUIPMENT: Yashica Eletro TLS 35 **LENS**: 400 mm **FILM**: RE-2475 - 1000 ASA **APERTURE**: ?

SHUTTERSPEED: ? **HEIGHT**: ? **DISTANCE**: ? **APPARENTSIZE**: ? **SPEED**: ?

TRAJECTORY: ? **COLOR**: ? **FORM**: Undefined **DESCRIPTION**: Luminous body

PHOTOGRAPER AND/OR OBSERVERS: Military team from A2 – I COMAR

OBSERVATIONS: Luminous UFO in Baía do Sol with no defined shape.

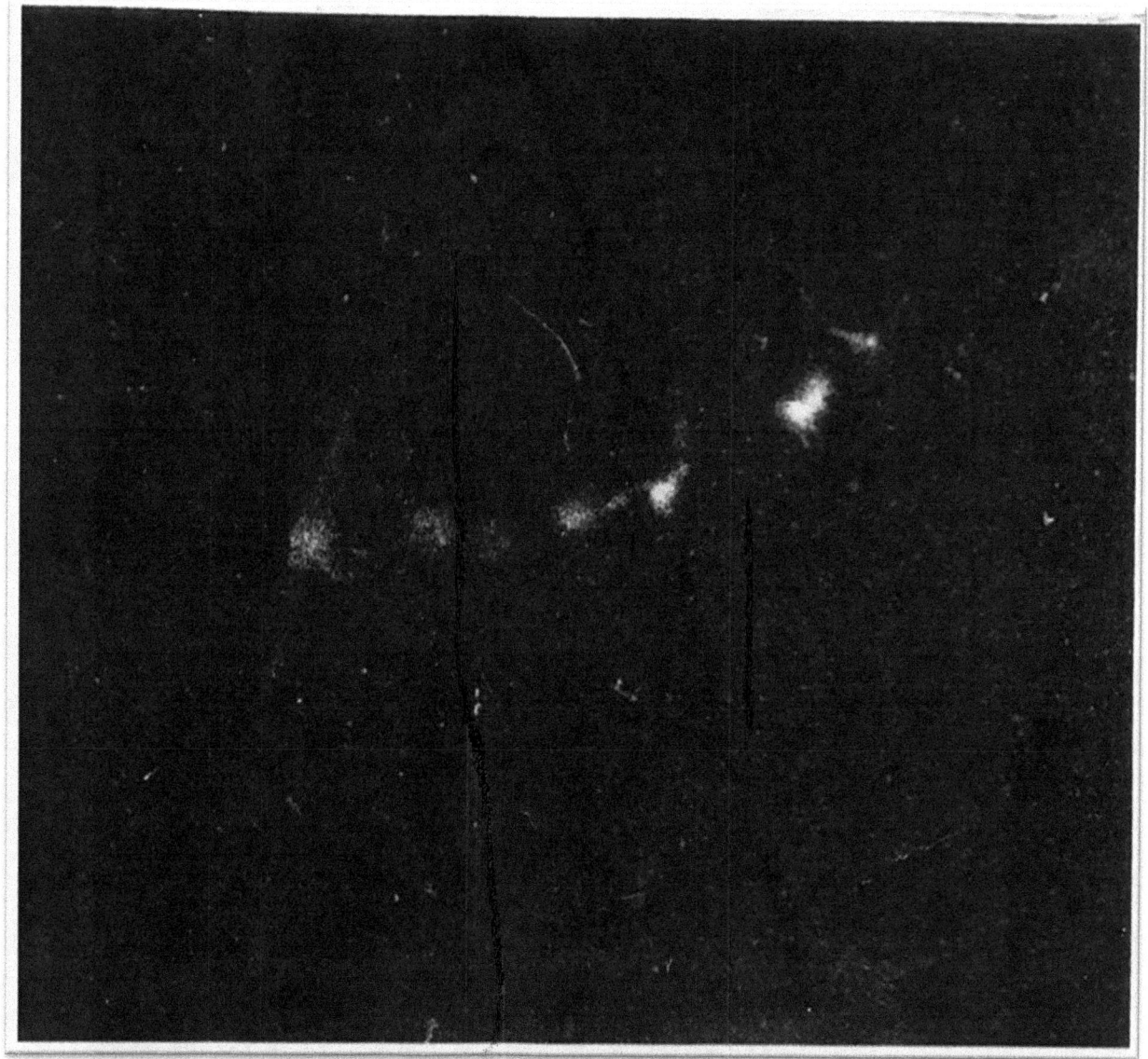

MILITARY OPERATION: Operation Saucer – I COMAR - FAB MILITARY **REGISTRATION**: ?

LOCATION: Baía do Sol, Mosqueiro – Pará **DATE**: December 1977 **TIME**: ?

EQUIPMENT: Yashica Eletro TLS 35 **LENS**: 400 mm **FILM**: RE-2475 - 1000 ASA **APERTURE**: ?

SHUTTERSPEED: ? **HEIGHT**: ? **DISTANCE**: ? **APPARENTSIZE**: ? **SPEED**: ?

TRAJECTORY: ? **COLOR**: ? **SHAPE**: Undefined **DESCRIPTION**: Luminous body

PHOTOGRAPER AND/OR OBSERVERS: Military team from A2 – I COMAR

OBSERVATIONS: Luminous UFO in Baía do Sol with no defined shape.

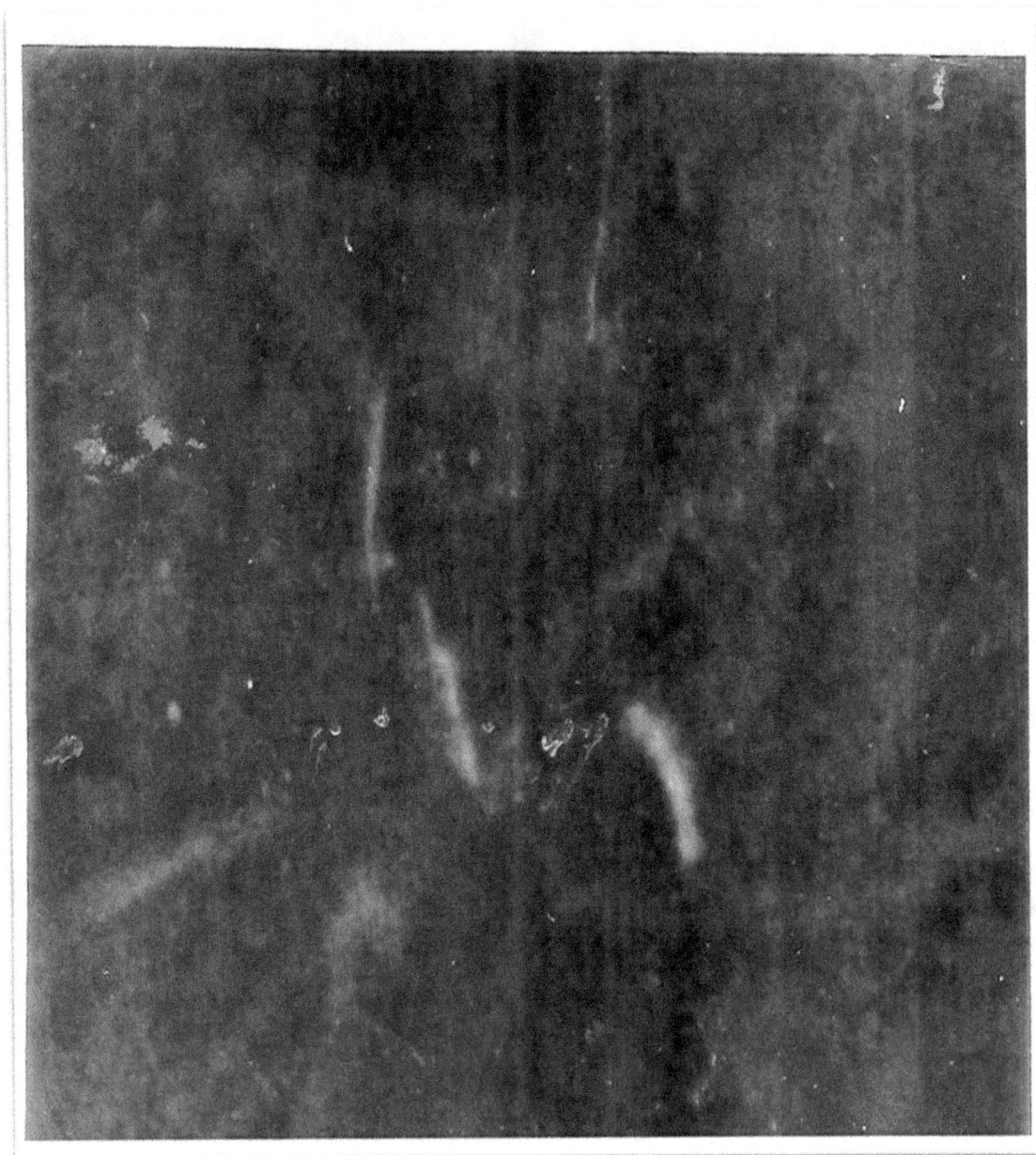

MILITARY OPERATION: Operation Saucer – I COMAR - FAB MILITARY **REGISTRATION**: ?

LOCATION: Baía do Sol – Pará **DATE**: November/December 1977 **TIME**: ?

EQUIPMENT: Yashièa Eletro TLS 35 **LENS**: 400 mm **FILM**: RE-2475 - 1000 ASA **APERTURE**: ?

SHUTTERSPEED: ? **HEIGHT**: ? **DISTANCE**: ? **APPARENTSIZE**: ? **SPEED**: ?

TRAJECTORY: ? **COLOR**: ? **SHAPE**: Circular **DESCRIPTION**: Luminous body

PHOTOGRAPER AND/OR OBSERVERS: Military team from A2 – I COMAR

OBSERVATIONS: Luminous UFO in Baía do Sol. Sequence of 3 photographs.

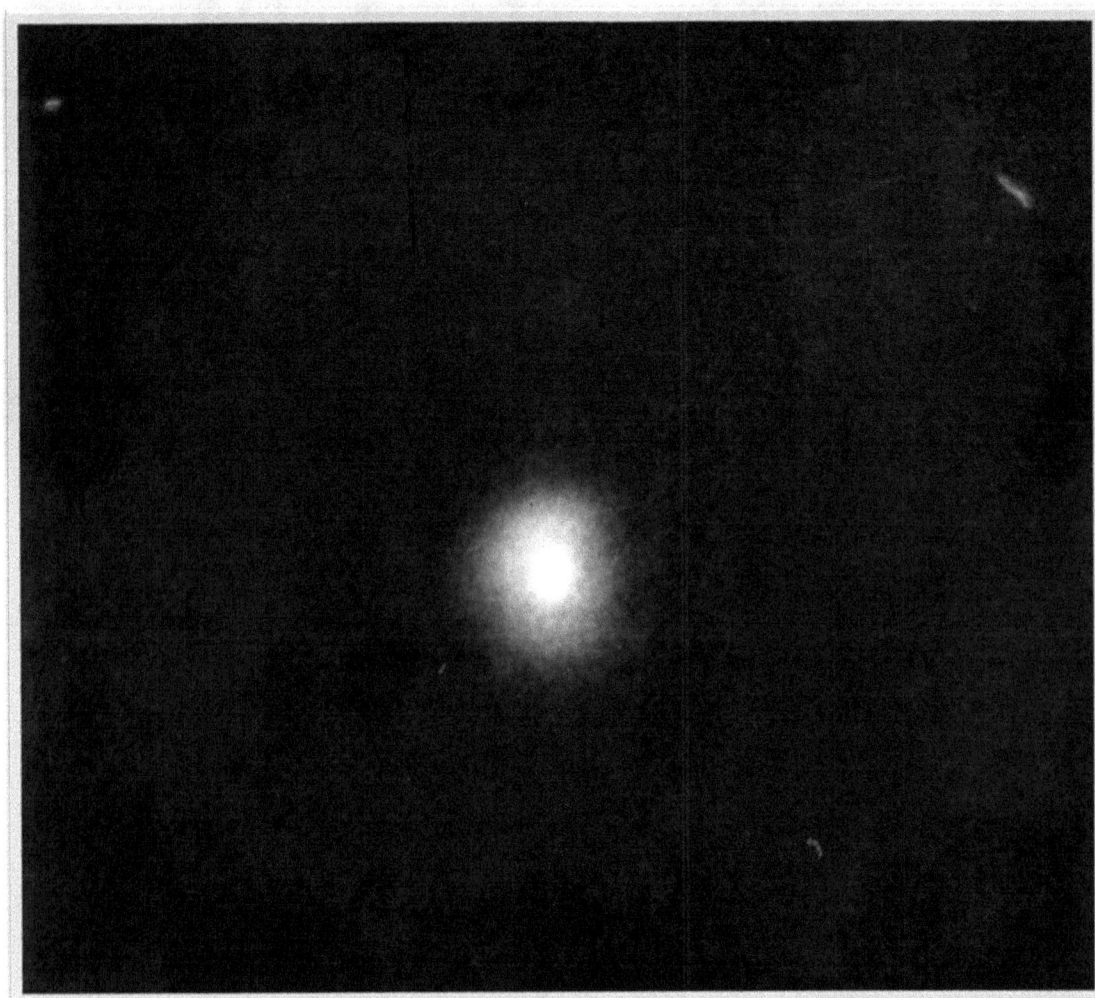

MILITARY OPERATION: Operation Saucer – I COMAR - FAB MILITARY **REGISTRATION**: ?

LOCATION: Colares – Pará **DATE**: November 1977 **TIME**: ?

EQUIPMENT: Yashica Eletro TLS 35 **LENS**: 400 mm **FILM**: RE-2475 - 1000 ASA **APERTURE**: ?

SHUTTERSPEED: ? **HEIGHT**: ? **DISTANCE**: ? **APPARENTSIZE**: ? **SPEED**: ?

TRAJECTORY: ? **COLOR**: ? **SHAPE**: Undefined **DESCRIPTION**: Luminous body

PHOTOGRAPER AND/OR OBSERVERS: Military team from A2 – I COMAR

OBSERVATIONS: Luminous UFO in the village of Colares with no defined shape and no noise.

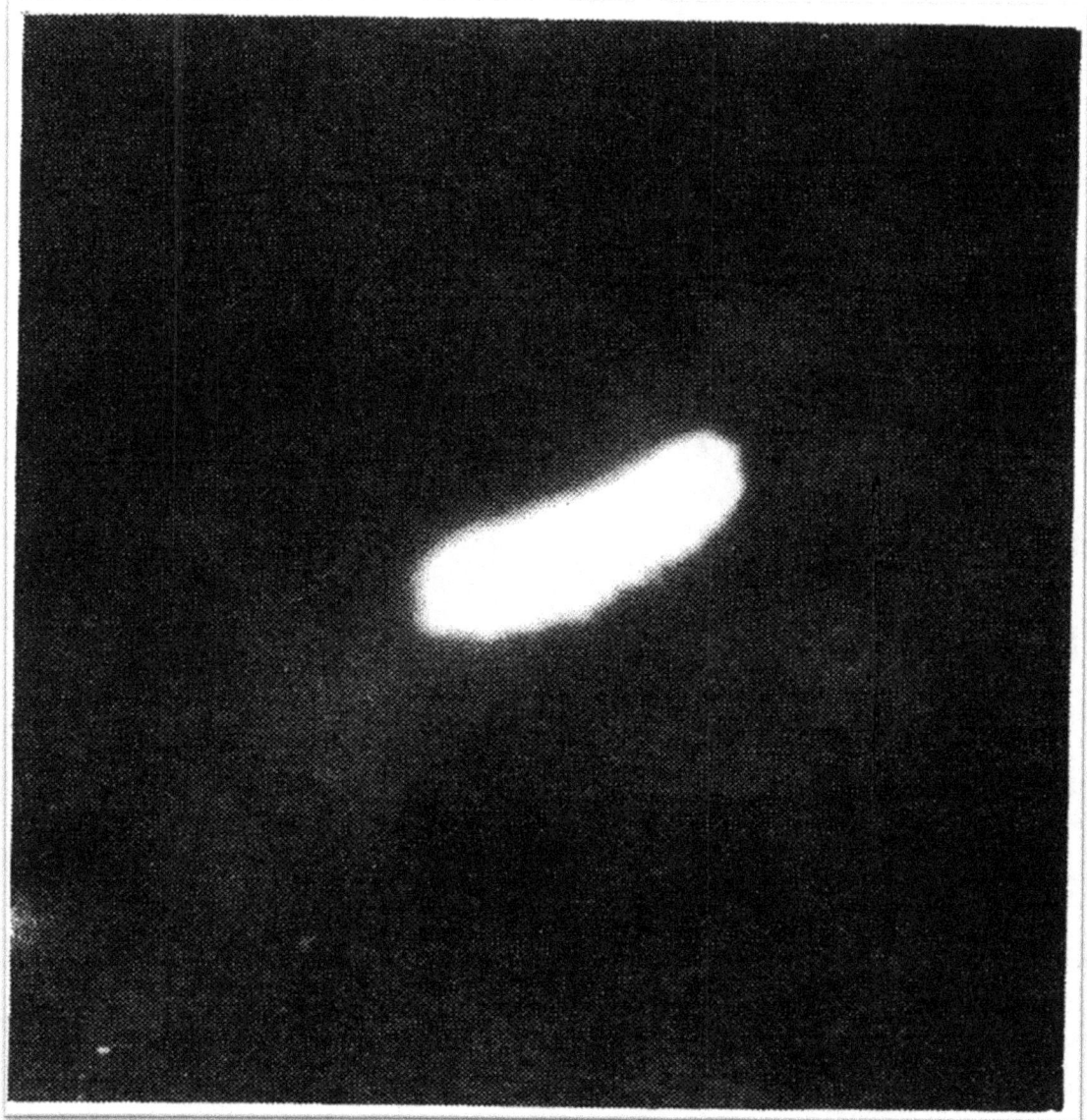

MILITARY OPERATION: Operation Saucer – I COMAR - FAB MILITARY **REGISTRATION**: ?

LOCATION: Guajará, Ananindeua River – Pará **DATE**: December 1977 **TIME**: ?

EQUIPMENT: CANON 814 (X4) Super 8 Camcorder **LENS**: ? **FILM**: G160 Ektaèrome (160 ASA) **APERTURE**:

? **SHUTTERSPEED**: ? **HEIGHT**: ? **DISTANCE**: ? **APPARENTSIZE**: ?

SPEED: ? **TRAJECTORY**: ? **COLOR**: ? **SHAPE**: Undefined **DESCRIPTION**: Luminous body

PHOTOGRAPHER AND/OR OBSERVERS: Military team from A2 – I COMAR

OBSERVATIONS: Enlarged copy of a frame from the video.

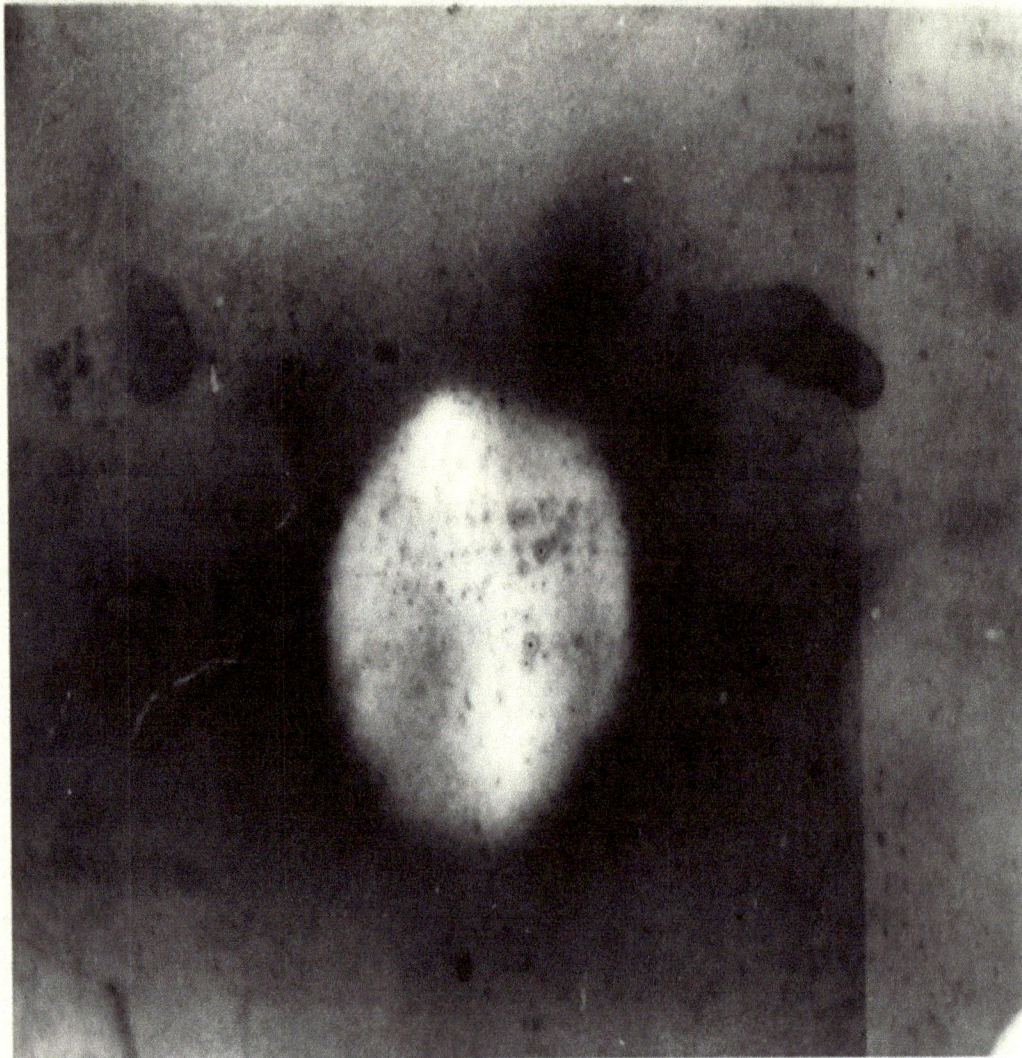

Extended video frame shots. Military Registry - FAB (unknown).
Guarajá River, Ananindeua, Pará state. December, 1977. Time:
(unknown). Photographer: A2 military team (credit: I COMAR)

MILITARY OPERATION: Operation Saucer – I COMAR - FAB MILITARY **REGISTRATION**: ?

LOCATION: Benevides – Pará **DATE**: July 1978 **TIME**: ? **EQUIPMENT**: Nikon **LENS**: 400 mm

FILM: 1000 ASA **APERTURE**: ? **SHUTTERSPEED**: ? **HEIGHT**: ? **DISTANCE**: ?

APPARENTSIZE: ? **SPEED**: ? **TRAJECTORY**: ? **COLOR**: ? **SHAPE**: Undefined

DESCRIPTION: Luminous body

PHOTOGRAPER AND/OR OBSERVERS: Military team from A2 – I COMAR

OBSERVATIONS: Large luminous body over the city of Benevides. At a certain point, the larger UFO released small flying objects.

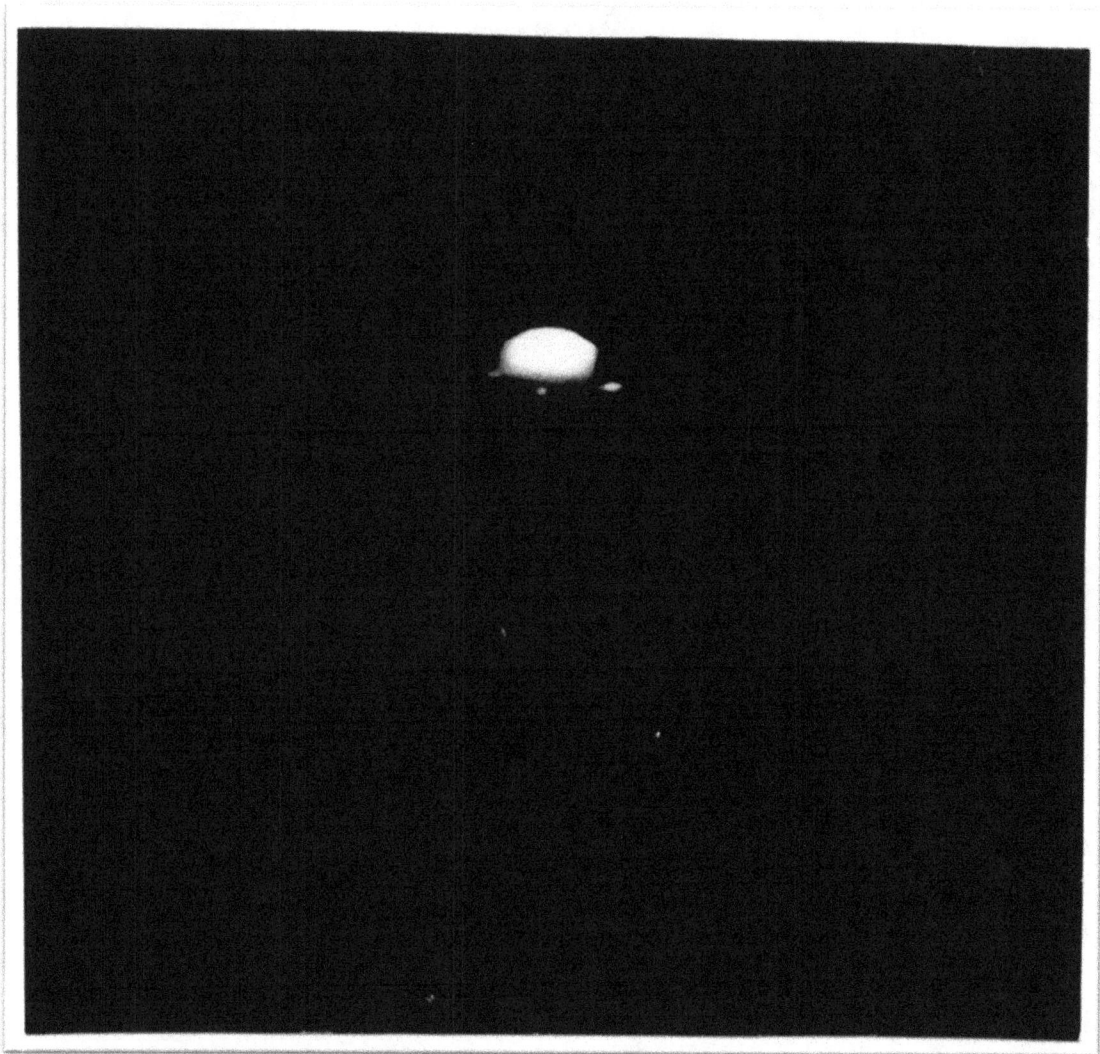

UFO Attacks in Brazil

MILITARY OPERATION: Operation Saucer – I COMAR - FAB MILITARY **REGISTRATION**: ?

LOCATION: Baía do Sol – Pará **DATE**: December 1977 **TIME**: ? **EQUIPMENT**: Minolta **LENS**: ?

FILM: 1000 ASA **APERTURE**: ? **SHUTTERSPEED**: ? **HEIGHT**: ? **DISTANCE**: ?

APPARENTSIZE: ? **SPEED**: ? **TRAJECTORY**: ? **COLOR**: ? **SHAPE**: Undefined

DESCRIPTION: Luminous body

PHOTOGRAPER AND/OR OBSERVERS: Military team from A2 – I COMAR

OBSERVATIONS: UFO of undefined shape in Baía do Sol.

MILITARY OPERATION: Operation Saucer – I COMAR - FAB MILITARY **REGISTRATION**: ?

LOCATION: Baía do Sol – Pará **DATE**: December 1977 **TIME**: ? **EQUIPMENT**: Minolta **LENS**: ? **FILM**: 1000 ASA **APERTURE**: ? **SHUTTERSPEED**: ? **HEIGHT**: ? **DISTANCE**: ?

APPARENTSIZE: ? **SPEED**: ? **TRAJECTORY**: ? **COLOR**: ? **SHAPE**: Undefined

DESCRIPTION: Luminous body

PHOTOGRAPER AND/OR OBSERVERS: Military team from A2 – I COMAR

OBSERVATIONS: UFO of undefined shape in Baía do Sol.

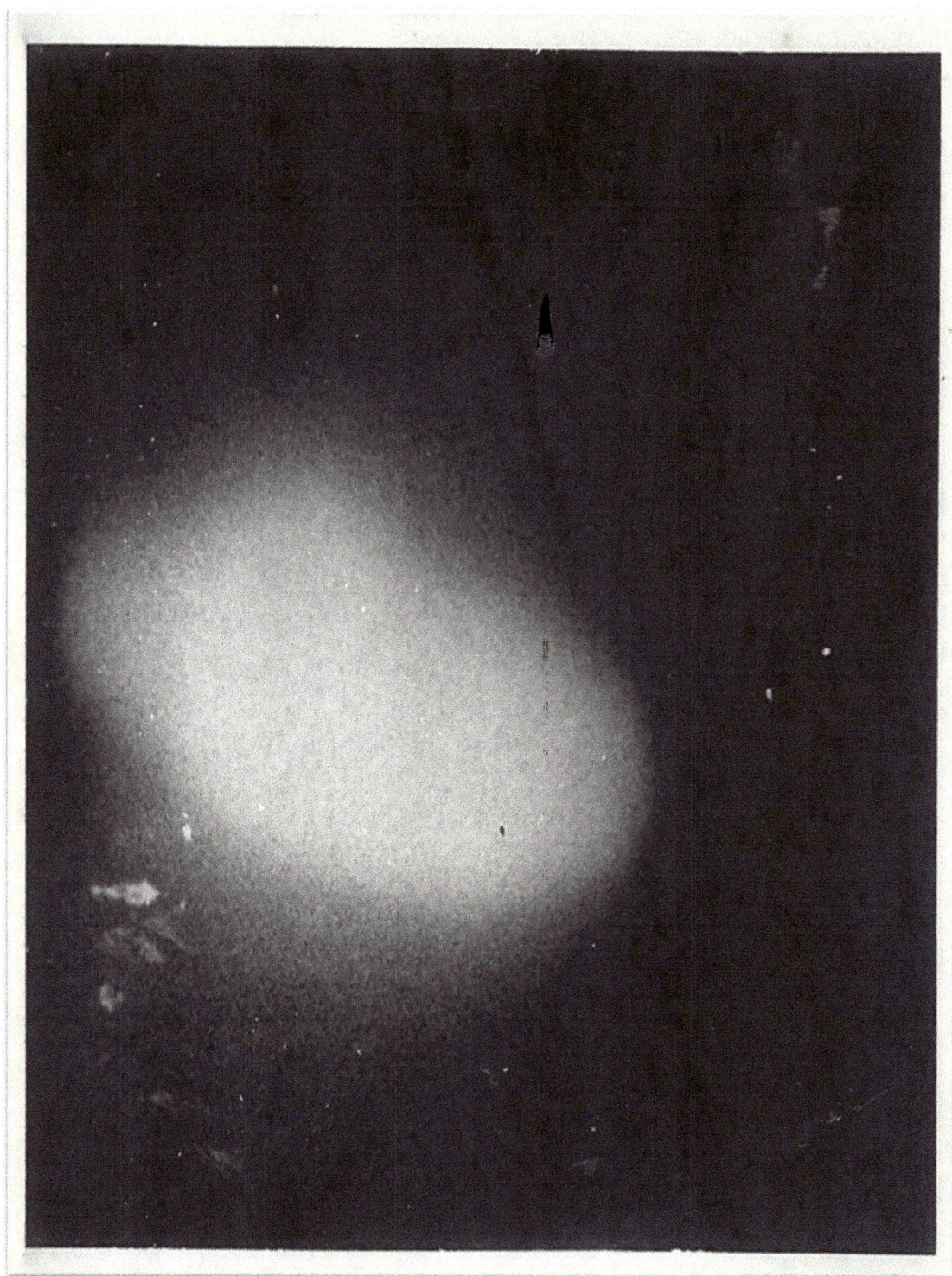

MILITARY OPERATION: Operation Saucer – I COMAR - FAB MILITARY **REGISTRATION**: ?

LOCATION: Baía do Sol – Pará **DATE**: December 1977 **TIME**: ? **EQUIPMENT**: Minolta LENS: ? **FILM**: 1000

ASA **APERTURE**: ? **SHUTTERSPEED**: ? **HEIGHT**: ? **DISTANCE**: ?

APPARENTSIZE: ? **SPEED**: ? **TRAJECTORY**: ? **COLOR**: ? **SHAPE**: Undefined

DESCRIPTION: Luminous body

PHOTOGRAPER AND/OR OBSERVERS: Military team from A2 – I COMAR

OBSERVATIONS: UFO of undefined shape in Baía do Sol.

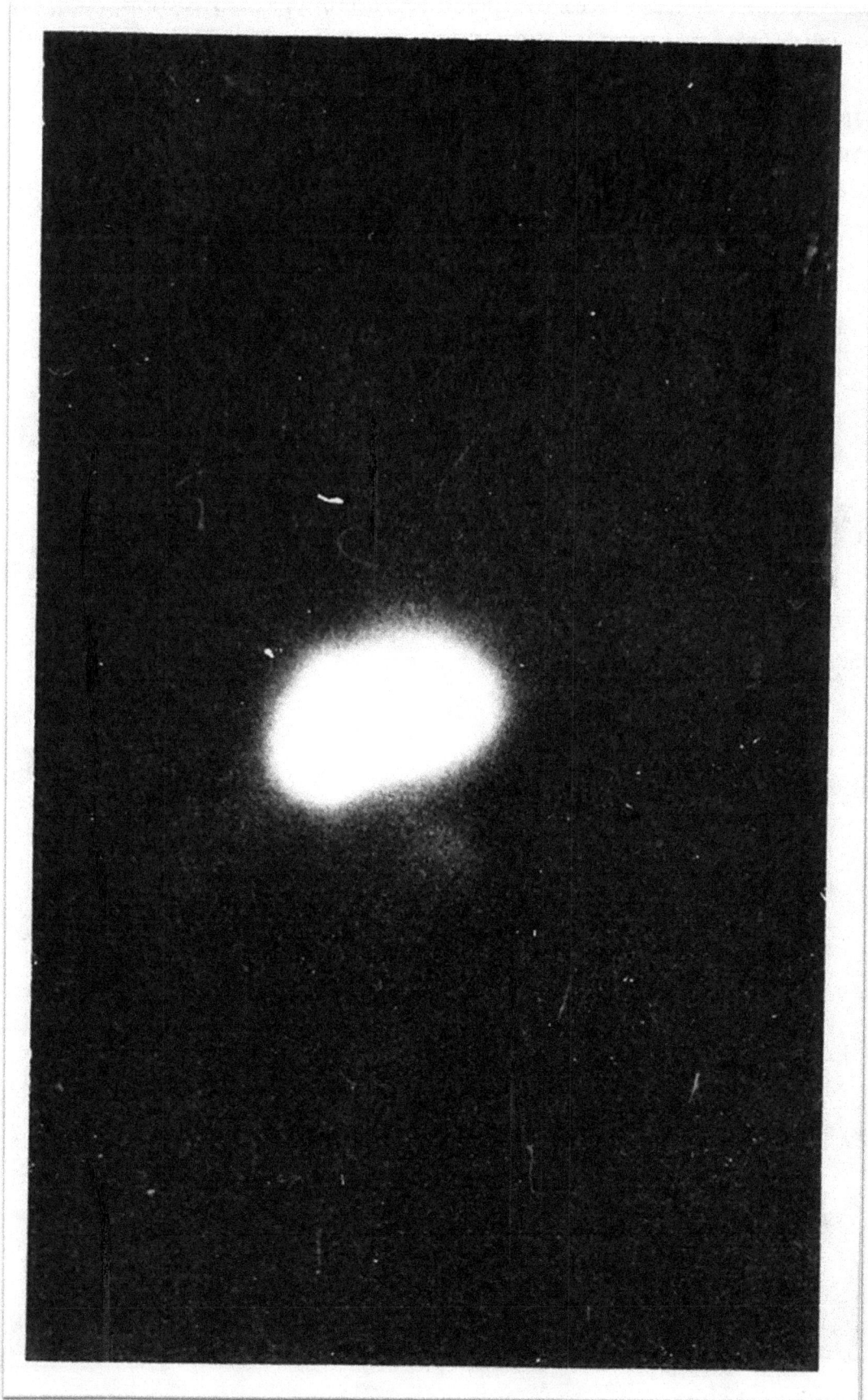

UFO Attacks in Brazil

MILITARY OPERATION: Operation Saucer – I COMAR - FAB MILITARY **REGISTRATION**: ?

LOCATION: Baía do Sol – Pará **DATE**: June 1978 **TIME**: ? **EQUIPMENT**: Nikon LENS: ?

FILM: 1000 ASA **APERTURE**: ? **SHUTTERSPEED**: ? **HEIGHT**: ? **DISTANCE**: ?

APPARENTSIZE: ? **SPEED**: ? **TRAJECTORY**: ? **COLOR**: ? **SHAPE**: Undefined

DESCRIPTION: Luminous body

PHOTOGRAPER AND/OR OBSERVERS: Military team from A2 – I COMAR and civilian JOSÉ RIBAMAR PRAZERES **OBSERVATIONS**: Luminous body over Baía do Sol.

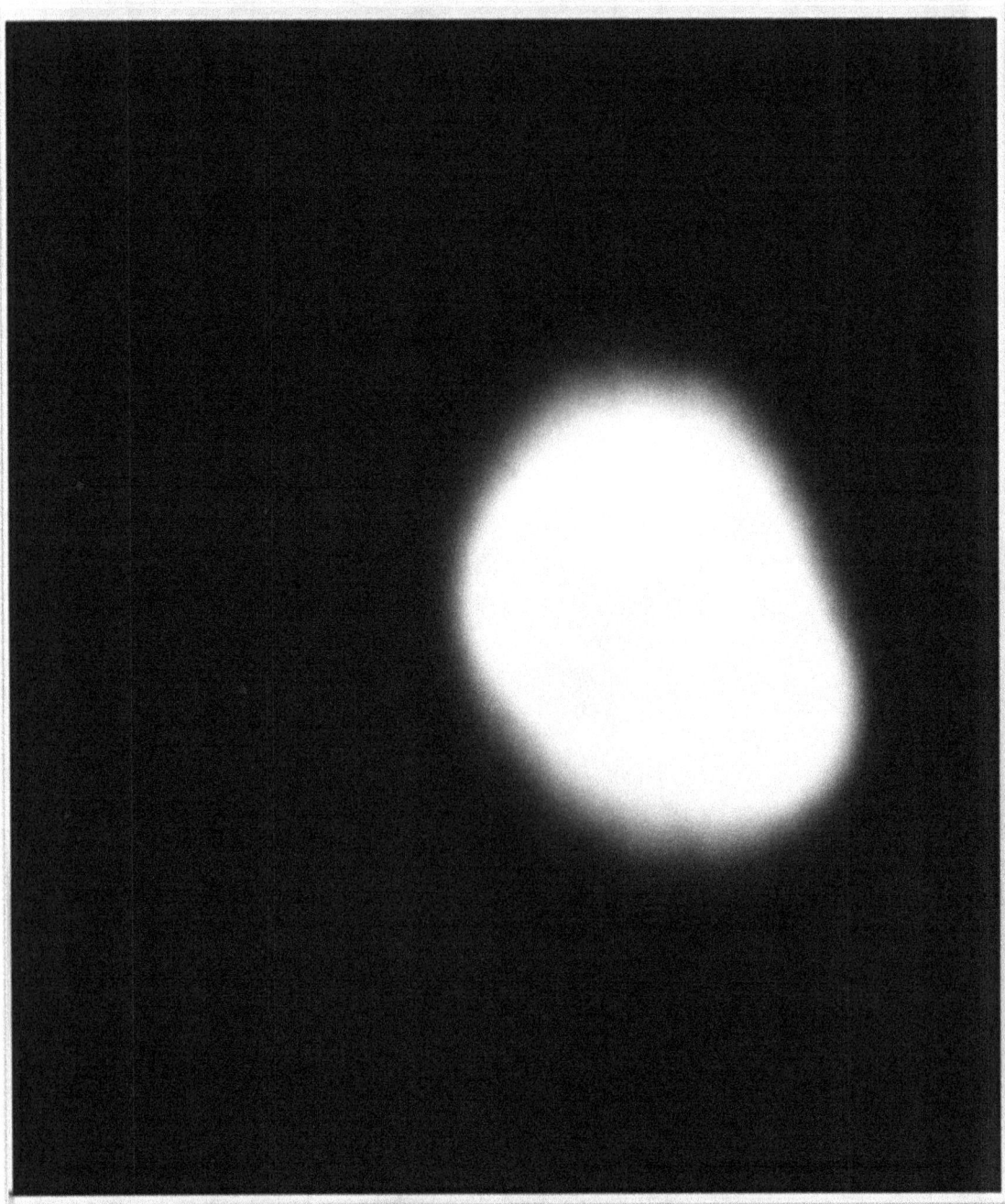

MILITARY OPERATION: Operation Saucer – I COMAR - FAB MILITARY **REGISTRATION**: ?

LOCATION: Guajará River, Ananindeua – Pará **DATE**: December 1977 **TIME**: ?

EQUIPMENT: Super 8 Camcorder **LENS**: ? FILM: 160 ASA **APERTURE**: ?

SHUTTERSPEED: ? **HEIGHT**: ? **DISTANCE**: ? **APPARENTSIZE**: ? **SPEED**: ?

TRAJECTORY: ? **COLOR**: Reddish yellow **SHAPE**: Oval **DESCRIPTION**: Luminous body

PHOTOGRAPER AND/OR OBSERVERS: Military team from A2 – I COMAR

OBSERVATIONS: Enlarged copy of a frame from the video.

UFO Attacks in Brazil

MILITARY OPERATION: Operation Saucer – I COMAR - FAB MILITARY **REGISTRATION**: 075

LOCATION: Highway PA-47 – Jejú/Piçarreira Farm – São Domingos do Capim – Pará

DATE: Friday, December 16, 1977 **TIME**: 23:50**EQUIPMENT**: Minolta SRT 101

LENS: ROKKOR ZOOM 100/200 mm**FILM**: RE-2475 - 1000 ASA **APERTURE**: 5.6

SHUTTERSPEED: 1/30 sec **HEIGHT**: 4,000 m **DISTANCE**: 2,500 m

APPARENTSIZE: 2 to 3 cm **SPEED**: Medium (over 300 km/h)

TRAJECTORY: Straight and descending **COLOR**: Light yellow, bright **SHAPE**: Undefined **DESCRIPTION**: Luminous body

PHOTOGRAPER AND/OR OBSERVERS: Military team from A2 – I COMAR **OBSERVATIONS**: Composed of 6 photographs in the military report. The UFO followed in a Southeast/Northwest direction. From a distance, the UFO looked like a very large and bright star. When it leveled off (300 m), it was covered by tall grass. The team moved on foot to get a better look.

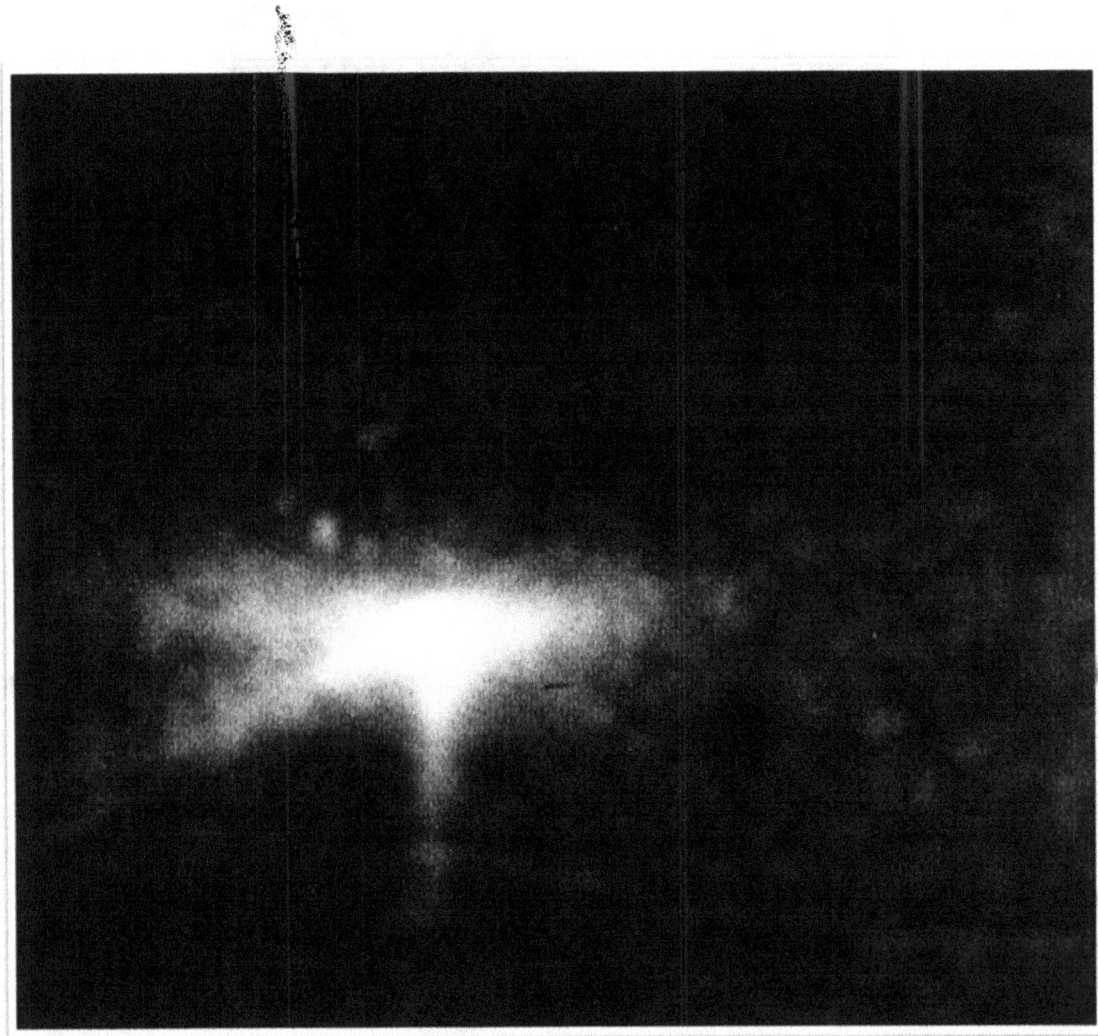

MILITARY OPERATION: Operation Saucer – I COMAR - FAB MILITARY **REGISTRATION**: ?

LOCATION: Guajará River, Ananindeua – Pará **DATE**: December 1977 **TIME**: ?

EQUIPMENT: Yashica LENS: ? **FILM**: 1000 ASA **APERTURE**: ? **SHUTTERSPEED**: ?

HEIGHT: ? **DISTANCE**: ? **APPARENTSIZE**: ? **SPEED**: ? **TRAJECTORY**: ? **COLOR**: ?

SHAPE: Undefined **DESCRIPTION**: Luminous body **PHOTOGRAPER AND/OR OBSERVERS**: Military team from A2 – I COMAR **OBSERVATIONS**: Enlarged copy of photo of UFO near Guajará River.

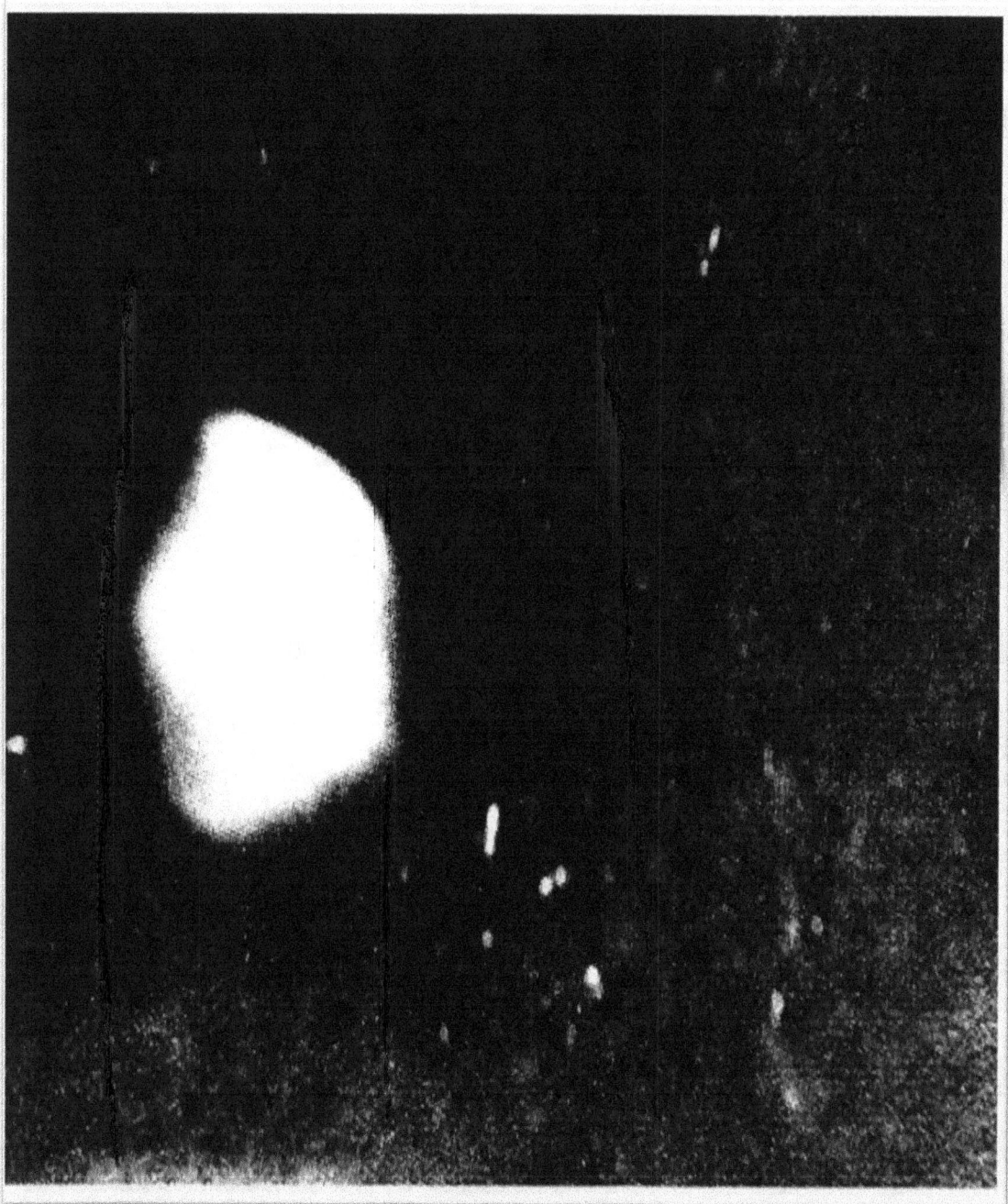

UFO Attacks in Brazil

MILITARY OPERATION: Operation Saucer – I COMAR - FAB **MILITARYREGISTRATION**: ?

LOCATION: Baía do Sol – Pará **DATE**: June 1978 **TIME**: ? **EQUIPMENT**: Nikon **LENS**: ?

FILM: 1000 ASA **APERTURE**: ? **SHUTTERSPEED**: ? **HEIGHT**: ? **DISTANCE**: ?

APPARENTSIZE: ? **SPEED**: ? **TRAJECTORY**: ? **COLOR**: ? **SHAPE**: Undefined

DESCRIPTION: Luminous body **PHOTOGRAPER AND/OR OBSERVERS**: Team of military personnel from A2 – I COMAR and civilian JOSÉ RIBAMAR PRAZERES **OBSERVATIONS**: Luminous body over Baía do Sol. Photograph featured on the cover of the newspaper *O ESTADO DO PARÁ*, issue 16,545, dated June 29th, 1978.

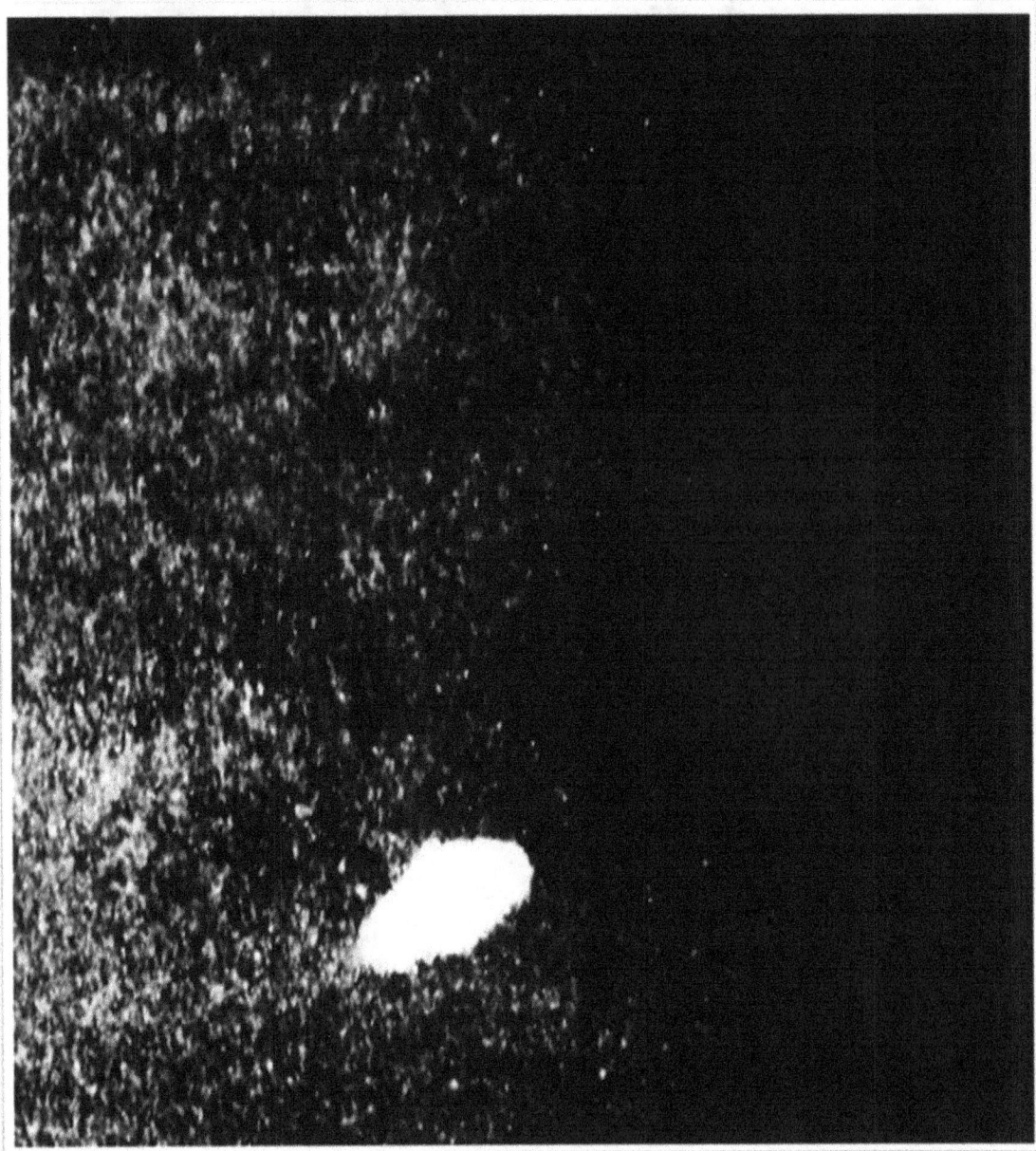

MILITARY OPERATION: Operation Saucer – I COMAR - FAB MILITARY **REGISTRATION**: ?
LOCATION: Baía do Sol – Pará **DATE**: June 1978 **TIME**: ? **EQUIPMENT**: Nikon **LENS**: ?
FILM: 1000 ASA **APERTURE**: ? **SHUTTERSPEED**: ? **HEIGHT**: ? **DISTANCE**: ?
APPARENTSIZE: ? **SPEED**: ? **TRAJECTORY**: ? **COLOR**: ? **SHAPE**: Undefined **DESCRIPTION**: Luminous
body **PHOTOGRAPER AND/OR OBSERVERS**: Team of military personnel from A2 – I COMAR and civilian
JOSÉ RIBAMAR PRAZERES **OBSERVATIONS**: Elongated luminous body over Baía do Sol. Photograph
published in the newspaper O ESTADO DO PARÁ, No. 16,545, on June 29, 1978, on page 14. After analysis,
it was found to be only a hole and a scratch in the negative, that is, the negative used in the newspaper article
was damaged.

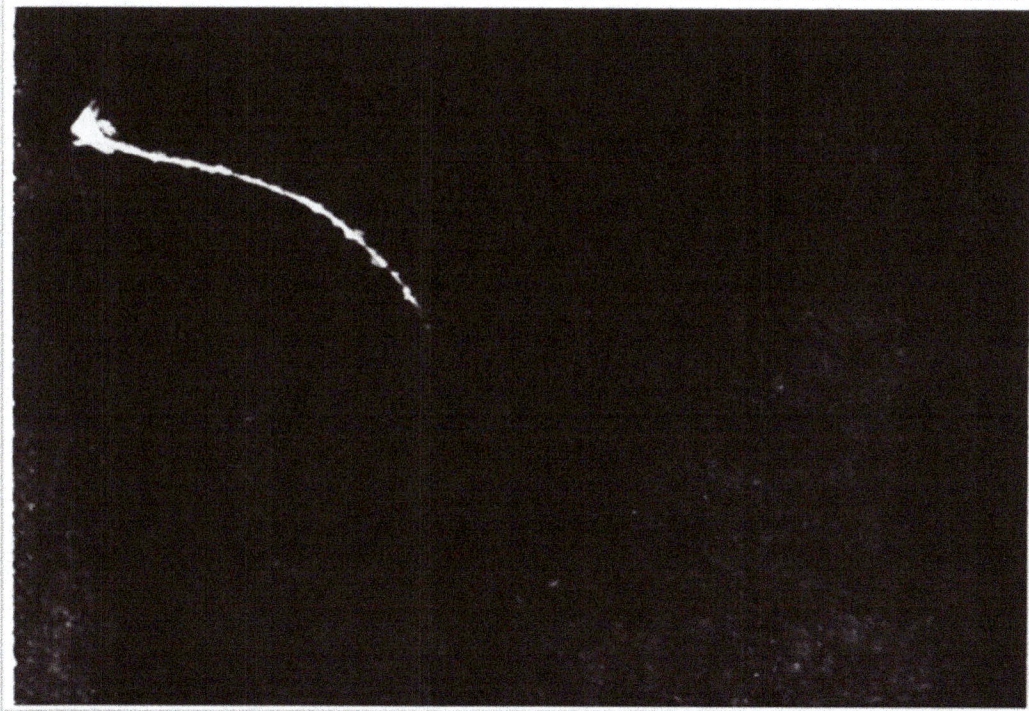

Chapter 23

Press publications

Figure 130. "Flying Saucer in Icoaraci". Newspaper O Estado do Pará. November 14th, 1977

Figure 131. "Flying Saucer Route". Newspaper O Estado do Pará.

A luz da morte também apareceu em Marapanim

De Marapanim vem notícias de pânico entre a população, alarmada com o que os vigienses chamam de "chupa-chupa": um estranho objeto voador, com fachos de luz que paralisam e, mais que isso, aterrorizam a população. Na Vigia, novamente se registram casos, indo o delegado do Município proceder diligências para saber o que, realmente está acontecendo no interior. (Pag. 2)

Figure 132. The light of death also appeared in Marapanim.

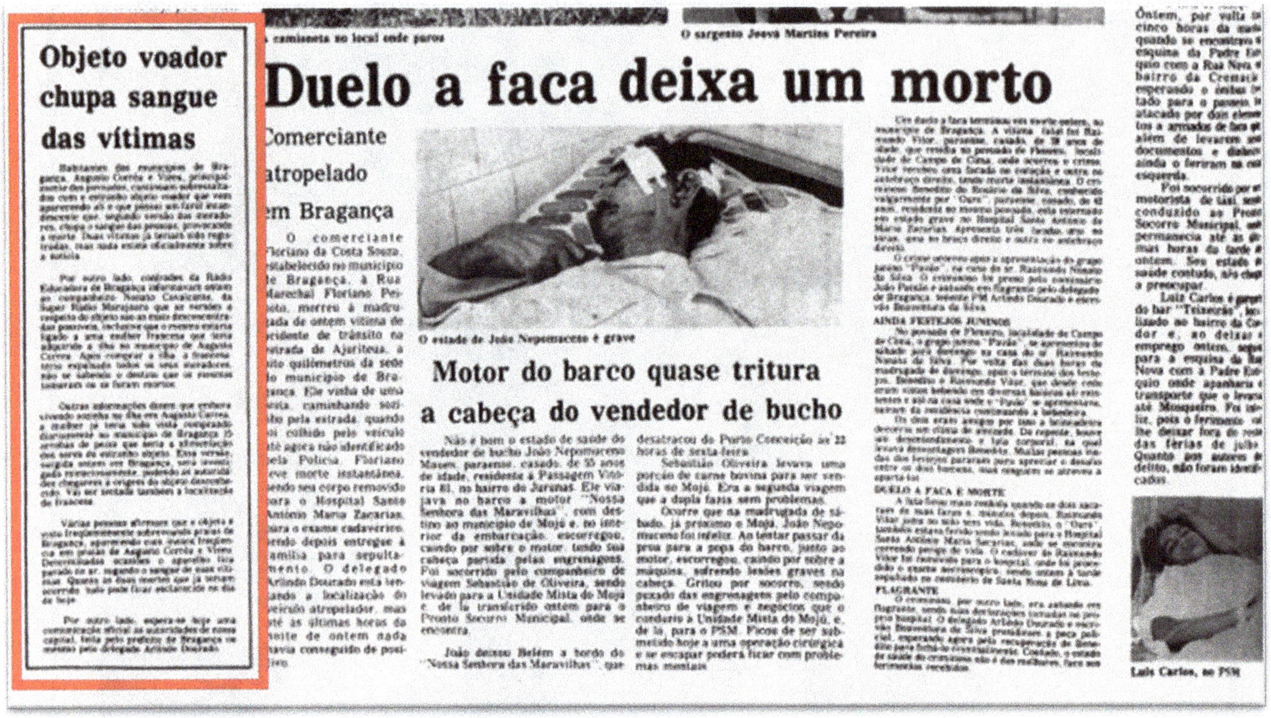
Objeto voador chupa sangue das vítimas

camiseta no local onde parou

O sargento Jeová Martins Pereira

Duelo a faca deixa um morto

Comerciante atropelado em Bragança

O estado de João Nepomuceno é grave

Motor do barco quase tritura a cabeça do vendedor de bucho

Luiz Carlos, no PSM

Figure 133. Flying object sucks blood from victims

A Província do Pará

IRRADIADOR DOS DIÁRIOS ASSOCIADOS: ASSIS CHATEAUBRIAND

ss O - Nº 25.610   Belém - Quinta-feira, 20 de outubro de 1977   Cr$ 4,00

Tempo

Diretório Nacional mantém expulsão de Paz

Remo vence América no Mineirão

O Clube do Remo se reabilizou da derrota de domingo, diante do Atlético Mineiro, e salvou o futebol paraense de novo vexame total na segunda rodada da Copa Brasil, ao derrotar o América por 1 x 0, ontem à noite, no Mineirão, com gol de Bira, aos 31 minutos do segundo tempo. Por sua vez, o Paissandu levou de 4 x 0 do Santos Futebol Clube, na Vila Belmiro, no reaparecimento do goleiro Reginaldo, que na sua última partida pelo clube, sofreu cinco gols diante do Clube do Remo, pelo Ciplo-76. O Remo não jogou um futebol de primeira porque o adversário procurou marcar em cima e o gramado estava bastante escorregadio, dificultando as manobras azulinas. Entretanto, foi a equipe que criou as melhores oportunidades. No primeiro tempo, Leônidas fez uma boa jogada, aos sete minutos, e perdeu um gol certo, o mesmo acontecendo aos 13, com Mesquita. No segundo tempo, Bira desperdiçou outra oportunidade aos 25, mas depois o América cresceu e só não marcou porque a estréia de Edson foi auspiciosa, tanto pela sua técnica como pela sua sorte, pois duas bolas chutadas por Marção, acertaram os dois postes da sua trave. O gol de Bira surgiu numa investida de Marinho pela direita, falhando a zaga americana. Em Belém, iniciando a decisão do voleibol, o Remo derrotou a Tuna no feminino, por três "sets" a uma, e no masculino, a Tuna venceu os remistas por três "sets" a zero. A decisão será amanhã, no "Serra Freire". (Esportes)

Telefoto e Estado de Minas

Zé Maurício, que demonstrou intranqüilidade, impede a queda de seu arco num dos ataques do Remo

Objeto não identificado assusta a Vigia

A população da Vigia assistiu assustada, anteontem, um espetáculo insólito: um objeto não identificado irradiando uma forte luz amarela, sobrevoou por várias vezes a cidade, deixando todos os habitantes apavorados. Há cerca de dois meses que o aparecimento de um Objeto Voador Não Identificado-Oval, vem deixando em polvorosa os moradores do povoado de Ubiteúba, distante 20 quilômetros da sede do município. Anteontem, as denúncias foram confirmadas, inclusive, pelo próprio prefeito José Idasse Favacho, que presenciou o acontecimento. Em Santo Antônio do Tauá, o Ovni foi também provocado tanto em várias pessoas. Na ramal do Triunfo, a senhora Denorith Pereira Oliveira depois de ter visto o aparelho, ficou completamente alienada. Em Campo Cerrado, o Ovni planou cerca de 5 minutos em frente à casa de um caboco, que está com os membros superiores paralisados em consequência do raio irradiado pelo objeto. (Polícia)

SEFA ratifica multa contra Cooperativa

A multa de 25 milhões de cruzeiros imposta à Cooperativa Agrícola de Tomé-Açu, foi considerada procedente pela Secretaria da Fazenda do Estado, e por isso será mantida. A Cooperativa já foi comunicada e deverá depositar o valor da multa dentro de 8 dias, ou então, apresentar recurso voluntário para o Conselho de Contribuintes do Estado. A multa, a maior até hoje aplicada a um contribuinte do Estado, tomou por base a Lei 5.484/76, que considera irrelevante a isenção concedida pela Sefa à Cooperativa, e exigiu o pagamento da diferença do ICM não depositado, a partir de 1972. Reunidos, ontem, com o governador Aloysio Chaves, os dirigentes da Cooperativa, receberam a promessa de que o "problema do ato que determinou a multa seria convenientemente estudado para posterior solução". (Pág. 7)

A Comissão de Ética do Diretório Nacional do MDB, reunida, ontem em Brasília, sob a presidência do deputado João Menezes, resolveu manter, por unanimidade, a decisão de expulsão do vereador Alvaro Paz do Nascimento, aplicada pelo Diretório Regional. O relator do processo foi o senador Ivandro Cunha Lima, que em seu parecer disse que as denúncias contra o presidente da Câmara são procedentes, "notadamente no que concerne ao comportamento ético de Alvaro Paz do Nascimento, cuja incontinência de linguagem, refratária à lealdade partidária, desserve a causa do partido, na medida em que faz do MDB, alvo de candentes críticas da imprensa." O recurso impetrado por Alvaro Paz, contra a decisão do Diretório Regional, foi indeferido tomando por base a gravidade das faltas que ele cometeu. Os deputados Getúlio Dias, Nélson Maculan, Argilano Dário, Ruy Lino, Carlos Costa e o senador Evandro Carrera, mantiveram a decisão. (Pág. 3)

TELEFOTO A-32

Figure 134. Unidentified object scares watchman. Newspaper A Província do Pará**, October 20th, 1977.**

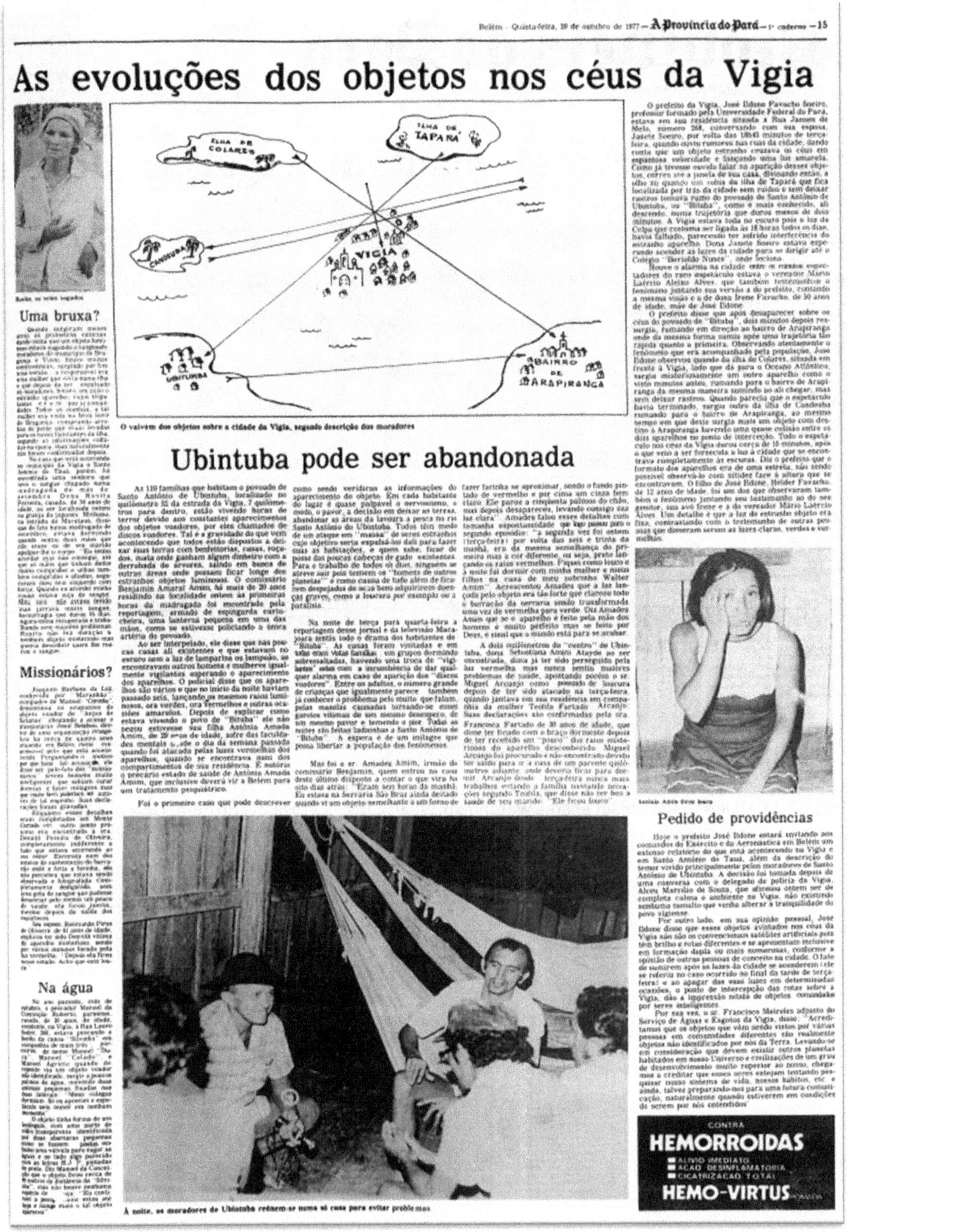

Belém · Quinta-feira, 20 de outubro de 1977 — A Província do Pará — 1º caderno — 15

As evoluções dos objetos nos céus da Vigia

Uma bruxa?

O valvém dos objetos sobre a cidade da Vigia, segundo descrição dos moradores

Ubintuba pode ser abandonada

Missionários?

Na água

À noite, os moradores de Ubintuba reúnem-se numa só casa para evitar problemas

Pedido de providências

Figure 135. "The evolution of objects in the skies of Vigia". "Ubintuba may be abandoned". "A witch?". "Missionaries?". "Request for Action". A *Província do Pará* newspaper. October 20th, 1977.

Uma estranha nave sobrevoando um terreiro.
Uma arma que não funcionou. Um homem atingido por uma luz e com um lado do corpo
paralisado. O seio sugado pela mão de unhas compridas. Duas mulheres que perderam o juízo.
A evolução de dois objetos não identificados sobre a cidade da Vigia e
que foi assistida pelo prefeito. Uma população que ameaça abandonar
o vilarejo. A luz misteriosa chega a Santo Antônio do Tauá.

LUZ E PAVOR NAS NOITES VIGIENSES

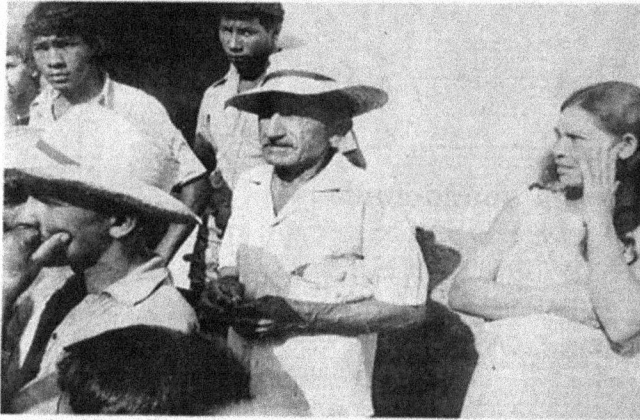

Aterrorizados, os habitantes do Vigia esperam uma explicação para o fenômeno da luz misteriosa

Manoel "Coronha" (foto acima) demonstra como enfrentou os tripulantes do disco. Ao lado, a reprodução do incidente a partir da narração feita por ele.

Manoel "Coronha" tentou atirar no disco. A arma não funcionou

Figure 136. "Light and terror of the Vigia night. A strange ship flying over a yard. A weapon that didn't work. A man hit by a light and with one side of his body paralyzed. A breast sucked by a hand with long nails. Two women who lost their minds. The evolution of two unidentified objects over the city of Vigia and which were seen by the mayor. A population that threatens to abandon the village. The mysterious light arrives in Santo Antônio do Tauá." "Manoel Coronha tried to shoot the disk. The weapon didn't work".

351

Figure 137. "The appearance of flying saucers denied". Newspaper A Província do Pará**. November 5ᵗʰ, 1977.**

Figure 138. "1ˢᵗ COMAR claims that UFO in Vigia was pure optical illusion".

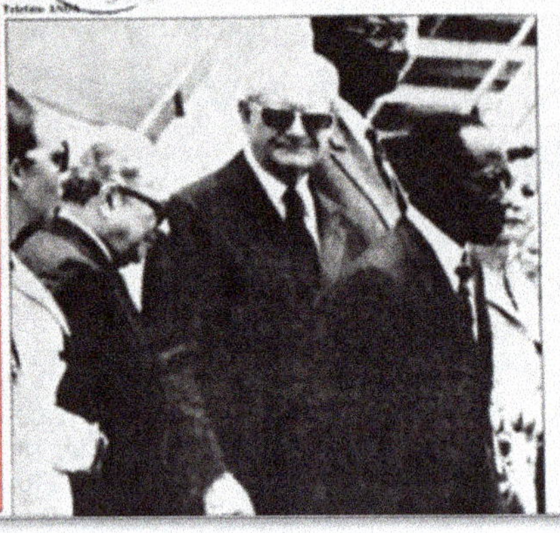

LUZ MISTERIOSA APAVORA VISEU

Em pequenas localidades próximas à cidade de Viseu, seus moradores vivem apreensivos e apavorados. Não saem mais de suas casas após às 18 horas nem para visitar os vizinhos mais próximos porque têm "medo duma lanterna com luz forte que vôa pelo céu e vem sugar o sangue da gente até deixar morto".

Se o fato é verídico ou apenas fantasia do nosso caboclo bragantino ninguém sabe explicar, desmentir ou confirmar. A verdade é que a conversa anda de "boca em boca" e já chegou até a pequenos locais do litoral maranhense próximo à zona bragantina.

A tal "lanterna com luz forte" para muitos que sabem contar a estória, sem nada de concretismo, é "disco voador dos russos" que estão pousando "aí pros lados de Viseu, Curupati, Urumajó e Itaçu".

Isto é o que muita gente sabe informar. Mas ninguém se atreve a dizer que viu, ouviu. Apenas de "ouvi falar".

Em meio a todo esse mistério de "disco voador" e "lanterna com luz forte" há sempre a citação de uma "suíça ou americana loura" que vive numa ilha, cujo nome é desconhecido, mas que para um pescador "é a do Cajueiro". Ela é bastante conhecida em Bragança, porém ninguém sabe seu nome, nem o que é o que faz efetivamente. A curiosidade da população de Bragança até passou a se aperceber de que ela é "hippie", vive sozinha nessa ilha e quando vem à cidade, que também não sabem informar por quais meios, sempre adquire "duzentos quilos de peixe". Por isso todos se perguntam: "para que a loura hippie quer tanto peixe, se vive sozinha?".

No aparecimento de mulher em toda essa estória há o depoimento da sra. Margarida, enteada do administrador do Aeroporto de Bragança (um campo de pouso cuja pista é gramada).

"Ela ia por uma capoeira e de repente apareceu uma mulher muito bonita. Estava toda de preto, da cabeça aos pés. Seus braços eram cobertos por uma roupa folgada até os punhos e usava luvas. As pernas também estavam cobertas até os tornozelos. Usava um negócio na cabeça. Ela se aproximou da Margarida e perguntou quantos filhos tinha, o que fazia, se não tinha medo de andar por ali sozinha. Depois desapareceu misteriosamente. A Margarida chegou em casa com uma dor de cabeça que não tinha mais tamanho".

O mesmo Américo (administrador do campo de pouso) é quem volta a falar, sempre com o "porronca" à boca e apagado:

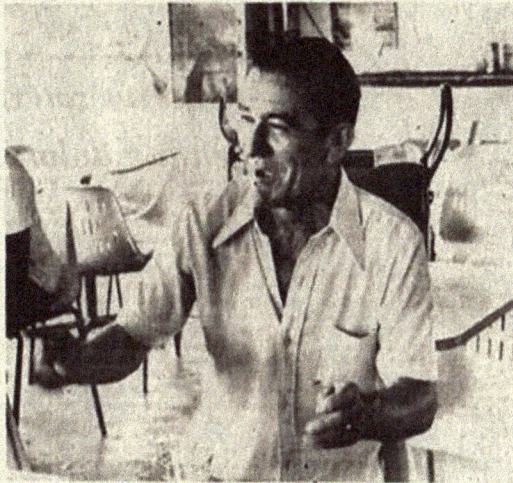

Para o deputado tudo não passa de fantasia.

O delegado acredita, com reservas.

"Ainda na quarta-feira a minha filha (mostrou a filha) estava aqui na frente de casa. Eram 10 da noite. De repente passou aquele clarão forte duma estrela que se movimentava. Eu contei aqui para os meus vizinhos e eles disseram que era o disco voador que estava vindo de Vizeu".

O delegado de Bragança, sargento Arlindo Dourado, ouviu falar no caso, mas foi como o próprio disse "foi conversa de bar, de gente que estava bebendo e logo não merecia crédito". O próprio Arlindo ainda comentava que depois tomara conhecimento de que duas pessoas tinham falecido em Vizeu depois de terem "o sangue chupado por uma luz que voava".

Muito calmo estava o deputado João Motta, que preferiu desmentir tudo:

"Tudo isso, meu amigo, não passa de mentira, boato. É gente que não tem o que fazer e vive inventando coisas. Eu já estive no Curupati de onde surgiu essa conversa tola e procurei lhes mostrar que tudo não passava de invenção".

O deputado ainda conta que soube das mortes, mas em Vizeu, ninguém sabia informar a identidade de quem tinha sido atingido pela "luz forte da lanterna".

Aliás, é o próprio João Motta quem conta uma outra estória surgida em Vizeu, em que um monstro, que eles apelidaram de "Ataíde", estava vindo em direção de Bragança. Por onde passava suas "patas deixavam tudo queimado. Não ficava ninguém pra contar a estória".

No "disse me disse" e "ouvir contar assim e assado" apareceu Carmen Lúcia Venanço Ribeiro. Ela é cozinheira da lanchonete do Terminal Rodoviário de Bragança. A mãe da dona da casa onde mora teve seu filho atingido pela luz e passou quase um mês no hospital:

"Foi lá no Itaçu, bem próximo de Viseu. O filho da mãe da mulher onde moro foi caçar. Passaram-se dois dias e ele não voltava para casa. Os amigos então foram procurá-lo e como sabiam onde era a "espera" dele o encontraram desfalecido. Trouxeram para o hospital e aqui ele ficou quase um mês. Ele contou que já era mais de meia noite quando viu aquela luz se aproximar da "espera". À medida que se aproximava de sua rede ia ficando como que embriagado, mas não de todo inconsciente. No braço direito sentiu aquela picada. . Depois não se lembra de nada. Disseram os médicos que estava quase sem uma gota de sangue no corpo".

Toda essa estória da "luz da lanterna" é conhecida em Bragança em especial pelos motoristas de táxi, ônibus e caminhão, e pelos habituais freqüentadores dos bares. Todos eles contam a estória sem ter um pingo de certeza.

João Martinho de Almeida Filho, por exemplo, informava que um cidadão de Viseu contou-lhe que uma mulher saiu da tal luz. Estava com roupa toda prateada, aproximou-se dum morador e o manietou. Como ele gritasse e aparecesse gente, "de repente desapareceu a luz e a mulher". É ele quem informa também que quem não tinha arma nessas localidades próximas de Viseu passou a usá-las: "tem gente até que está dormindo com pedaço de pau embaixado da cama com medo desse mistério".

Figure 139. "Mysterious light terrifies Viseu"."For the Congressman, it's all just a fantasy." "The sheriff believes, with reservations."

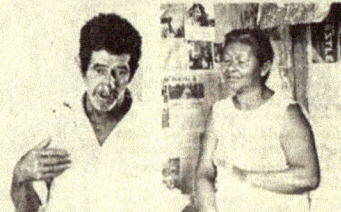

Figure 140. "Mysterious light still causes fear in Viseu". "The end of the world". "Figment of their imagination". "More unbelievers". "Teacher saw". "Child was about to die". "The case of the fishermen". "Alone with the light". "The Island of Bats".

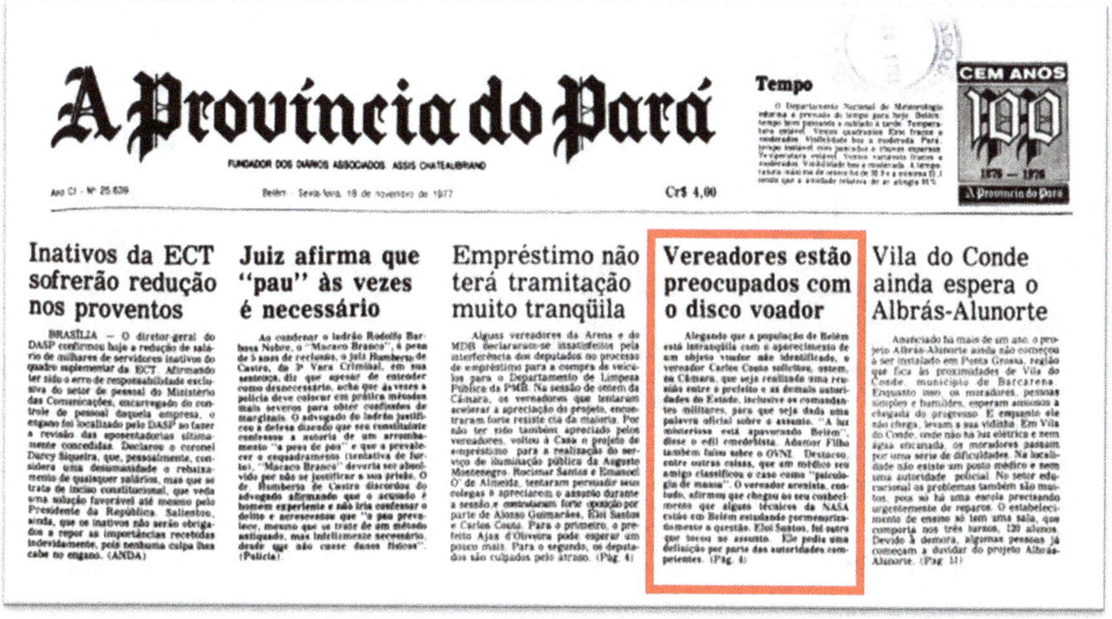

A Província do Pará

FUNDADOR DOS DIÁRIOS ASSOCIADOS ASSIS CHATEAUBRIAND

Tempo

CEM ANOS

Inativos da ECT sofrerão redução nos proventos

Juiz afirma que "pau" às vezes é necessário

Empréstimo não terá tramitação muito tranqüila

Vereadores estão preocupados com o disco voador

Vila do Conde ainda espera o Albrás-Alunorte

Figure 141. "Councilmen are worried about the flying saucer". Newspaper A Província do Pará**. November 18th, 1977.**

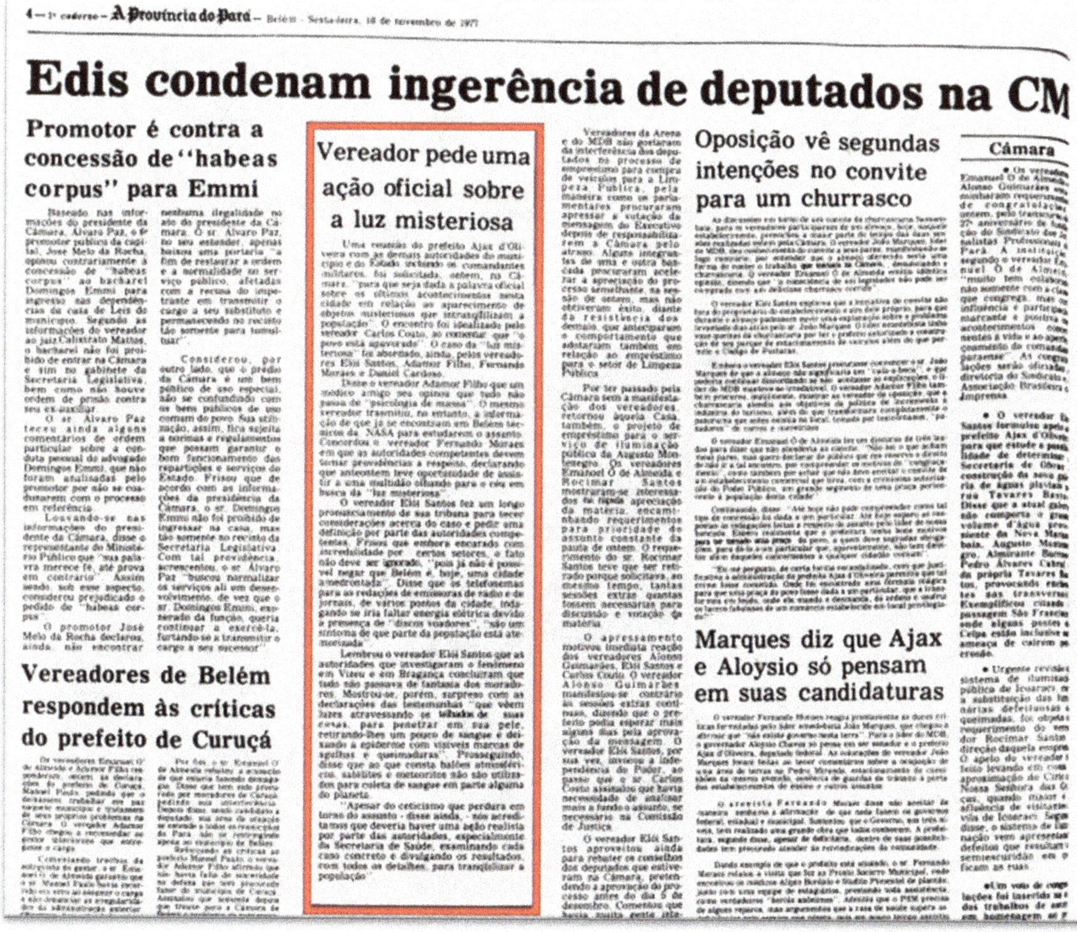

Edis condenam ingerência de deputados na CM

Promotor é contra a concessão de "habeas corpus" para Emmi

Vereador pede uma ação oficial sobre a luz misteriosa

Oposição vê segundas intenções no convite para um churrasco

Câmara

Vereadores de Belém respondem às críticas do prefeito de Curuçá

Marques diz que Ajax e Aloysio só pensam em suas candidaturas

Figure 142. "Councilman calls for official action on mysterious light". Newspaper A Província do Pará**. November 18th, 1977.**

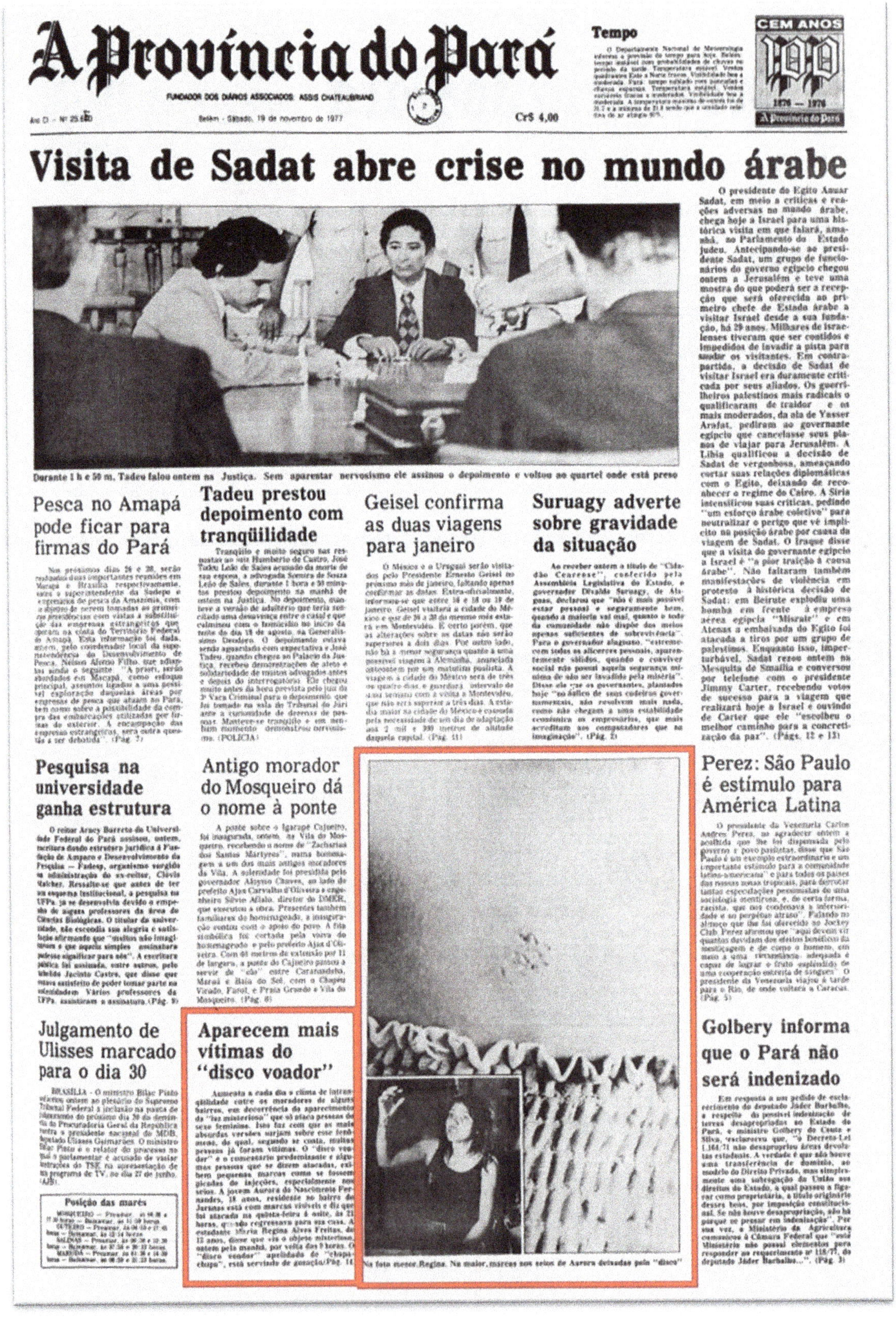

A Província do Pará

FUNDADOR DOS DIÁRIOS ASSOCIADOS ASSIS CHATEAUBRIAND

Belém - Sábado, 19 de novembro de 1977 — Cr$ 4,00

CEM ANOS 1876 — 1976

Visita de Sadat abre crise no mundo árabe

O presidente do Egito Anuar Sadat, em meio a críticas e reações adversas no mundo árabe, chega hoje a Israel para uma histórica visita em que falará, amanhã, no Parlamento do Estado judeu. Antecipando-se ao presidente Sadat, um grupo de funcionários do governo egípcio chegou ontem a Jerusalém e teve uma mostra do que poderá ser a recepção que será oferecida ao primeiro chefe de Estado árabe a visitar Israel desde a sua fundação, há 29 anos. Milhares de israelenses tiveram que ser contidos e impedidos de invadir a pista para saudar os visitantes. Em contrapartida, a decisão de Sadat de visitar Israel era duramente criticada por seus aliados. Os guerrilheiros palestinos mais radicais o qualificaram de traidor e os mais moderados, da ala de Yasser Arafat, pediram ao governante egípcio que cancelasse seus planos de viajar para Jerusalém. A Líbia qualificou a decisão de Sadat de vergonhosa, ameaçando cortar suas relações diplomáticas com o Egito, deixando de reconhecer o regime do Cairo. A Síria intensificou suas críticas, pedindo "um esforço árabe coletivo" para neutralizar o perigo que vê implícito na posição árabe por causa da viagem de Sadat. O Iraque disse que a visita do governante egípcio a Israel é "a pior traição à causa árabe". Não faltaram também manifestações de violência em protesto à histórica decisão de Sadat: em Beirute explodiu uma bomba em frente à empresa aérea egípcia "Misrair" e em Atenas a embaixada do Egito foi atacada a tiros por um grupo de palestinos. Enquanto isso, imperturbável, Sadat rezou ontem na Mesquita de Smailia e conversou por telefone com o presidente Jimmy Carter, recebendo votos de sucesso para a viagem que realizará hoje a Israel e ouvindo de Carter que ele "escolheu o melhor caminho para a concretização da paz". (Págs. 12 e 13)

Durante 1 h e 50 m, Tadeu falou ontem na Justiça. Sem aparentar nervosismo ele assinou o depoimento e voltou ao quartel onde está preso

Pesca no Amapá pode ficar para firmas do Pará

Tadeu prestou depoimento com tranqüilidade

Geisel confirma as duas viagens para janeiro

Suruagy adverte sobre gravidade da situação

Perez: São Paulo é estímulo para América Latina

Pesquisa na universidade ganha estrutura

Antigo morador do Mosqueiro dá o nome à ponte

Julgamento de Ulisses marcado para o dia 30

Posição das marés

Aparecem mais vítimas do "disco voador"

Aumenta a cada dia o clima de intranqüilidade entre os moradores de alguns bairros, em decorrência do aparecimento da "luz misteriosa" que só ataca pessoas do sexo feminino. Isso faz com que as mais absurdas versões surjam sobre esse fenômeno, do qual, segundo se conta, muitas pessoas já foram vítimas. O "disco voador" é o comentário predominante e algumas pessoas que se dizem atacadas, exibem pequenas marcas como se fossem picadas de injeções, especialmente nos seios. A jovem Aurora do Nascimento Fernandes, 18 anos, residente no bairro do Jurunas está com marcas visíveis e diz que foi atacada na quinta-feira à noite, às 21 horas, quando regressava para sua casa. A estudante Maria Regina Alves Freitas, de 12 anos, disse que viu o objeto misterioso, ontem pela manhã, por volta das 9 horas. O "disco voador" apelidado de "chupa-chupa", está servindo de gozação.(Pág. 14)

Na foto menor, Regina. Na maior, marcas nos seios de Aurora deixadas pelo "disco"

Golbery informa que o Pará não será indenizado

Figure 143. "More victims of flying saucer appear". Newspaper A Província do Pará**. November 19th, 1977.**

O LIBERAL/1o. Caderno   Belém, quinta-feira, 17 de novembro de 1977 — Página 21

Fenômeno da luz intranqüiliza a cidade

Um clima de intranqüilidade e insegurança está se registrando e se ampliando em toda a cidade, em decorrência de estranhas e contraditórias notícias sobre uma misteriosa luz que ataca as pessoas, sugando-lhes o sangue. Isso é o que contam dezenas de pessoas, talvez na falta de uma explicação plausível para os acontecimentos.

Em nossa redação, são freqüentes os telefonemas recebidos informando sobre aparições do objeto, geralmente definido como sendo uma luz que passa a noite em direção de suas vítimas, deixando-as paralisadas. Mas têm sido inúteis os esforços da reportagem em encontrar simultaneamente, ou logo após os mirabolantes ataques aqueles que se dizem vítimas do "foco paralisador". A cidade está cheia de boatos. E são muitas as pessoas que se dizem vítimas.

Maria Augusta também foi atingida pelo "chupa-chupa"

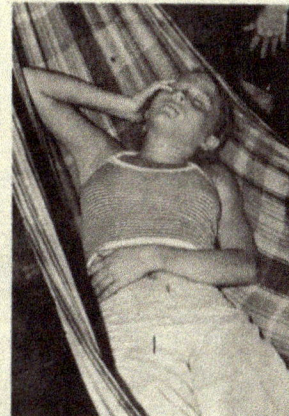

Carmem do Socorro recebe constantes visitas

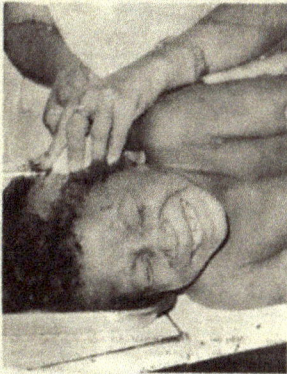

O menino Tadeu de Jesus teria sido atingido pelo "fogo misterioso".

Maria de Belém sofreu a violência do irmão atacado pela luz

Muitos casos, mas até agora nenhuma prova concreta

Provocando reações diferentes, versões nem sempre idênticas da mesma história, contadas num misto de exaltado espanto ou riso incrédulos, telefonemas aqui, e sem deixar nenhuma prova física de sua "presença", a luz misteriosa apareceu também na Travessa 9 de Janeiro, próximo à Oliveira Belo, na noite de terça-feira passada, e na tarde de ontem.

As reações diferentes, porém, não podem ser atribuídas apenas à imaginação dos que viram a luz: nas duas casas visitadas na 9 de Janeiro também os cachorros reagiram de forma diversa. Dona Neuza Marinho, que mora na casa no. 77, tem um cachorro grande que é o terror de quem porventura entra na casa desautorizado. Uma vez, além da luz deixado nenhum sinal ou marca física, visível, nas pessoas que atacou, a luz misteriosa parece ter esquecido também o cachorro da casa no. 13.

UM, DOIS, TRÊS...

Ontem à tarde surgiu a notícia de que a luz havia aparecido na 9 de Janeiro, de tarde. Mas a aparição havia começado na madrugada, por volta de 1,30 horas.

Naquela hora Oberlando de Almeida Teixeira, de 18 anos, estudante do Curso de Mineração na Escola Técnica, estudava sozinho em casa, à 9 de Janeiro, 13. [...]

A Província do Pará

FUNDADOR DOS DIÁRIOS ASSOCIADOS: ASSIS CHATEAUBRIAND

Belém - Domingo, 20 de novembro de 1977 — Cr$ 5,00

CEM ANOS 100 1876 — 1976 A Província do Pará

Médico examina vítimas e explica "Chupa-chupa"

"As visões observadas por algumas pessoas atacadas pelo "vampiro extraterreno", são frutos do estado d'alma, em sintonia com o inconsciente produzindo uma excitação psicomotora". A afirmação é do médico Orlando Zoghbi que, ontem, em companhia da reportagem de A Província do Pará, esteve visitando várias pessoas que se dizem atacadas pelo popular "Chupa-chupa". No que tange às lesões verificadas em algumas pessoas, disse: "Elas são decorrentes das reações de horror, ocasionadas por choque adrenérgico". Quanto ao caso da jovem Aurora que ficou com marcas no seio depois de ter visto uma luz estranha, Zoghbi explicou que "as mulheres instintivamente, num ato de proteção, levam as mãos aos seios e a ação motora contraindo as mãos em garras, ocasionam as lesões nas glândulas mamárias". (Polícia)

Figure 145. "Doctor examines victims and explains Chupa-Chupa". Newspaper A Província do Pará**. November 20th, 1977.**

"Vampiro interplanetário" só gosta de mulher

(O relato daqueles que já estiveram face a face com o "Vampiro", popularmente conhecido por "Chupa-Chupa")

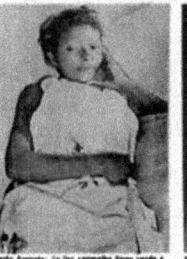

CIRCUITO FORENSE

Juiz nega ordem de habeas-corpus para o homicida

O fenômeno, duas opiniões

As estórias que o povo anda contando

Pastor Firmino: "os homens vão desmaiar de terror"

Frei Paulino: "não há base científica"

Figure 146. "Interplanetary vampire only likes women". **Newspaper** A Província do Pará. **November 19th, 1977.**

359

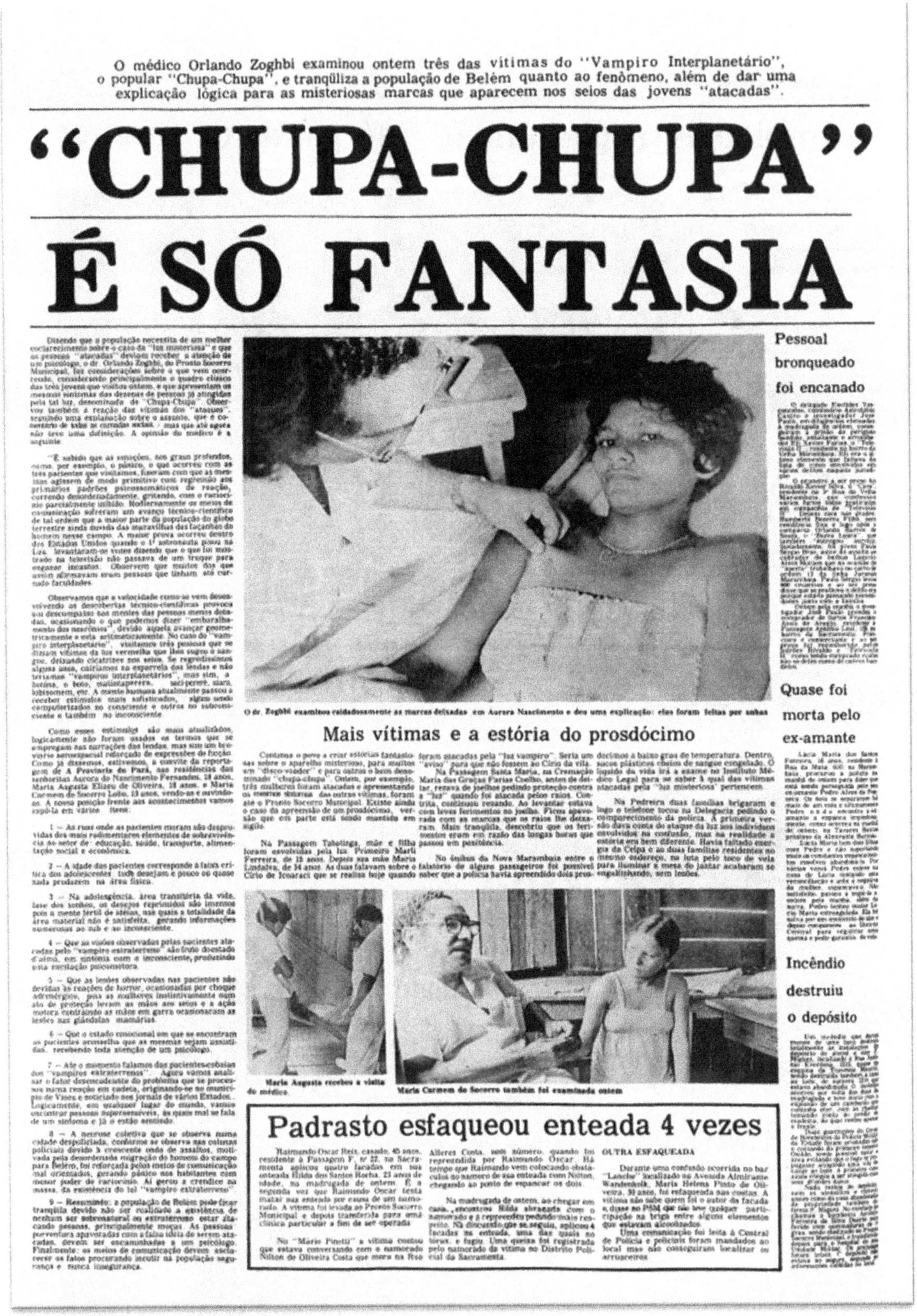

O médico Orlando Zoghbi examinou ontem três das vítimas do "Vampiro Interplanetário", o popular "Chupa-Chupa", e tranqüiliza a população de Belém quanto ao fenômeno, além de dar uma explicação lógica para as misteriosas marcas que aparecem nos seios das jovens "atacadas".

"CHUPA-CHUPA" É SÓ FANTASIA

Pessoal bronqueado foi encanado

Quase foi morta pelo ex-amante

Incêndio destruiu o depósito

O dr. Zoghbi examinou cuidadosamente as marcas deixadas em Aurora Nascimento e deu uma explicação: elas foram feitas por unhas

Mais vítimas e a estória do prosdócimo

Maria Augusta recebeu a visita do médico

Maria Carmem do Socorro também foi examinada ontem

Padrasto esfaqueou enteada 4 vezes

Figure 147. "Chupa-Chupa is just fantasy". "Yesterday, doctor Orlando Zoghbi examined three victims of the Interplanetary Vampire, the popular Chupa-Chupa, and reassured the population of Belém about the phenomenon, in addition to giving a logical explanation for the mysterious marks that appear on the breasts of the young women who were attacked".

Figure 148. "Flying object terrifies Umbituba". O Liberal newspaper. October 16th, 1977.

Figure 149. "Flying saucer attacks in Mosqueiro". O Estado do Pará newspaper. November 1st, 1977.

O ESTADO DO PARÁ

Na localidade Tapiapanema, às margens do rio Pratiquara, na Ilha do Mosqueiro, os moradores estão tomados de grande temor devido ao estranho fato ocorrido no sábado passado, dia 29 de outubro, e que se repetiu no dia de ontem: um aparelho desconhecido, circulando no céu em grande velocidade e emitindo uma luz esverdeada, está atacando as casas locais e fazendo vítimas. O alarmante fato já foi comunicado ao Agente Municipal do Mosqueiro, ao Delegado local.

Em Belém a notícia chegou reservadamente à Central de Polícia, mas quase nenhuma providência foi tomada.

O pavor continua grassando, e já começa a se estender para o centro da Vila balneária. A reportagem de "O Estado do Pará" foi até o local dos acontecimentos documentar o testemunho de várias pessoas, procurando dar um melhor espelho da angústia dos poucos moradores de Tapiapanema bem como de outros moradores da Ilha.

Sílvia e Benedito moram nesta casa

O casal atingido e que está hospitalizado

Disco voador ataca mulher.
Pavor na ilha do Mosqueiro

Sílvia deitou em uma rede como esta. O foco chegou pela fresta.

Raimunda foi uma testemunha e vive traumatizada.

O objeto estranho passou a aparecer por trás das árvores.

Eles querem fugir de suas casas, temerosos.

O cachorro não late mais. Foi atingido pelo foco

Sílvia pode perder a criança

Familiares tomam providências

O foco esverdeado fez várias vítimas

A mãe de Sílvia é a mais abalada.

A cadela não lateu mais.

Figure 150. "Flying saucer attacks woman. Fear on Mosqueiro Island". "Dog no longer barks. It was attacked by the fire". "Sílvia may lose her child". "Family members take measures". "The greenish fire attacked several victims". O Estado do Pará **newspaper. November 1st, 1977.**

Figure 151. "Flying saucer in Belém. Appeared in Pedreira". "Colares was attacked. Victims in Belém". "In Mosqueiro the atmosphere is tense. Residents armed". O Estado do Pará newspaper. November 2nd, 1977.

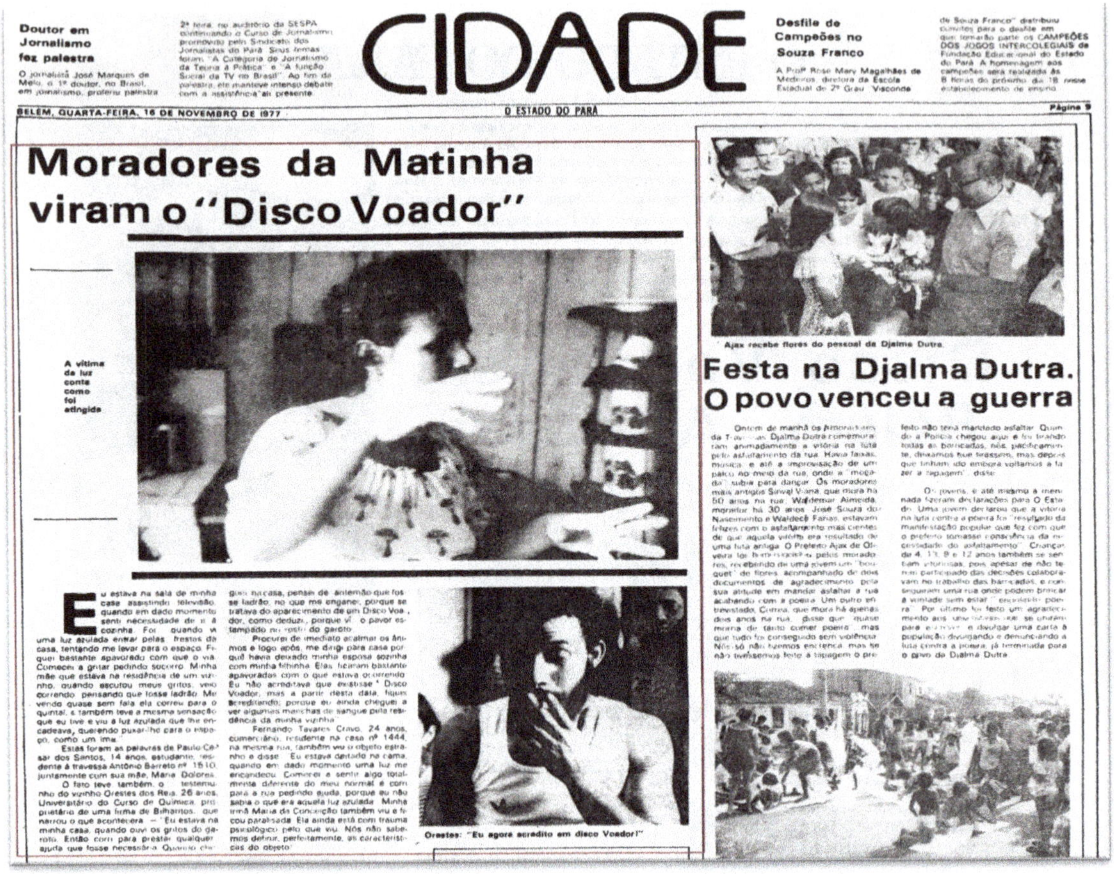

Doutor em Jornalismo faz palestra

CIDADE

Desfile de Campeões no Souza Franco

BELÉM, QUARTA-FEIRA, 16 DE NOVEMBRO DE 1977 — O ESTADO DO PARÁ — Página 9

Moradores da Matinha viram o "Disco Voador"

A vítima da luz conta como foi atingida

Orestes: "Eu agora acredito em disco Voador!"

Ajax recebe flores do pessoal da Djalma Dutra

Festa na Djalma Dutra. O povo venceu a guerra

Figure 152. "Matinha Residents Saw Flying Saucer". O Estado do Pará newspaper. November 16th, 1977.

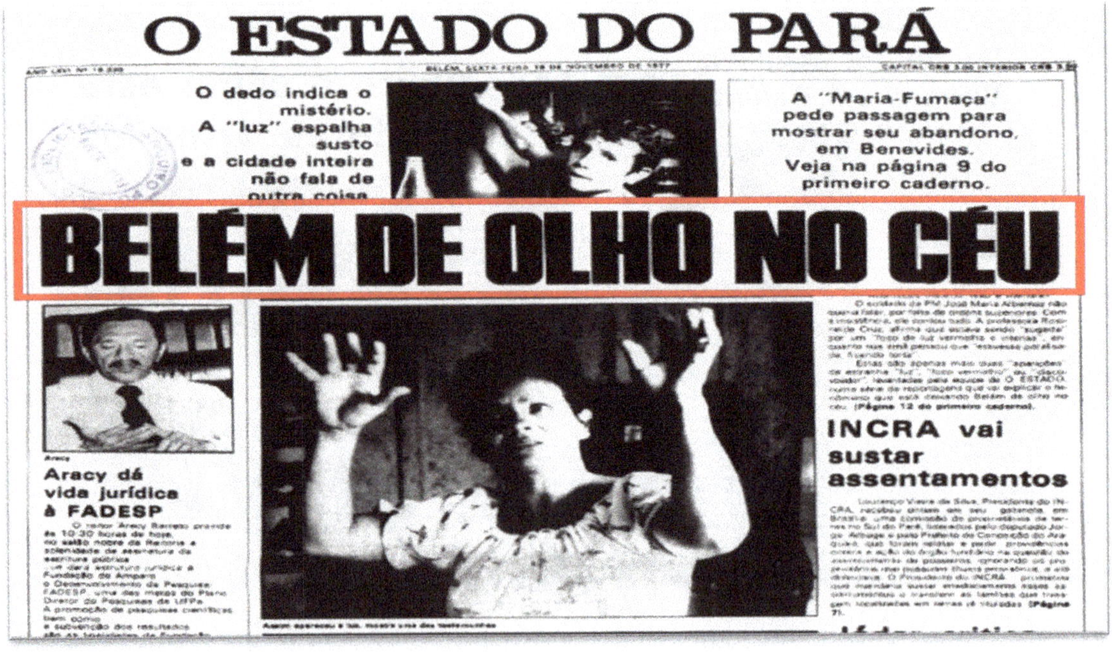

O ESTADO DO PARÁ

O dedo indica o mistério. A "luz" espalha susto e a cidade inteira não fala de outra coisa

A "Maria-Fumaça" pede passagem para mostrar seu abandono, em Benevides. Veja na página 9 do primeiro caderno.

BELÉM DE OLHO NO CÉU

Aracy dá vida jurídica à FADESP

INCRA vai sustar assentamentos

Figure 153. "Belém with an eye on the sky". O Estado do Pará newspaper. November 18th, 1977.

O ESTADO DO PARÁ

"Discos voadores"

Professora e policial, as novas vítimas do estranho foco

A cada dia que passa, as vítimas do misterioso "chupa-chupa", "disco voador", "foco do diabo", "luz estranha", "arrebatador" e muitos outros nomes dados pelo povo, ao inexplicável fenômeno do espaço, aumentam assustadoramente. Ontem, mais de seis pessoas foram atingidas em diversos bairros de nossa cidade.

Na Cremação

Na Cremação, a senhora Maria Ruth Santos Braz, residente na Conceição, passagem São José, 47, estava na sala assistindo televisão com seu esposo, quando por volta das 23.00 horas, resolveu fechar a janela do quarto, ao fazê-lo, virou-se para o corredor; nesse momento, sentiu um clarão no seu lado direito, que esquentou todo o seu corpo. Então ela gritou, e o clarão foi em cima de sua filha de meses que estava na cama. Quando seu esposo correu para socorrê-la, o clarão vermelho, segundo ela sumiu, e durante todo o dia, de vez em quando sente seu corpo esquentar como se estivesse com febre.

No Comércio

No comércio, a professora Rosineide Cruz, de 23 anos de idade e sua irmã, Ruth Maria de 18 anos, residentes na travessa Frutuoso Guimarães, 722, estavam estudando na sala, quando por volta de 1:00 hora da madrugada, Rosineide, viu uma estrela, bem defronte à janela (era a única). Então chamou sua mãe para ver e antes que sua genitora chegasse, a estrela sumiu, e voltou rapidamente, vindo das imediações do ed. Manuel Pinto da Silva, tocando uma luz vermelha branca, bem à mesa onde estavam. Nesse instante, Rosineide, sentiu que estava sendo atraída e sugada, enquanto que sua irmã, Ruth Maria pensou que estivesse paralisada, completamente, pois sentia que estava ficando torta e sua cabeça foi virada sem que ela a movimentasse. Segundo as meninas, o "foco" ia e voltava e cada vez que fazia isso, elas sentiam que estava sendo sugadas. Foi quando sua mãe apareceu e em fração de segundos tudo desapareceu. Agora, elas estão agindo normalmente.

Em Icoaraci no conjunto da COHAB, na casa IW 5), onde reside a senhora Dionéia da Silva Santana, o "foco misterioso", atacou por volta das 20:00 horas. Dona Dionéia, estava vendo a novela, juntamente com o seu esposo, que em dado momento foi até a cozinha buscar palito de dente, deixando sua esposa, sozinha, na sala, de repente, dona Dionéia, viu um clarão que iluminava toda a frente de sua casa, mas não se importou. Foi quando desceu em sua direção um raio vermelho, que quanto mais se aproximava, mais forte ficava. Sem ação, Dionéia ficou tremendo sentindo uma fervura por

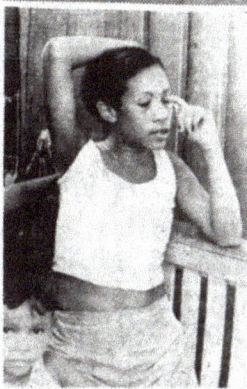

Maria José: "Fui perseguida pelo 'foco'".

Ruth Maria: "Eu estava ficando torta".

todo o corpo, quando percebeu que algo queria entranhar-se em sua carne. Gritou por seu esposo, que ao correr para acudi-la, não encontrou mais nada, pois o "disco voador", já havia desaparecido.

Na Estrada Nova

No bairro da Estrada Nova, na Padre Eutíquio, com a passagem São Cristóvão, 14, Maria José Santos Silva foi perseguida pelo "foco". Maria, vinha de uma festa dançante, na Conceição, e ao chegar no canto da Padre Eutíquio, sentiu seu corpo esquentar e viu um clarão nas suas costas. Olhou pela trás, e constatou que uma luz muito forte, aproximando, vinha em sua direção. Saiu correndo até sua residência e a luz continuou a segui-la. Já em sua casa, Maria que só tem uma filha de 2 anos, tomou um copo de água e gaguejando, contou à empregada o que tinha ocorrido; depois foi se deitar. De madrugada, aproximadamente, às 3:30 horas, ela acordou com o clarão no canto de seu quarto. Gritou por socorro, e a vizinhança correu para acudi-la. Desde então, sente, constantemente, uma quentura no corpo.

Ruth Santos: "O clarão veio lá do canto do telhado".

Soldado Ferraz: "Aconteceu mesmo, não é mentira".

Soldado também viu

"Eram quase 00:30 minutos. Eu, como já é de costume, faço a ronda pela cidade na minha bicicleta. Quando ia tranquilamente pela rua de frente da delegacia, vi uma luz muito forte que vinha na minha direção. Comecei a sentir um frio e pensei logo na tal "luz misteriosa", então fiz a curva rapidamente e pedalei mais apressado. O clarão passou por cima de mim, foi quando senti uma dor de cabeça forte". Assim contou o soldado PM, José Maria Albernaz, que está servindo no destacamento do município de Benevides. De início ele não queria contar o que tinha lhe acontecido, por não ter permissão de seu superior. Depois de pensar um pouco, exclamou: "Aconteceu mesmo não é mentira", e começou a contar o que tinha ocorrido, demonstrando estar bastante temeroso com o que viu.

Disse à reportagem que muita gente naquele município já foi atingida pelo "foco", mas não dizem nada pra reportagem (só para familiares) temendo serem atacados novamente, já que souberam, pelo rádio que se alguém continuar dizendo que foi atingido será procurado outra vez pela "coisa".

Os discos-voadores fazem parte da história desde que o mundo é mundo. Agora em Belém, uma luz forte que dizem sugar o sangue das pessoas substituiu os objetos metálicos.

Luz pode atacar, mas não tira vida de ninguém

Crendice ou não, estórias de discos-voadores, homens verdes, focos de luz, bola de fogo, etc acompanham o homem desde que o mundo é mundo. Não é de uma simples questão de mais ou de menos cultura das pessoas diretamente ligadas a essas aparições, pois muitos depoimentos têm sido prestados por gente às vezes de considerável respeitabilidade no mundo social. Contudo, há períodos em que essas manifestações coletivas se tornam acentuadas, predispondo considerações de todo o tipo, nas quais se entremeiam gozações e pitadas de seriedade.

Antes de Cristo

Documentos históricos, alguns de caráter científico, e até livros religiosos, como a Bíblia, sugerem em suas páginas, na linguagem simbólica que os caracteriza, a presença de seres de outros planetas e objetos estranhos vistos nos céus. Interpretada dentro de uma visão científica, a destruição de Sodoma e Gomorra teria sido obra de seres extraterrestres, que indignados com a atividade pecaminosa dos habitantes dessas duas cidades, teriam atirado bombas atômicas, aniquilando-as. As sete pragas do Egito também teriam

sido, nada mais nada menos, que uma contaminação bacteriológica das lavouras por entidades vindas de outros mundos. Mas não é somente de destruição do patrimônio terrestre que vivem as páginas da história. A construção das pirâmides do Egito, por exemplo, hoje se inserem na mente de alguns cientistas como uma obra que dificilmente, para ser erguida, teria sido construída por seres humanos, sendo produto de alta engenharia espacial, empregada por um civilização evoluída.

À medida em que os séculos foram passando, a evolução do pensamento científico não conseguiu apagar a idéia de que tudo eram coisas de mentalidades ainda adormecidas pela fome de saber.

Muito pelo contrário, hoje em dia, quando as maiores conquistas espaciais nos campos da física, química, cibernética e até na mente humana se acentuam o privilégio dessas grandes descobertas, ainda é mentido que se sempre em sigilo por meia dúzia de homens. Enquanto isso, o povo, volta-se para suas próprias preocupações, adicionando a elas uma que não possui nenhuma vinculação com as manifestações cotidianas e as manifestações

extraterrestres, que sempre existiram mas que ainda, guardadas as devidas proporções causam rebuliço.

Sol azul

Justamente por ser um assunto em que o único requisito "a priori" é a própria imaginação, os discos-voadores, para a maioria, ainda fazem parte de um universo de ficção cinematográfica. No entanto, antes do surgimento do cinema, mais precisamente, na segunda metade do século XIX, os jornais, as emissoras de rádio e os livros, citaram a maior proliferação desses objetos que a história já registrou. É muito mais Na Europa e nos Estados Unidos, chuvas de rãs, aguaceiros de sangue, monstros espaciais, sóis azuis, seres minúsculos vindos em discos do tamanho de um pires, desfilaram por muito tempo nas rodas de bar, em conferências em universidades, em teses científicas e, principalmente, nos lares. Naquela época, não faltaram comentários do tipo "o fim do mundo está próximo" ou "isso é um castigo de Deus". No âmbito científico, nem certos cientistas de reputação, escaparam às fórmulas apressadas para salvar o planeta em perigo, havendo um, Carl Sanders, que chegou a uma conclusão original

pará evitar um possível bombardeamento por discos-voadores: um grupo de cientistas deixaria a Terra. Ele não disse como isso seria possível, mas mas sem dúvida, pela fama que possuía, foi menos ridicularizado que um sujeito qualquer que afirmasse ter visto uma luz estranha.

Triângulo de morte

A mais recente dúvida da ciência é sobre o que de fato acontece com aviões e navios que cruzam o mar do Caribe, e, vez por outra, desaparecem. Para uns é um vácuo no espaço que leva a outra dimensão. Para outros, tudo não passa de uma atração magnética da Terra que recolhe objetos que passam em sua zona de ação. Expedições científicas já foram enviadas mas até hoje nenhum relatório foi tornado público, o que mostra a precaução excessiva dos estudiosos para um problema até há bem pouco tempo de fácil avaliação.

No Brasil, País privilegiado em vários aspectos, inclusive o imaginativo, aparições de OVNIs, luzes misteriosas, anões do espaço, etc, já formam um considerável rol de dúvidas para as autoridades ligadas ao assunto, e a facilidade que existe para se difundir um

boato. Se alguém viu um objeto qualquer no céu, imediatamente outra pessoa comunicada sobre o assunto o deturpa, querendo às vezes provocar temor em alguém que a escuta. É claro que declarações de peso, feitas por pessoas sérias, não podem ser incluídas na enxurrada de depoimentos, a maioria carentes de substância e povoados de fantasias. Porém, quando um fato ou algo assim revestido corre na boca do povo, difícil se torna saber qual a verdadeira estória.

Precauções

As aparições do foco em Belém ganharam força a partir dos episódios narrados por moradores de municípios da Vigia e Mosqueiro. Em pouco tempo, a luz começou a rondar a cercanias da cidade, aparecendo em Icoaraci. Agora, não há mais um local fixo, tanto pode ser no Jurunas, na Marambaia, como no pacato bairro Comercial. Os boatos que engordam as manhãs das narrativas do belenense, à noite se transformam em temerosas conversas de vizinhos e familiares. Uma senhora, residente no bairro da Cremação, afirma que ainda não viu o foco "chupa-chupa", mas tem muito medo que ele apareça em sua casa.

Amanhã: A invasão dos marcianos

Figure 154. "Teacher and police officer, the new victims of the strange focus". "Light can attack, but it does not take anyone's life". "Belém with an eye on the sky". O Estado do Pará newspaper. November 18th, 1977.

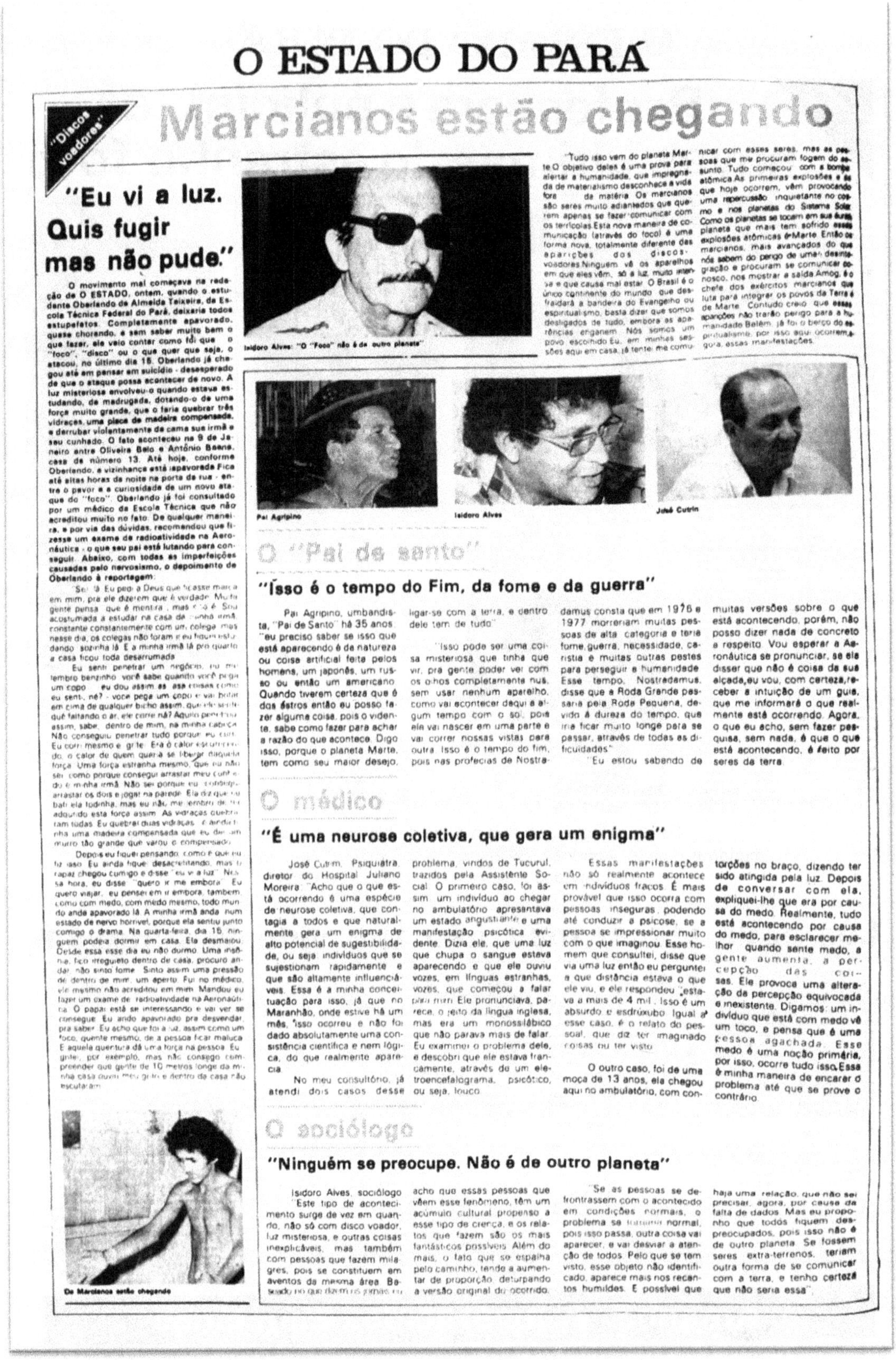

O ESTADO DO PARÁ

"Discos voadores"

Marcianos estão chegando

"Eu vi a luz. Quis fugir mas não pude."

O movimento mal começava na redação de O ESTADO, ontem, quando o estudante Oberlando de Almeida Teixeira, da Escola Técnica Federal do Pará, deixaria todos estupefatos. Completamente apavorado, quase chorando, e sem saber muito bem o que fazer, ele veio contar como foi que o "foco", "disco" ou o que quer que seja, o atacou, no último dia 15. Oberlando já chegou até em pensar em suicídio - desesperado de que o ataque possa acontecer de novo. A luz misteriosa envolveu-o quando estava estudando, de madrugada, dotando-o de uma força muito grande, que o faria quebrar três vidraças, uma placa de madeira compensada, e derrubar violentamente da cama sua irmã e seu cunhado. O fato aconteceu na 9 de Janeiro entre Oliveira Belo e Antônio Baena, casa de número 13. Até hoje, conforme Oberlando, a vizinhança está apavorada. Fica até altas horas da noite na porta da rua - entre o pavor e a curiosidade de um novo ataque do "foco". Oberlando já foi consultado por um médico da Escola Técnica que não acreditou muito no fato. De qualquer maneira, e por via das dúvidas, recomendou que fizesse um exame de radioatividade na Aeronáutica - o que seu pai está lutando para conseguir. Abaixo, com todas as imperfeições causadas pelo nervosismo, o depoimento de Oberlando à reportagem:

"Sei lá. Eu pedi a Deus que ficasse marca em mim, pra ele dizerem que é verdade. Muita gente pensa que é mentira, mas não é. Sou acostumada a estudar na casa de minha irmã, constante constantemente com um colega, mas nesse dia, os colegas não foram e eu fiquei estudando sozinha lá. E a minha irmã lá pro quarto, a casa ficou toda desarrumada.

Eu senti penetrar um negócio, eu me lembro benzinho, você sabe quando você pega um copo ... eu dou assim as asa coisas como eu senti, né? - você pega um copo e vai botar em cima de qualquer bicho assim, que ele sente que faltando o ar, ele corre né? Aquilo penetrou assim, sabe, dentro de mim, na minha cabeça. Não conseguiu penetrar tudo porque eu corri. Eu corri mesmo e gritei. Era o calor esquentando, o calor de quem queria se liberar daquela força. Uma força estranha mesmo, que eu não sei como porque consegui arrastar meu cunhado e minha irmã. Não sei porque eu consegui arrastar os dois e jogar na parede. Ela diz que eu bati ela todinha, mas eu não me lembro de ter adquirido esta força assim. As vidraças quebraram todas. Eu quebrei duas vidraças, e ainda tinha uma madeira compensada que eu dei um murro tão grande que varou o compensado.

Depois eu fiquei pensando como é que eu fiz isso. Eu ainda fiquei desacreditando, mas o rapaz chegou cumigo e disse "eu vi a luz". Nessa hora, eu disse "quero ir me embora". Eu quero viajar, eu pensei em ir embora, também como com medo, com medo mesmo, todo mundo anda apavorado lá. A minha irmã anda num estado de nervo horrível, porque ela sentiu junto comigo o drama. Na quarta-feira, dia 15, ninguem podeia dormir em casa. Ela desmaiou. Desde essa esse dia eu não durmo. Uma insônia, fico irrequieto dentro de casa, procuro andar, não sinto fome. Sinto assim uma pressão de dentro de mim, um aperto. Fui no médico, ele mesmo não acreditou em mim. Mandou eu fazer um exame de radioatividade na Aeronáutica. O papai está se interessando e vai ver se consegue. Eu ando apavorado pra desvendar, pra saber. Eu acho que foi a luz, assim como um foco, quente mesmo, de a pessoa ficar maluca. E aquela quentura dá uma força na pessoa. Eu gritei, por exemplo, mas não consego compreender que gente de 10 metros longe da minha casa ouviu meu grito e dentro da casa não escutaram.

Os Marcianos estão chegando

Isidoro Alves: "O "Foco" não é de outro planeta"

"Tudo isso vem do planeta Marte. O objetivo deles é uma prova para alertar a humanidade, que impregnada de materialismo desconhece a vida fora da matéria. Os marcianos são seres muito adiantados que querem apenas se fazer comunicar com os terrícolas. Esta nova maneira de comunicação (através do foco) é uma forma nova, totalmente diferente das aparições dos discos-voadores. Ninguem vê os aparelhos em que eles vêm, só a luz, muito intensa e que causa mal estar. O Brasil é o único continente do mundo que desfraldará a bandeira do Evangelho ou espiritualismo, basta dizer que somos desligados de tudo, embora as aparências enganem. Nós somos um povo escolhido. Eu, em minhas sessões aqui em casa, já tentei me comunicar com esses seres, mas as pessoas que me procuram fogem do assunto. Tudo começou com a bomba atômica. As primeiras explosões e as que hoje ocorrem, vêm provocando uma repercussão inquietante no cosmo e nos planetas do Sistema Solar. Como os planetas se tocam em sua áuras, planeta que mais tem sofrido essas explosões atômicas é Marte. Então os marcianos, mais avançados do que nós sabem do perigo de uma desintegração e procuram se comunicar conosco, nos mostrar a saída. Amog, é o chefe dos exércitos marcianos que luta para integrar os povos da Terra e de Marte. Contudo creio que essas aparições não trarão perigo para a humanidade. Belém já foi o berço do espiritualismo, por isso aqui ocorrem, agora, essas manifestações.

Pai Agripino

Isidoro Alves

José Cutrin

O "Pai de santo"

"Ísso é o tempo do Fim, da fome e da guerra"

Pai Agripino, umbandista, "Pai de Santo" há 35 anos: **"eu preciso saber se isso que está aparecendo é da natureza ou coisa artificial feita pelos homens, um japonês, um russo ou então um americano. Quando tiverem certeza que é dos ástros então eu posso fazer alguma coisa, pois o vidente, sabe como fazer para achar a razão do que acontece.** Digo isso, porque o planeta Marte, tem como seu maior desejo, ligar-se com a terra, e dentro dele tem de tudo".

"Isso pode ser uma coisa misteriosa que tinha que vir, pra gente poder ver com os olhos completamente nus, sem usar nenhum aparelho, como vai acontecer daqui a algum tempo com o sol, pois ele vai nascer em uma parte e vai correr nossas vistas para outra. Isso é o tempo do fim, pois nas profecias de Nostradamus consta que em 1976 e 1977 morreriam muitas pessoas de alta categoria e teria fome, guerra, necessidade, carestia e muitas outras pestes para perseguir a humanidade. Esse tempo, Nostradamus, disse que a Roda Grande passaria pela Roda Pequena, devido à dureza do tempo, que iria ficar muito longe para se passar, através de todas as dificuldades".

"Eu estou sabendo de muitas versões sobre o que está acontecendo, porém, não posso dizer nada de concreto a respeito. Vou esperar a Aeronáutica se pronunciar, se ela disser que não é coisa da sua alçada, eu vou, com certeza, receber a intuição de um guia, que me informará o que realmente está ocorrendo. Agora, o que eu acho, sem fazer pesquisa, sem nada, é que o que está acontecendo, é feito por seres da terra.

O médico

"É uma neurose coletiva, que gera um enigma"

José Cutrim, Psiquiátra, diretor do Hospital Juliano Moreira: "Acho que o que está ocorrendo é uma espécie de neurose coletiva, que contagia a todos e que naturalmente gera um enigma de alto potencial de sugestibilidade, ou seja, indivíduos que se sujestionam rapidamente e que são altamente influenciáveis. Essa é a minha conceituação para isso, já que no Maranhão, onde estive há um mês, isso ocorreu e não foi dado absolutamente uma consistência científica e nem lógica, do que realmente aparecia.

No meu consultório, já atendi dois casos desse problema, vindos de Tucuruí, trazidos pela Assistênte Social. O primeiro caso, foi assim: um indivíduo ao chegar no ambulatório apresentava um estado angustiante e uma manifestação psicótica evidente. Dizia ele, que uma luz que chupa o sangue estava aparecendo e que ele ouviu vozes, em línguas estranhas, vozes, que começou a falar para mim. Ele pronunciava, parece, o jeito da língua inglesa, mas era um monossilábico que não parava mais de falar. Eu examinei o problema dele, e descobri que ele estava francamente, através de um eletroencefalograma, psicótico, ou seja, louco.

Essas manifestações não só realmente acontece em indivíduos fracos. É mais provável que isso ocorra com pessoas inseguras, podendo até conduzir a psicose, se a pessoa se impressionar muito com o que imaginou. Esse homem que consultei, disse que via uma luz então eu perguntei a que distância estava o que ele viu, e ele respondeu "estava a mais de 4 mil". Isso é um absurdo e esdrúxubo. Igual a esse caso, é o relato do pessoal, que diz ter imaginado coisas ou ter visto.

O outro caso, foi de uma moça de 13 anos, ela chegou aqui no ambulatório, com contorções no braço, dizendo ter sido atingida pela luz. Depois de conversar com ela, expliquei-lhe que era por causa do medo. Realmente, tudo está acontecendo por causa do medo, para esclarecer melhor quando sente medo, a gente aumenta a percepção das coisas. Ele provoca uma alteração da percepção equivocada e inexistente. Digamos: um indivíduo que está com medo vê um toco, e pensa que é uma pessoa agachada. Esse medo é uma noção primária, por isso, ocorre tudo isso. Essa é minha maneira de encarar o problema até que se prove o contrário.

O sociólogo

"Ninguém se preocupe. Não é de outro planeta"

Isidoro Alves, sociólogo: "Este tipo de acontecimento surge de vez em quando, não só com disco voador, luz misteriosa, e outras coisas inexplicáveis, mas também com pessoas que fazem milagres, pois se constituem em aventos da mesma área. Baseado no que dizem os jornais, eu acho que essas pessoas que vêem esse fenômeno, têm um acúmulo cultural propenso a esse tipo de crença, e os relatos que fazem são os mais fantásticos possíveis. Além do mais, o fato que se espalha pelo caminho, tende a aumentar de proporção, deturpando a versão original do ocorrido.

"Se as pessoas se defrontassem com o acontecido em condições normais, o problema se tornaria normal, pois isso passa, outra coisa vai aparecer, e vai desviar a atenção de todos. Pelo que se tem visto, esse objeto não identificado, aparece mais nos recantos humildes. É possível que haja uma relação, que não sei precisar, agora, por causa da falta de dados. Mas eu proponho que todos fiquem despreocupados, pois isso não é de outro planeta. Se fossem seres extra-terrenos, teriam outra forma de se comunicar com a terra, e tenho certeza que não seria essa".

Figure 155. "The Martians are coming". "I saw the light. I wanted to run away, but I couldn't". O Estado do Pará **newspaper. November 19th, 1977.**

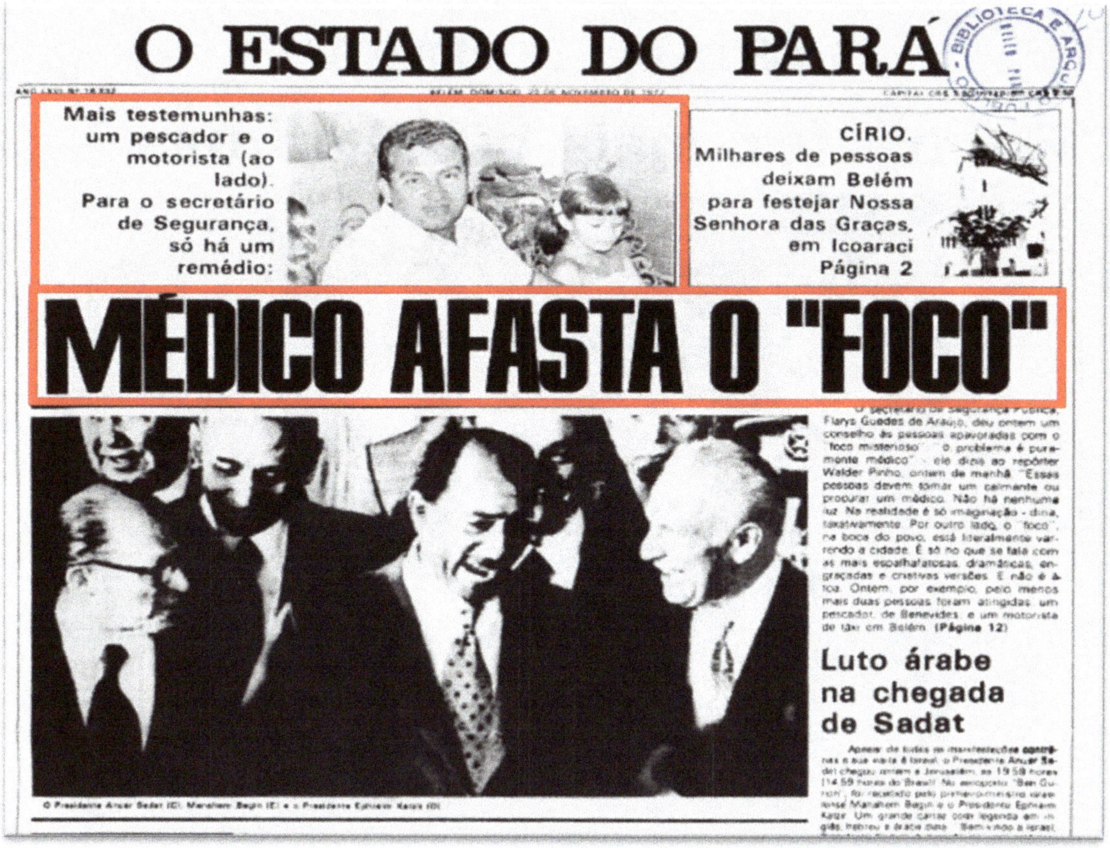

O ESTADO DO PARÁ

Mais testemunhas: um pescador e o motorista (ao lado). Para o secretário de Segurança, só há um remédio:

CÍRIO. Milhares de pessoas deixam Belém para festejar Nossa Senhora das Graças, em Icoaraci Página 2

MÉDICO AFASTA O "FOCO"

O secretário de Segurança Pública, Flarys Guedes de Araújo, deu ontem um conselho às pessoas apavoradas com o "foco misterioso": "o problema é puramente médico" - ele disse ao repórter Walder Pinho, ontem de manhã. "Essas pessoas devem tomar um calmante ou procurar um médico. Não há nenhuma luz. Na realidade é só imaginação - disse, taxativamente. Por outro lado, o "foco", na boca do povo, está literalmente varrendo a cidade. É só no que se fala com as mais espalhafatosas, dramáticas, engraçadas e criativas versões. E não é à toa. Ontem, por exemplo, pelo menos mais duas pessoas foram atingidas: um pescador, de Benevides, e um motorista de táxi em Belém. (Página 12)

Luto árabe na chegada de Sadat

Figure 156. "Doctor removes the focus". More witnesses: a fisherman and the driver. For the Secretary of Security, there is only one remedy". O Estado do Pará newspaper. November 20th, 1977.

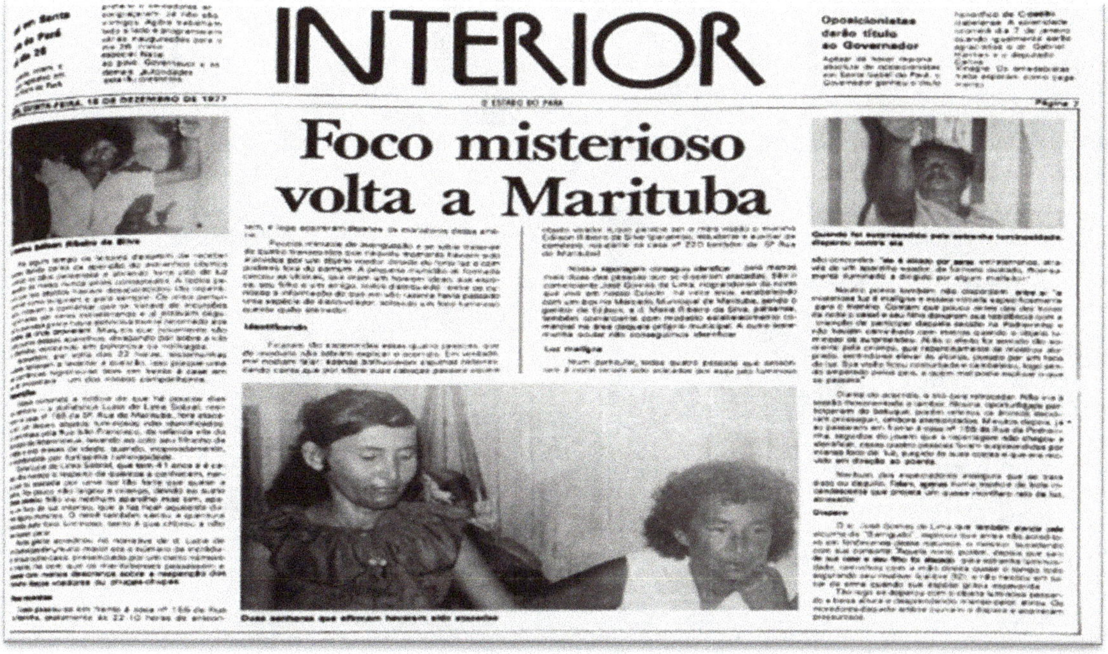

INTERIOR

Oposicionistas darão título ao Governador

O ESTADO DO PARÁ

Foco misterioso volta a Marituba

Figure 157. "Mysterious focus returns to Marituba". O Estado do Pará newspaper. December 15th, 1977.

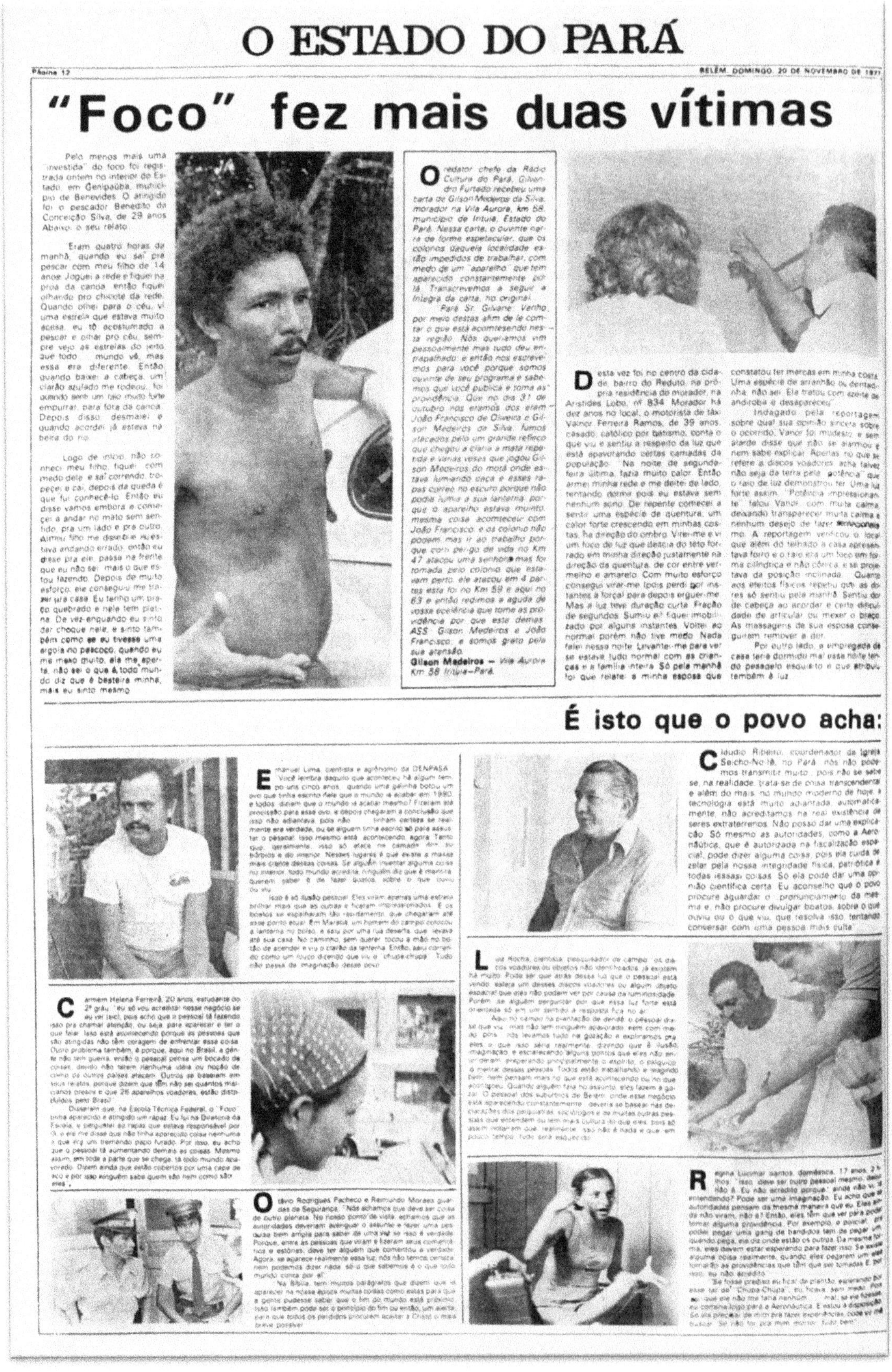

O ESTADO DO PARÁ

Página 12 — BELÉM, DOMINGO, 20 DE NOVEMBRO DE 1977

"Foco" fez mais duas vítimas

Pelo menos mais uma "investida" do foco foi registrada ontem no interior do Estado, em Genipaúba, município de Benevides. O atingido foi o pescador Benedito da Conceição Silva, de 29 anos. Abaixo, o seu relato.

"Eram quatro horas da manhã, quando eu saí pra pescar com meu filho de 14 anos. Joguei a rede e fiquei na proa da canoa, então fiquei olhando pro chicote da rede. Quando olhei para o céu, vi uma estrela que estava muito acesa, eu tô acostumado a pescar e olhar pro céu, sempre vejo as estrelas do jeito que todo mundo vê, mas essa era diferente. Então quando baixei a cabeça, um clarão azulado me rodeou, foi quando senti um raio muito forte empurrar, para fora da canoa. Depois disso desmaiei e quando acordei já estava na beira do rio.

Logo de início, não conheci meu filho, fiquei com medo dele e saí correndo, tropecei e caí, depois da queda é que fui conhecê-lo. Então eu disse vamos embora e comecei a andar no mato sem sentido, pra um lado e pra outro. Aí meu filho me disse que eu estava andando errado, então eu disse pra ele, passa na frente que eu não sei mais o que estou fazendo. Depois de muito esforço, ele conseguiu me trazer pra casa. Eu tenho um braço quebrado e nele tem platina. De vez enquando eu sinto dar choque nele, e sinto também como se eu tivesse uma argola no pescoço, quando eu me mexo muito, ela me aperta, não sei o que é, todo mundo diz que é besteira minha, mais eu sinto mesmo.

O redator chefe da Rádio Cultura do Pará, Gilvandro Furtado recebeu uma carta de Gilson Medeiros da Silva, morador na Vila Aurora, km 58, município de Intuia, Estado do Pará. Nessa carta, o ouvinte narra de forma espetacular, que os colonos daquela localidade estão impedidos de trabalhar, com medo de um "aparelho" que tem aparecido constantemente por lá. Transcrevemos a seguir a íntegra da carta, no original.

Pará Sr. Gilvane Venho, por meio destas afim de le comtar o que está acomtesendo nesta região. Nós queriamos vim pessoalmente mas tudo deu entrapalhado e então nos escrevemos para você porque somos ouvinte de seu programa e sabemos que você publica e toma as providência. Que no dia 31 de outubro nos eramos dos erem João Francisco de Oliveira e Gilson Medeiros da Silva, fumos atacados pelo um grande reflecço que chegou a clariá a mata repetida e varias veses que jogou Gilson Medeiros do mutá onde estava lumiando caça e esses rapas correo no escuro porque não podia lumia a sua lanterna porque o aparelho estava muito. mesma coisa acomteceu com João Francisco. e os colonio não podem mas ir ao trabalho porque corn perigo de vida no Km 47 atacou uma senhora mas foi tomada pelo colonio que estavam perto. ele atacou em 4 partes esta foi no Km 59 e aqui no 63 e então redimos a aguda de vossa ecelência que tome as providência por que este demas. ASS Gilson Medeiros e João Francisco. e somos grato pela sua atensão.

Gilson Medeiros — *Vila Aurora Km 58 Intuia—Pará.*

Desta vez foi no centro da cidade, bairro do Reduto, na própria residência do morador, na Aristides Lobo, nº 834. Morador há dez anos no local, o motorista de táxi Valnor Ferreira Ramos, de 39 anos, casado, católico por batismo, conta o que viu e sentiu a respeito da luz que está apavorando certas camadas da população: "Na noite de segunda-feira última, fazia muito calor. Então armei minha rede e me deitei de lado, tentando dormir pois eu estava sem nenhum sono. De repente comecei a sentir uma espécie de quentura, um calor forte crescendo em minhas costas, na direção do ombro. Virei-me e vi um foco de luz que descia do teto forrado em minha direção justamente na direção da quentura, de cor entre vermelho e amarelo. Com muito esforço consegui virar-me (pois perdi por instantes a força) para depois erguer-me. Mas a luz teve duração curta. Fração de segundos. Sumiu e fiquei imobilizado por alguns instantes. Voltei ao normal porém não tive medo. Nada falei nessa noite. Levantei-me para ver se estava tudo normal com as crianças e a família inteira. Só pela manhã foi que relatei à minha esposa que constatou ter marcas em minha costa. Uma espécie de arranhão ou dentadinha, não sei. Ela tratou com azeite de andiroba e desapareceu".

Indagado pela reportagem sobre qual sua opinião sincera sobre o ocorrido, Valnor foi modesto e sem alarde disse que não se alarmou e nem sabe explicar. Apenas, no que se refere a discos voadores, acha talvez não seja da terra pela "potência" que o raio de luz demonstrou ter. Uma luz forte assim. "Potência impressionante" falou Valnor com muita calma, deixando transparecer muita calma e nenhum desejo de fazer sensacionalismo. A reportagem verificou o local que além do telhado a casa apresentava forro e o raio era um foco em forma cilíndrica e não cônica, e se projetava da posição inclinada. Quanto aos efeitos físicos repetiu que as dores só sentiu pela manhã. Sentiu dor de cabeça ao acordar e certa dificuldade de articular ou mexer o braço. As massagens de sua esposa conseguiram remover a dor.

Por outro lado, a empregada de casa teria dormido mal essa noite tendo pesadelo esquisito e que atribuiu também à luz.

É isto que o povo acha:

Emanuel Lima, dentista e agrônomo da DENPASA: "Você lembra daquilo que aconteceu há algum tempo uns cinco anos, quando uma galinha botou um ovo que tinha escrito nele que o mundo ia acabar em 1980, e todos diziam que o mundo ia acabar mesmo? Fizeram até procissão para esse ovo, e depois chegaram à conclusão que isso não adiantava, pois não tinham certeza se realmente era verdade, ou se alguém tinha escrito só para assustar o pessoal. Isso mesmo está acontecendo agora. Tanto que, geralmente, isso só ataca na camada dos subúrbios e do interior. Nesses lugares é que existe a massa mais crente dessas coisas. Se alguém inventar alguma coisa no interior, todo mundo acredita, ninguém diz que é mentira. Querem saber é de fazer [illegible] sobre o que ouviu ou viu.

Isso é só ilusão pessoal. Eles viram apenas uma estrela brilhar mais que as outras e ficaram impressionados. E os boatos se espalharam tão rapidamente, que chegaram até esse ponto atual. Em Marabá, um homem do campo colocou a lanterna no bolso, e saiu por uma rua deserta, que levava até sua casa. No caminho, sem querer tocou a mão no botão de acender e viu o clarão da lanterna. Então, saiu correndo como um louco dizendo que viu o "chupa-chupa". Tudo não passa de imaginação desse povo".

Cláudio Ribeiro, coordenador da Igreja Seicho-No-Ie, no Pará: "nós não podemos transmitir muito, pois não se sabe se, na realidade, trata-se de coisa transcendental e além do mais, no mundo moderno de hoje, a tecnologia está muito adiantada, automaticamente, não acreditamos na real existência de seres extraterrenos. Não posso dar uma explicação. Só mesmo as autoridades, como a Aeronáutica, que é autorizada na fiscalização espacial, pode dizer alguma coisa, pois ela cuida de zelar pela nossa integridade física, patriótica e todas (essas) coisas. Só ela pode dar uma opinião científica certa. Eu aconselho que o povo procure aguardar o pronunciamento da mesma e, não procure divulgar boatos sobre o que ouviu ou o que viu, que resolva isso tentando conversar com uma pessoa mais culta".

Carmem Helena Ferreira, 20 anos, estudante do 2º grau: "eu só vou acreditar nesse negócio se eu ver (sic), pois acho que o pessoal tá fazendo isso pra chamar atenção, ou seja, para aparecer e ter o que falar. Isso está acontecendo porque as pessoas que são atingidas não têm coragem de enfrentar essa coisa. Outro problema também, é porque, aqui no Brasil, a gente não tem guerra, então o pessoal pensa um bocado de coisas, devido não terem nenhuma idéia ou noção de como os outros países atacam. Outros se baseiam em seus relatos, porque dizem que têm não sei quantos marcianos presos e que 26 aparelhos voadores estão distribuídos pelo Brasil".

Disseram que, na Escola Técnica Federal, o "Foco" tinha aparecido e atingido um rapaz. Eu fui na Diretoria da Escola, e perguntei ao rapaz que estava responsável por lá, e ele me disse que não tinha aparecido coisa nenhuma e que era um tremendo papo furado. Por isso, eu acho que o pessoal tá aumentando demais as coisas. Mesmo assim, em toda a parte que se chega, tá todo mundo apavorado. Dizem ainda que estão cobertos por uma capa de aço e por isso ninguém sabe quem são nem como são eles".

Luiz Rocha, cientista, pesquisador de campo: "os discos voadores ou objetos não identificados já existem há muito. Pode ser que atrás dessa luz que o pessoal está vendo, esteja um desses discos voadores ou algum objeto espacial que eles não podem ver por causa da luminosidade. Porém, se alguém perguntar por que essa luz forte está orientada só em um sentido a resposta fica no ar.

Aqui no campo na plantação de dendê o pessoal disse que viu, mas não tem ninguém apavorado, nem com medo, pois nós levamos tudo na gozação e explicamos pra eles o que isso seria realmente, dizendo que é ilusão, imaginação, e esclarecendo alguns pontos que eles não entenderam, [illegible] principalmente o espírito, o psíquico e mental dessas pessoas. Todos estão trabalhando e [illegible] bem, nem pensam mais no que está acontecendo ou no que aconteceu. Quando alguém fala no assunto, eles fazem a gozar. O pessoal dos subúrbios de Belém, onde esse negócio está aparecendo constantemente, deveria se basear nas declarações dos psiquiatras, sociólogos e de muitas outras pessoas que entendem ou tem mais cultura do que eles, pois só assim notariam que, realmente, isso não é nada e que em pouco tempo, tudo será esquecido".

Otávio Rodrigues Pacheco e Raimundo Moraes, guardas de Segurança: "Nós achamos que deve ser coisa de outro planeta. No nosso ponto de vista, achamos que as autoridades deveriam averiguar o assunto e fazer uma pesquisa bem ampla para saber de uma vez se isso é verdade. Porque, entre as pessoas que viram e fizeram seus comentários e estórias, deve ter alguém que comentou a verdade. Agora, se aparece realmente essa luz, nós não temos certeza nem podemos dizer nada, só o que sabemos é o que todo mundo conta por aí".

"Na Bíblia, tem muitos parágrafos que dizem que iria aparecer na nossa época muitas coisas como estas para que a gente pudesse saber que o fim do mundo está próximo. Isso também pode ser o princípio do fim ou então, um alerta para que todos os perdidos procurem aceitar a Cristo o mais breve possível".

Regina Lucimar Santos, doméstica, 17 anos, 3 filhos: "Isso deve ser outro pessoal mesmo, daqui não é. Eu não acredito porque ainda não vi, tá entendendo? Pode ser uma imaginação. Eu acho que as autoridades pensam da mesma maneira que eu. Eles ainda não viram, não é? Então, eles têm que ver para poder tomar alguma providência. Por exemplo, o policial, pra poder pegar uma gang de bandidos tem de pegar um, quando pega, ele diz onde estão os outros. Da mesma forma, eles devem estar esperando para fazer isso. Se existe alguma coisa realmente, quando eles pegarem um, eles tomarão as providências que têm que ser tomadas. E por isso, eu não acredito.

"Se fosse preciso eu ficar de plantão esperando por esse tal de "Chupa-Chupa", eu ficava, sem medo. Pois eu sei que ele não me faria nenhum mal, se ele fosse eu corria logo para a Aeronáutica. E estou à disposição. Se ela precisar de mim pra fazer experiências, pode vir me buscar. Se não for pra mim morrer, tudo bem".

Figure 158. "Focus makes two more victims. O Estado do Pará **newspaper. November 20th, 1977.**

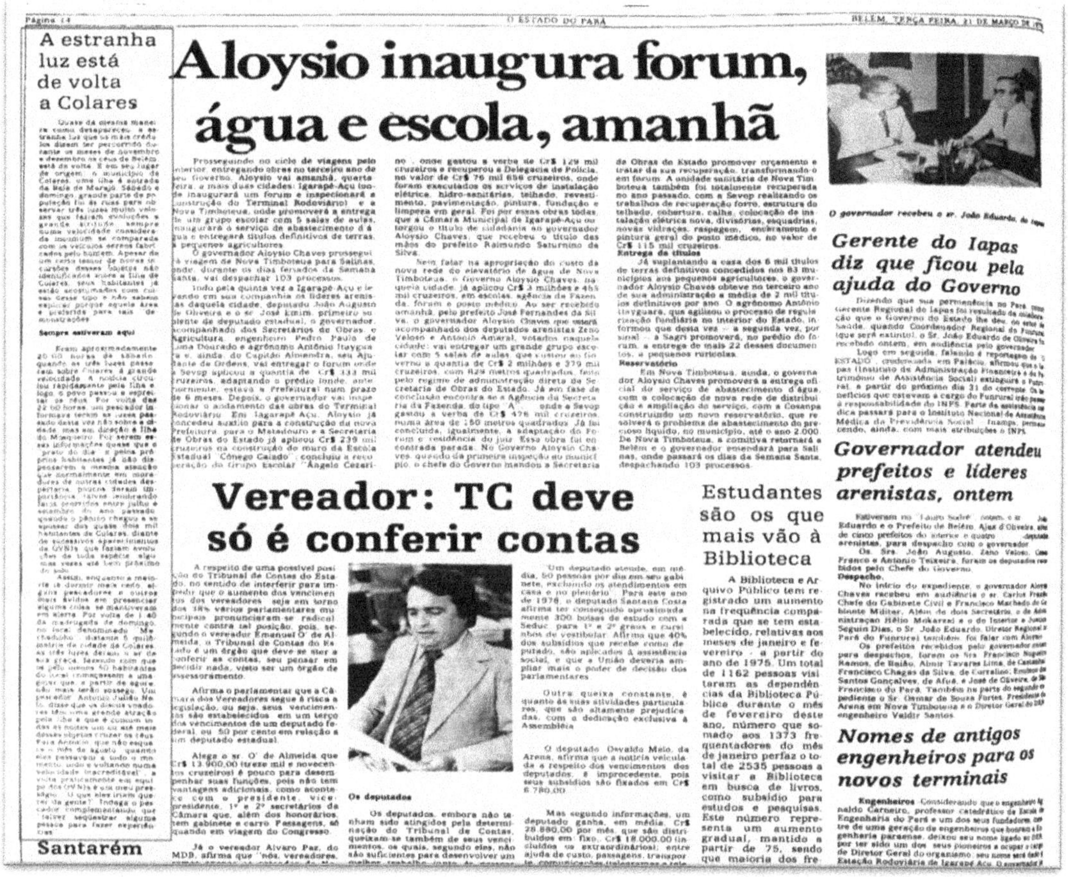
A estranha luz está de volta a Colares

Aloysio inaugura forum, água e escola, amanhã

Vereador: TC deve só é conferir contas

Estudantes são os que mais vão à Biblioteca

Gerente do Iapas diz que ficou pela ajuda do Governo

Governador atendeu prefeitos e líderes arenistas, ontem

Nomes de antigos engenheiros para os novos terminais

Santarém

Figure 159. "The strange light is back in Colares". O Estado do Pará **newspaper. March 21**th**, 1978.**

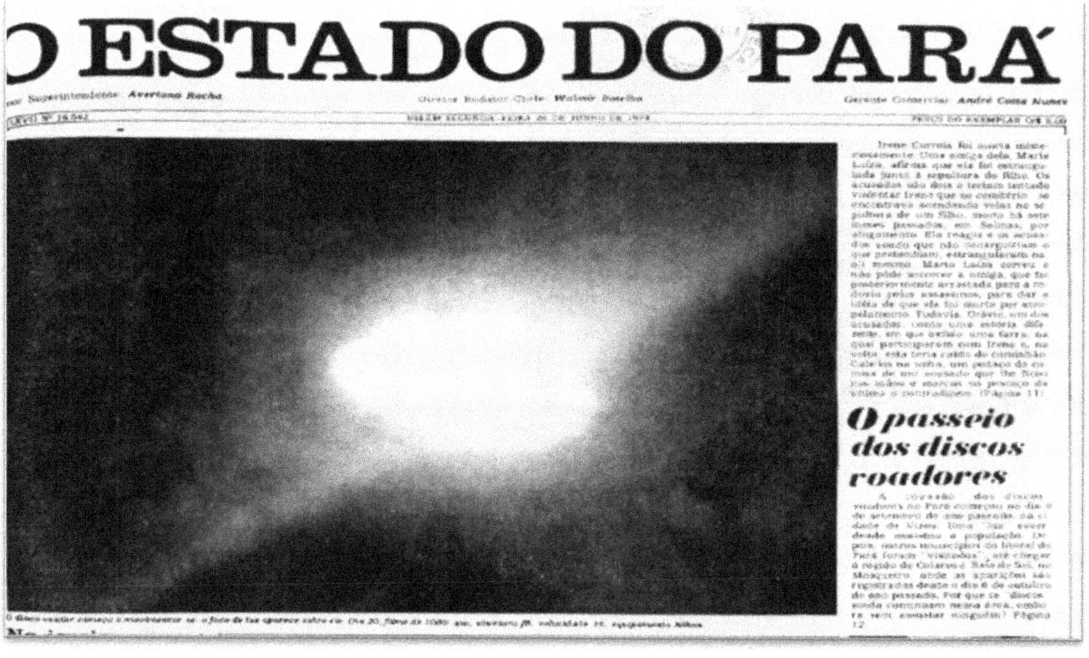
O ESTADO DO PARÁ

O passeio dos discos voadores

Figure 160. "The flying saucers' tour". O Estado do Pará newspaper. June 26th**, 1978.**

ARX. 184, p. 153/160

RELATÓRIO OVNI

CONFIDENCIAL

IMPRENSA LOCAL

JORNAL — O ESTADO DO PARÁ

DATA — 25 Jun 1978

O ESTADO DO PARÁ

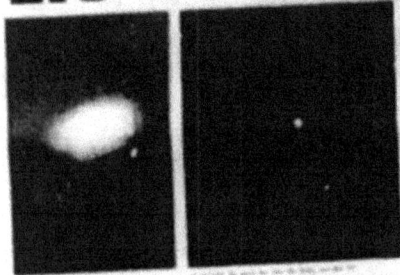

EIS O "CHUPA-CHUPA"

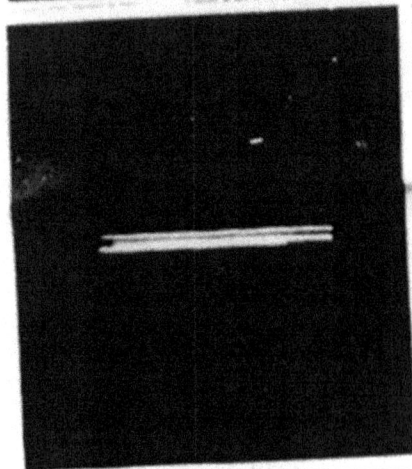

Figure 161. "Behold the Chupa-Chupa". O Estado do Pará newspaper. June 25th, 1978.

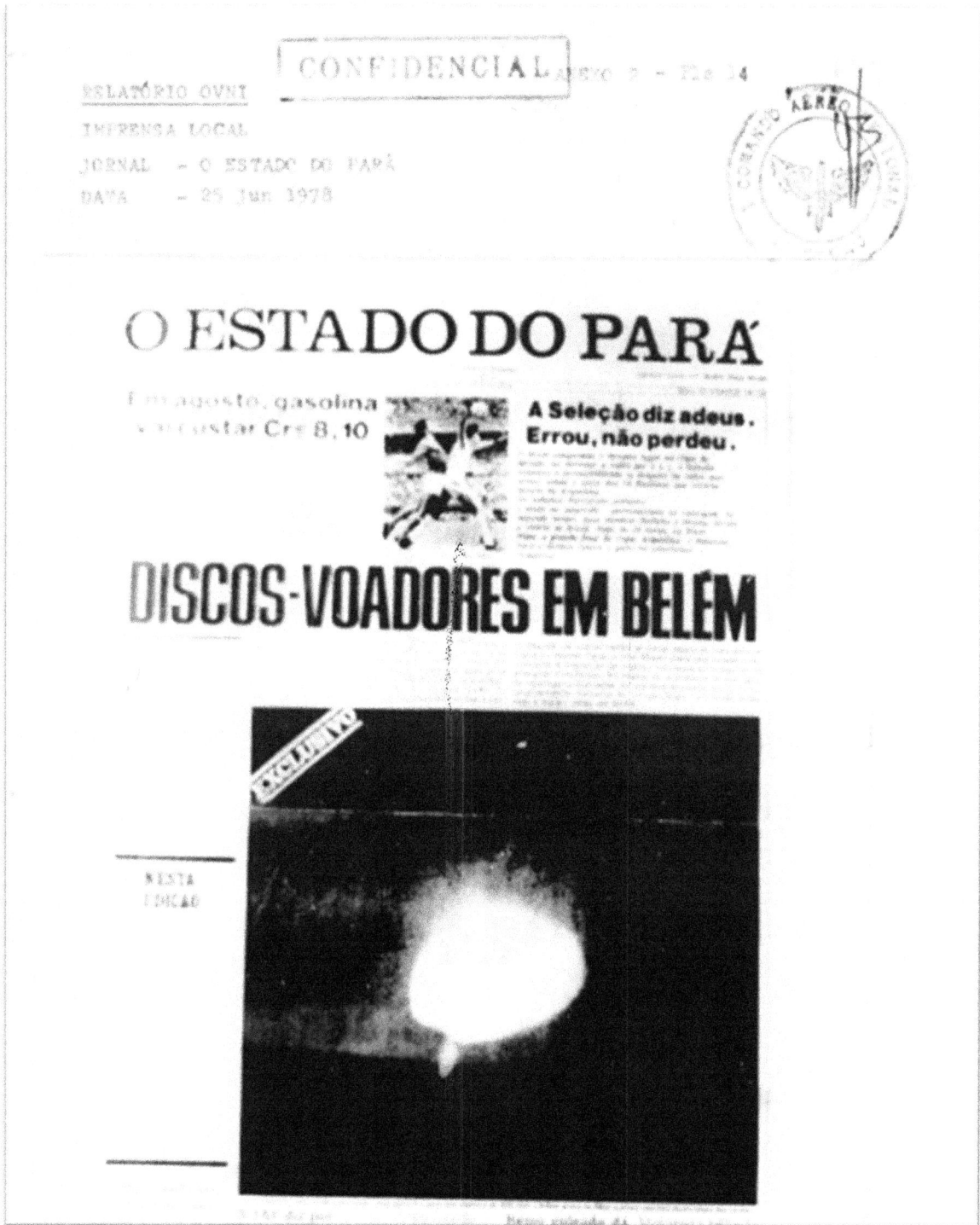

Figure 162. "Flying saucers in Belém". O Estado do Pará **newspaper.June 25th, 1978.**

O ESTADO DO PARÁ

ROTA DOS DISCOS-VOADORES

A rota dos discos-voadores no Pará acompanha a faixa litorânea. O primeiro município visitado pelos estranhos aparelhos foi Vizeu, precisamente no dia 9 de setembro do ano passado. Durante uma semana, fatos inacreditáveis aconteceram, deixando verdadeiramente apavorados moradores de diversos lugarejos pertencentes ao município.

Seguindo a costa paraense, os objetos voadores não identificados assustaram Bragança durante nove dias. Pescadores, além de pequenos colonos, avistaram inúmeras vezes a estranha luz esverdeada cruzar o céu em vertiginosa velocidade. Tracuateua, um dos povoados do município, recebeu visitas periódicas; seus moradores até hoje não esquecem das fantásticas aparições.

Ainda na primeira quinzena de setembro, os OVNIs singraram, rapidamente, o céu de Capanema, durante alguns dias. Em pouco tempo estavam em Ourém, onde também foram para São Miguel do Guamá. Neste município, os aparelhos foram divisados em maior escala pelos moradores ribeirinhos. No povoado de Quatipuru, fantásticas versões surgiram sobre o aparecimento dos discos-voadores.

No final deste mês, Salinópolis foi a localidade escolhida pelos indescritíveis naves espaciais. Durante três dias seguidos, luzes até então nunca vistas, apareceram no céu e suscitaram as mais diversas interpretações.

Em outubro, na primeira quinzena, pequenos registros dos OVNIs foram feitos nos municípios de Timboteua, Santa Maria do Pará, Igarapé-Açu e Maracanã, Curuçá e Castanhal. A extraordinária luz neste período foi vista principalmente nas áreas rurais.

No dia 16 de outubro, os OVNIs abalariam Vigia. No bairro do Arapiranga, assim como no povoado de Bituca, fatos extraordinários aconteceram. Dezenas de pessoas, até hoje, juram que divisaram no céu luzes diferentes das estrelas. O assunto chegou a provocar pânico entre os moradores de Bituca, assim como em Colares, onde houve até romaria.

Aparecendo esporadicamente na costa do Pará os discos-voadores, em outubro, chegaram a Santa Isabel do Pará, Benevides e Ananindeua. No final do mês, descobriram Mosqueiro. Em quatro pontos da vila, os objetos voadores não identificados foram divisados. Na Baía do Sol foi registrada a maior incidência.

REALIDADE FANTÁSTICA

Em cada lugar em que apareceu, a desconhecida luz esverdeada deixou uma história para contar. Uma história, na verdade, fantástica. No quilômetro 29 da estrada Santa Isabel do Pará-Vigia, dona Amélia Sarmento quando dormia foi despertada por um foco luminoso vindo do telhado. Espantada, saiu da casa e viu, com absoluta nitidez, um estranho aparelho emitindo uma forte luz que tinha o poder de ultrapassar o teto da residência. Este é apenas um dos fatos contados por aqueles que já viram os discos-voadores.

Outro depoimento inacreditável é o de Antônio Santino, fiscal do mercado de Benevides. Ele, que foi inclusive alvo de pesquisas por órgãos oficiais, numa noite de luar viu um objeto voador não identificado no quintal de seu terreno, retirar um por um os pés de legumes de sua horta. A nave ficou parada a uns seis metros de altura do solo; abriu uma porta e emitiu uma luz que destruiu sua plantação.

Em Colares, um dos lugares preferidos pelos discos-voadores, sua população, constantemente visitada pelos estranhos objetos, ficou em pânico. As vilas de Jucaratuea, Jenipapo, Tupinambás e Fazenda, em polvorosa, organizaram ladainhas. Nestes povoados, como nos demais visitados pelos OVNIs, o incrível passou a ser um lugar-comum uma realidade fantástica das claras madrugadas.

Reportagem: Biamir Siqueira e José Ribamar Prazeres.
Texto final: Nelson Pantoja.

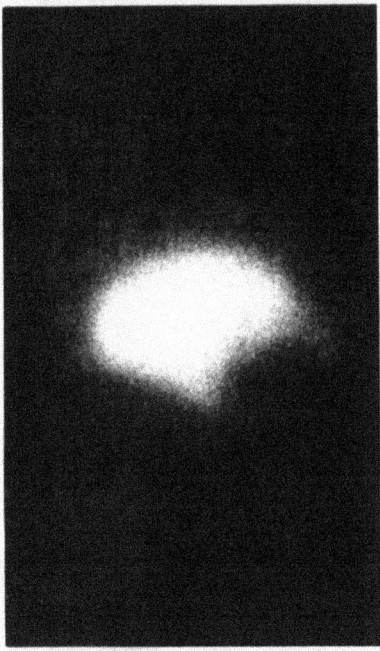

Dia 22, a nave, em forma de pião, é fotografada sobre a baía do Sol

Uma seqüência de fatos fantásticos

Madrugada de 7/06/78. No céu claro da Baía do Sol, inexplicavelmente, uma nuvem começa a se mover em direção contrária ao vento expelindo pequenos raios multicoloridos. Era a primeira vez que se distinguia no firmamento, durante o mês de junho, algo de inusitado. Eram precisamente 3h25min.

Minutos depois um espetáculo fantástico, literalmente inexplicável, seria presenciado pelas pessoas que estavam há três horas de plantão na praça principal da Vila. Quando todos ainda comentavam, perplexos, e tensos o que tinham visto momentos antes, dois objetos voadores não identificados cruzaram o céu em impressionante velocidade.

Com uma lâmpada giratória no topo de sua superfície, os indescritíveis aparelhos, sem o menor barulho, efetuaram diversas evoluções. Navegando lado a lado, os objetos voadores, sempre abaixo das nuvens, cruzavam o céu num vaivém impressionante. Depois de 10 minutos, inesperadamente, estancaram no espaço.

O tempo em que ficaram parados não foi anotado – todos abismados procuravam o melhor ângulo para distinguir sua forma. Quando menos se esperava, numa velocidade vertiginosa sumiram no firmamento.

Depois de um intervalo de cinco dias, no dia 13, foi reiniciada a reportagem. Os objetos voadores não identificados, nesta madrugada, passeram abaixo das nuvens durante 30 minutos. O fato novo: em determinado momento da nave mãe, através de uma abertura na parte lateral do aparelho, saem seis naves menores.

O que aconteceu depois foi extraordinário. A nave mãe dirigiu-se em alta velocidade e parou entre as estrelas. Os corpos estranhos que liberou, entretanto, passaram a cruzar o firmamento realizando inconcebíveis manobras. Momentos depois, desapareceram tão misteriosamente como tinham surgido.

A GRANDE MADRUGADA

Três dias depois, 16/06/78, às 4h35min, o grande acontecimento registrado pela teleobjetiva de 400 milímetros. Um corpo incandescido surge no céu e se aproxima da praça em alta velocidade e pára. Totalmente visível a olho nu, inesperadamente libera seis pequenos objetos voadores não identificados.

Enquanto os seis OVNIs começam a fazer uma série de evoluções nave-mãe, sempre incandescida, passa sobre a praça em moderada dade. De repente, pára e corre vertiginosamente para o firmamento

Foi a primeira vez que a nave ficou tão perto. Observou-se em tão, com bastante cuidado, todos os seus movimentos. Localizado no horizonte como se fosse o planeta Vênus inclusive em termos de diâmetro brilho, o aparelho, repentinamente desce e pára no espaço.

Depois de alguns minutos, a nave incandesce. A impressão que se tem é que ela fica envolta por uma luz concentrada em sua estrutura. Não expele, e isto foi atentamente observado, raios. Nesta forma, nave ficou exatamente, duas horas e trinta e sete minutos, até desaparecer no firmamento." Um detalhe: ninguém viu os aparelhos dispersados, serem recolhidos.

EM OUTROS PONTOS

Nas praias de Colares, Baía do Sol e perto do barrancu que fica em frente ao bairro do Arapiranga, na Vigia, pontos que registraram as maiores incidências dos OVNIs fatos conhecidos como e atento se, realmente, foi a distância em que ficou a nave mãe parada no espaço. Nesta madrugada além dos repórteres Biamir Siqueira e José Ribamar, presenciou o inusitado espetáculo da pracinha da Baía do Sol, o tabelião de Salinópolis. Raimundo Emir d'Oliveira.

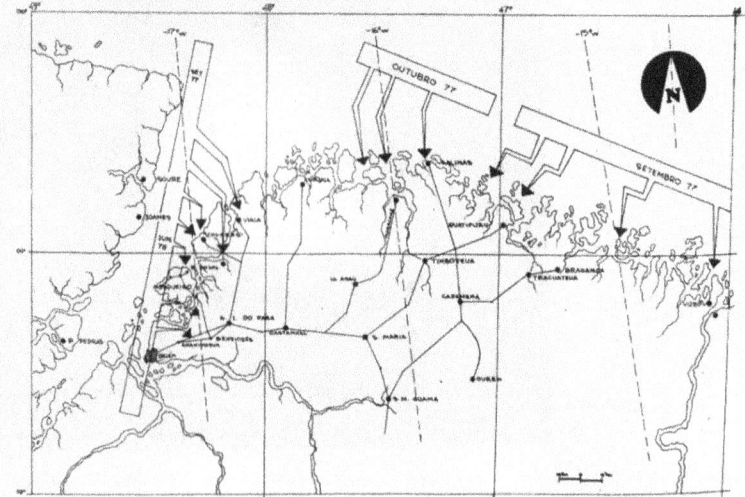

Um mapa inédito: a presença dos discos no Pará. As setas, à esquerda indicam as "visitas" recentes.

Amanhã: As estranhas "coincidências"

O ESTADO DO PARÁ

As estranhas luzes nas noites de Belém

Figure 164. "The strange lights in Belém's nights". O Estado do Pará newspaper. June 26th, 1978.

Acidente com morte depois da farra deixa rastro enigmático

Luz dos mistérios volta aos céus do Maranhão

Figure 165. "The light of mysteries returns to the skies of Maranhão". O Liberal Newspaper. July 14th, 1977.

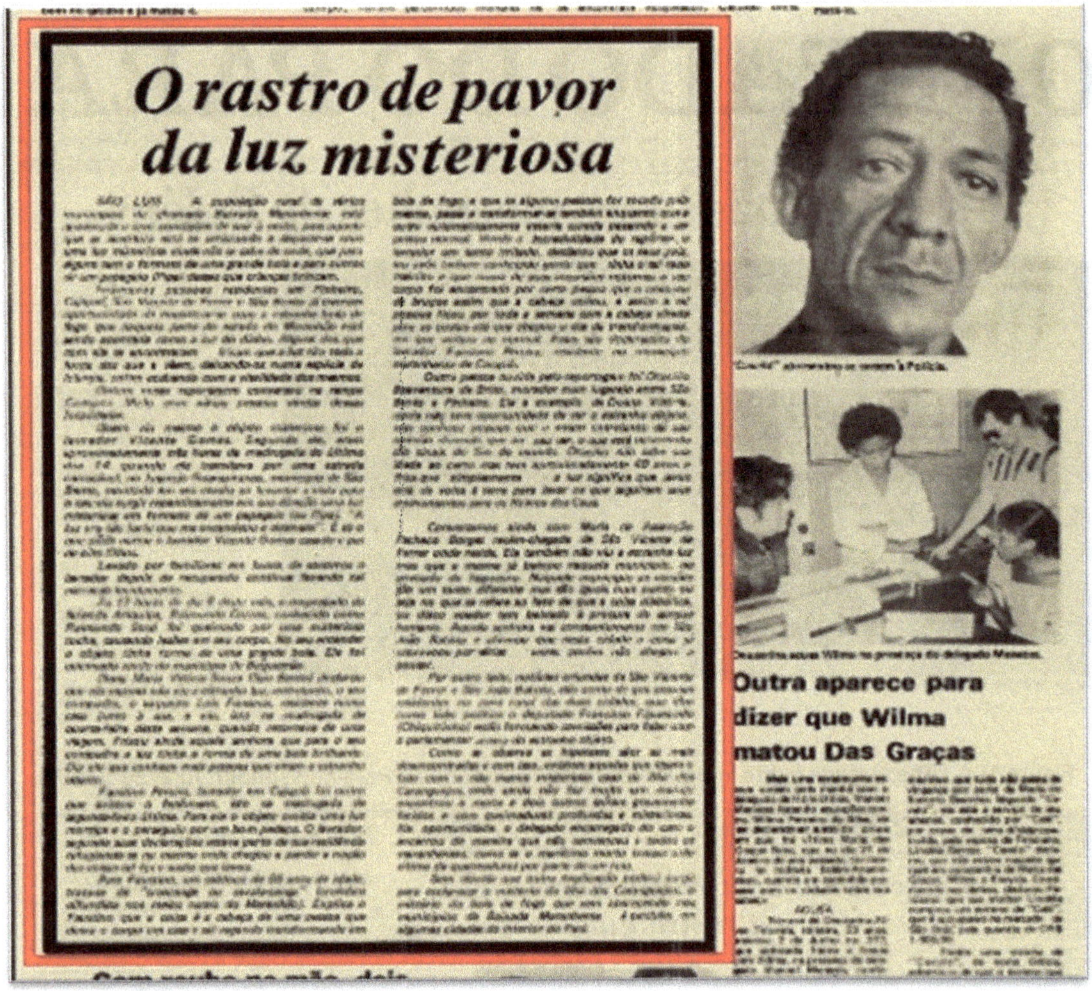
O rastro de pavor da luz misteriosa

Outra aparece para dizer que Wilma matou Das Graças

Figure 166. "The trail of terror of the mysterious light". O Liberal newspaper. July 16th, 1977.

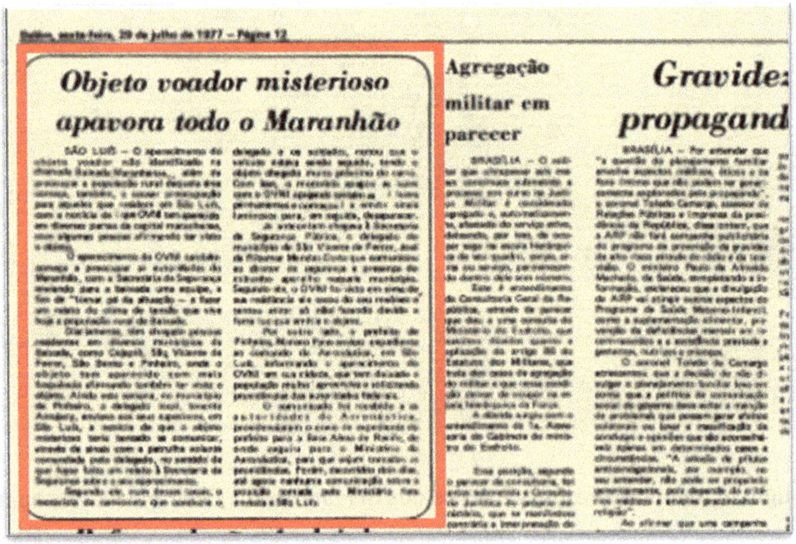
Objeto voador misterioso apavora todo o Maranhão

Agregação militar em parecer

Figure 167. "Mysterious object terrifies the entire Maranhão". O Liberal newspaper. July 29th, 1977.

Figure 168. "The sucking creature attacks women and men in the village of Vigia". "In Maracanã, a heat ray strikes down children". O Liberal **newspaper. October 8th, 1977.**

Chapter 24

Conclusion

Whatever happened to people in the cities of Pará and Maranhão states, and perhaps in other states, remains unexplained to this day, but everything leads us to believe that it was nothing from this world. Could they be extraterrestrial beings on their ships collecting human material? Could they be top-secret terrestrial experiments using a technology that has not yet been shown to us? We do not know. The "*Chupa-Chupa*" or "Vampire Lights" phenomenon is the largest record of ufological sightings and contacts in the entire history of Ufology.

Explanations come along with the fear of the experience they had with something unknown and the confrontation with their religious beliefs, or lack thereof. What did these people see and feel? They were not ordinary lights, the kind that crosses the sky at night from one point on the horizon to another, as occurs with meteorites, weather balloons, atmospheric phenomena, magnetic anomalies and space satellites. The lights were of various sizes, shapes and colors. In dozens of reports, they descended from the sky, chased people and, in some cases, after striking them with light beams, sucked their blood and then abandoned them, unconscious on any deserted street, country path or riverbank. The victims felt physically weak and the result of the attacks was revealed in marks on their skin, on their necks, arms, legs and breasts. Burns that, weeks later, in most cases, disappeared as if they had never existed.

What was it? Where did these lights come from? Were there human war experiments tested among major world powers, using these humble people as guinea pigs? What did they want from the people they attacked? Just to scare them, or to obtain something more? Were the victims guinea pigs in some sinister experiment? From this world or another? Questions without satisfactory answers to a phenomenon that is unparalleled anywhere in the world. A challenge for governments, secret research organizations and military intelligence services. Throughout this work the need to question the "certainties" that were presented by several scholars from Pará to explain the phenomenon will become very clear, as well as the need to doubt the doubt itself and accept other explanations in order to try to understand these ongoing apparitions.

Theories, some bizarre, others conspiratorial, involve stories and characters, mixing them in a kind of cauldron of rumors whose purpose is not to explain, but to confuse. This work also attempts to demystify some myths that had nothing to do with the appearance of the lights. Those who are unaware of the mysteries and legends of the Amazon and the way they have taken root in our culture, especially among the simplest people, have mixed things up to the point of transforming folklore into facts.

On social media, which in the 1970s was nothing more than science fiction, situations have been invented that have nothing to do with the attacks of the lights, trying to establish a causal link for bizarre theories. However, there is nothing that essentially disqualifies the witnesses of the apparitions.

And, where would the films, photos and medical records of the Chupa-Chupa victims be? We, Brazilian ufologists, do not know. There are four possibilities. The first is that these materials were destroyed by the military until the end of the Military Dictatorship in Brazil, in 1985. Fearing that this information would be used by the "democracy" that was returning to the country and that they would be accused of some crime of concealing evidence, they thought it best to destroy everything. The second possibility, and it would be quite "common" in Brazil, is that these materials deteriorated over time and due to lack of care in maintaining them, that is, films were lost, papers and photos became food for moths. The third hypothesis is that these documents were taken, delivered or stolen by agents from other countries who had a direct interest in knowing what was happening. Journalist Carlos Mendes claims to be certain that the CIA was involved in all of it. Witnesses said that there were people who spoke "foreign" in Colares at that time.

There is a video where Hall Puthoff, a researcher and laser specialist, stated that he studied the reports of the *"Chupa-Chupa"* attacks, classifying them as very good.

And of course, we cannot forget our Peruvian Juan, who infiltrated the editorial office of the newspaper *O Estado do Pará* with the intention, as he later confessed, of negotiating the journalists's photos with newspapers in the US and Europe.

And the fourth hypothesis, which is not absurd, is that the Brazilian military still keeps these films and photos in its secret archives. In 2013, I, as a member of the Brazilian Commission of Ufologists and other colleagues, had a meeting at the then Ministry of Defense, at the invitation of the Minister, to officially ask the representatives present from the three armed forces (Navy, Army and Air Force) what we wanted about the Brazilian UFO archives. The answer was that they had nothing, and the Air Force said that everything it possessed had already been released and was available to the public at the Brazilian National Archives.

Let us face it, which of these theories would be the "least bad"? One that shows the military's total disregard for preserving something so impactful and important to the human race? Or another one, which shows that in terms of national sovereignty, this does not exist, while we deal with countries that influence us?

Whatever the cause of these attacks, they were real.

The victims carry a deep pain in their stories. Although many do not know each other, they have similar stories and suffered similar after-effects: anemia (possibly caused by blood loss), coma and psychological disorders.

The origin of these objects may never be revealed, but this is without a doubt the greatest and most terrifying episode in the entire history of world Ufology.

Bibliography

AGHATOS, Stelio e OLIVEIRA, Daniela. *UFOs Rondam a Floresta Amazônica* (Campo Grande/MS, Brazil, Revista UFO, n° 39, p. 8-11, 1995);

ANICETO, Hélio Amado Rodrigues. *Corpos Luminosos. Uma operação militar em busca de respostas* (Niterói/RJ, Brazil, 2014);

ATHAYDE, Reginaldo de. *Ets – Santos e Demônios na Terra do Sol* (Campo Grande/MS, Brazil: Biblioteca UFO, 2000);

ATHAYDE, Reginaldo. *Extraterrestres atacam e matam no Nordeste* (Campo Grande/MS, Brazil: Revista UFO, n° 7, p.7-11, 1989.

ATHAYDE, Reginaldo. *Os ataques do Chupa-Chupa começaram no Ceará* (Campo Grande/MS, Brazil, Revista UFO n° 117, p. 22-23, 2005);

BESSA, Jorge. *Discos Voadores na Amazônia: A Operação Prato* (Limeira/SP, Brazil: Editora Conhecimento, 2016);

CAVALCANTE, Agildo Monteiro. *Ilha de Colares na Amazônia: Fenômeno do Prato Voador* (Belém/PA, Brazil: Editora Café, 2014);

CHAVES, Pepe. *Como as assombrações da Amazônia se tornaram as assombrações de um homem: Coronel Uyrange Hollanda* (Campo Grande/MS, Brazil, Revista UFO, n° 116, p. 30-36, 2005);

EQUIPE UFO. *Alienígenas Representam Perigo no Nordeste* (Campo Grande/MS, Brazil, Revista UFO, n° 86, p. 8-11, 2003);

EQUIPE UFO. *Aliens Rondam a Floresta Amazônica* (Campo Grande/MS, Brazil, Revista UFO, n° 101, p. 08-27, 2004);

EQUIPE UFO. *Dossiê Amazônia: Continua a busca de informações sobre as ações militares na região* (Campo Grande/MS, Brazil, Revista UFO, n° 115, p. 26-35, 2005);

EQUIPE UFO. *O Impressionante Depoimento da Médica que Atendeu as Vítimas do Chupa-chupa* (Campo Grande/MS, Brazil, Revista UFO, n° 116, p. 20-29, 2005);

GEVAERD, Ademar J. *Fotos de OVNIs da Força Aérea Brasileira (FAB)* (Campo Grande/MS, Brazil: Ufologia Nacional e Internacional magazine, n° 3, p. 10-11, 1985);

GEVAERD, Ademar J. *Coronel Rompe Silêncio sobre UFOs* (Campo Grande/MS, Brazil, Revista UFO, n° 54, p. 18-27, 1997);

GEVAERD, Ademar J. *Os Resultados da Operação Prato* (Campo Grande/MS, Brazil, Revista UFO, nº 55, p. 46-52, 1997);

GEVAERD, Ademar J. *A Profundidade dos Casos Registrados na Amazônia* (Campo Grande/MS, Brazil, Revista UFO nº 114, p. 10-13, 2005);

GEVAERD, Ademar. J. *Amazônia – Campo de Experimento de Seres Alienígenas* (Campo Grande/MS, Brazil, Revista UFO, nº 114, p. 14-16, 2005);

GEVAERD, Ademar J. *Na Selva, UFOs deslumbram e amedrontam com seus voos rasantes e ataques impiedosos* (Campo Grande/MS, Brazil, Revista UFO, nº 114, p. 16-29, 2005);

GEVAERD, Ademar J. *Ainda Há Muito a Se Pesquisar no Pará* (Campo Grande/MS, Brazil, Revista UFO, nº 114, p. 31-35, 2005);

GEVAERD, Ademar J. *Não cedi às pressões dos militares* (Campo Grande/MS, Brazil, Revista UFO nº 117, p. 24-31, 2005);

GIESE, Daniel Rebisso. *Vampiros Extraterrestres na Amazônia* (Belém/PA, Brazil, 1991);

GIESE, Daniel Rebisso. *O Fenômeno "Chupa-chupa" na Amazônia* (Campo Grande/MS, Brazil: Revista UFO nº 7, p.13-14, 1989);

GIESE, Daniel Rebisso. Observações ufológicas no Litoral Paraense (Campo Grande/MS, Brazil, Ufologia Nacional e Internacional magazine, nº 3, p. 11-12, julho/agosto 1985);

GIESE, Daniel. *O Fenômeno "Chupa-Chupa": OVNIs atemorizam o estado do Pará* (Campo Grande/MS, Brazil, Ufologia Nacional e Internacional magazine, nº 5, p. 09-15, 1985);

GIESE, Daniel. *Novidades no Fenômeno "Chupa-Chupa"* (Campo Grande/MS, Brazil, Ufologia Nacional e Internacional magazine, nº 7, p. 14-15, 1986);

GOMES, Evelin. *Atividades extraterrestres ainda são registradas em Colares após anos dos primeiros contatos* (Campo Grande/MS, Brazil, Revista UFO, nº 114, p. 26-27, 2005);

MAUSO, Pablo Villarubia. *Las Luces De La Muerte* (Madrid/Spain: EDAF, 2004);

MAUSO, Pablo Villarubia. *Mistérios do Brasil* (São Paulo/SP, Brazil: Editora Mercuryo, 1997);

MAUSO, Pablo Villarubia. *Quando os UFOs Atacam* (Curitiba/PR, Brazil. Biblioteca UFO, 2021);

MAUSO, Pablo Villarrubia. *O Mistério das Luzes Assassinas na Amazônia* (Campo Grande/MS, Brazil, Revista UFO, nº 86, p. 32-35, 2003);

MENDES, Carlos. *Luzes do Medo* (Curitiba/PR, Brazil: Biblioteca UFO, 2019);

PETIT, Marco Antônio. *UFOs no Brasil: É hora de nossos militares encararem a verdade* (Campo Grande/MS, Brazil, Revista UFO, nº 115, p. 16-22, 2005);

PETIT, Marco Antônio. *Dossiê Amazônia: O último depoimento de Uyrangê Hollanda fornece inspiração para reflexões* (Campo Grande/MS, Brazil, Revista UFO, nº 117, p. 14-20, 2005);

PETIT, Marco Antônio. *Ufos Arquivo Confidencial – Um mergulho na Ufologia Militar Brasileira* (Campo Grande/MS, Brazil: Biblioteca UFO, 2007);

PRATT, Bob. *Perigo Alienígena no Brasil* (Campo Grande/MS, Brazil: Biblioteca UFO, 2003);

PRATT, Bob. *UFO Danger Zone: Terror and death in Brazil – where next?*(Inner Light Publications & Global Communications, 1996);

ROMERO, Alberto. *Verdades que Incomodam* (Campo Grande/MS, Brazil: Biblioteca UFO, 1999);

SILVESTRE, Fabiana. *UFOs Rondam a Floresta Amazônica.* (Campo Grande/MS, Brazil, Revista UFO, nº 75, p. 10-18, 2000);

TICCHETTI, Thiago Luiz. *Guia da Tipologia dos UFOs* (Curitiba/PR, Brazil: Biblioteca UFO, 2018);

TICCHETTI, Thiago Luiz. *UFO Contacts in Brazil* (London/UK, Flying Disk Press, 2019);

TICCHETTI, Thiago Luiz. *The Chupa-Chupa action on Colares Island was much more serious than one imagines* (London/UK, UFO Truth magazine, nº 63, p. 19-23, 2023);

TICCHETTI, Thiago Luiz. *Mission Para - Operation Saucer, Chupa-Chupa & Unprecedented Cases* (London/UK, UFO Truth magazine, nº 65, p. 22-31, 2024);

Documents & Reports

Chronological Input Files [ACE 3370/83] – (Serviço Nacional de Informações – SNI);

Operação Prato – 01.01.01 – Mission Report 1 – (FAB – 1st COMAR);

Operação Prato – 01.01.02 – Mission Report 2 – (FAB – 1st COMAR);

Operação Prato – 01.02.00 – Operational Information 1 – (FAB – 1st COMAR);

Operação Prato – 01.03.00 – Occurrences' Illustrations – (FAB – 1stCOMAR);

Operação Prato – 02.00.00 – Chronological Summary – (FAB – 1stCOMAR);

Operação Prato – 03.01.01 – Case Report 1 – (FAB – 1stCOMAR);

Operação Prato – 03.01.02 – Case Report 2 – (FAB – 1stCOMAR);

Operação Prato – 03.01.03 – Case Report 3 – (FAB – 1º COMAR);

Operação Prato – 03.01.04 – Case Report 4 – (FAB – 1º COMAR);

Operação Prato – 03.02.01 – Case Report 5 – (FAB – 1º COMAR);

Operação Prato – 03.02.02 – Case Report 6 – (FAB – 1º COMAR);

Operação Prato – 03.02.03 – Case Report 7 – (FAB – 1º COMAR);

Operação Prato – 03.02.04 – Case Report 8 – (FAB – 1º COMAR);

Operação Prato – 03.02.05 – Case Report 9 – (FAB – 1º COMAR);

Operação Prato – 03.02.06 – Additional Report 1 – (FAB – 1º COMAR);

Operação Prato – 03.02.07 – Additional Report 2 – (FAB – 1º COMAR);

Operação Prato – 03.02.08 – Additional Report 3 – (FAB – 1º COMAR);

Operação Prato – 03.02.09 – Additional Report 4 – (FAB – 1º COMAR);

Operação Prato – 04.00.00 – Occurrence Sheet 1 – (FAB – 1º COMAR);

Operação Prato – 05.00.00 – Special Report 1 – (FAB – 1º COMAR);

Operação Prato – 06.00.00 – Information Bureaus 1 – (FAB – 1º COMAR);

Operação Prato – 07.01.00 – General Terms 1 – (FAB – 1º COMAR);

Operação Prato – 07.02.00 – General Report 1 – (FAB – 1º COMAR);

Report 1- Mission Report – Parte Informativa;

Report 2 – Mission Report – II – Parte Informativa;

Report 3 – Mission Report – Ubintuba;

Report 4 – Mission Report – II – Parte Operacional;

Report 5 – Fishermen Report;

Report 6 – Comentários e Aspecto Psico-Social e Econômico;

Report 6 – Comentários e Aspecto Psico-Social e Econômico;

Report 7 – Mission Report – I – Parte Informativa;

Report 8 – Mission Report – II – Parte Informativa;

Report 9 – Extra 01;

Report 10 – Extra 02;

Report 11 – Extra 03;

Report 12 – Extra 04;

Report 13 – Extra 05;

Report 14 – Extra 06;

Report 15 – Mission Report – Jeju Farm;

Report 16 – Agent's Report – Jeju Farm.

Report 17 – Mission Report – Partes Operacional e Informativa – Fazenda Jeju;

Report 18 – Agent's Report – Extra 07;

Report 19 – Agent's Report – Extra 08;

Report 20 – Agent's Report – Extra;

Report 21 – Agent's Report – Extra 09;

Report 22 – Agent's Report – Extra 10;

Report 23 – Agent's Report – Extra 11;

Report 24 – Random Reports;

Report 25 – Extra 12;

SUPER 8 FOOTAGE REPORT – www.operacaoprato.com

MEDICAL REPORT – www.operacaoprato.com

Websites

www.operacaoprato.com

www.ufo.com.br

www.fenomenum.com.br/caso-chupa-chupa-e-operacao-prato

ABOUT THE AUTHOR

Born on February 5th, 1975, in Santa Cruz, state of Rio de Janeiro. His father, Luiz Mauro Ticchetti, was a Brazilian Air Force colonel. Thiago Luiz Ticchetti has been researching the UFO phenomenon for more than 30 years, having investigated hundreds of cases and participated in dozens of sky watches.

In 1997, he attended the 1st World Forum of Ufology, held in Brasília, where he was invited by Roberto Affonso Beck to join the Brazilian Entity for Extraterrestrial Studies (EBE-ET). For more than 10 years, he took an active part of the organization, becoming vice president of it.

He has been part of Revista UFO magazine since 1997, where he started as a volunteer translator, becoming a consultant, international coordinator, and co-editor, and since 2023 he has edited the magazine. Ticchetti was president of the Brazilian Commission of Ufologists (CBU) for seven years and is now its International Director. In 2017, he was appointed as Assistant to the International Director of Mutual UFO Network (MUFON) in Brazil. Since 2021,he has been the National Director of MUFON in Brazil. Ticchetti is a MUFON certified field researcher and member of MUFON's ERT and Star Team. In 2024, he became a member of The Scientific Coalition for UAP Studies (SCU). Author of more than a hundred articles for Revista UFO magazine, MUFON Journal and other publications around the world, he has published 15 books so far: "*Quedas de UFOs*" (2002), "*Arquivos UFO: Casos Ufológicos - Volume I*" (2013), "*Guia da Tipologia Extraterrestre*" (2014), "*Quedas de UFOs II*" (2015), "*Universo Insólito: O Livro de Bordo Partes 1 e 2*" (2015) , "*Arquivos UFO: Casos Ufológicos Volume II*" (2015), "*Arquivos UFO: Casos Ufológicos Volume III*" (2017), "*Guia da Tipologia dos UFOs*" (2017), "*Guia da Tipologia Extraterrestre – 3rd edition*" (2019), "UFO Contacts in Brazil" (2019), "*Cartas para Claudeir Covo*"(2022), "Contacts OVNIs Au Brésil" (2021), and "*Guia da Tipologia Extraterrestre – 4th edition*" (2023).

Thiago Luiz also had the opportunity to interview the greatest researchers of the UFO community, such as Nick Redfern, John Alexander, Don Schmitt, David H. Childress, Don Ledger, Phillip Mantle, David Jacobs, Kevin Randle, Stanton Friedman, Nick Pope, Jerome Clark, Graham Birdsall, Wendelle Stevens, Jacques Vallée, Paolla Harris, Preston Dennett, Linda Zimmerman, Graeme Rendall, Mike Masters, Jeff Sortino, Robert Salas, Mike Bara, Bruce Maccabee and among others. He has participated in dozens of interviews for podcasts, TV, YouTube channels and TV programs on National Geographic, History Channel, Discovery, Netflix and many more. He was the first and only Brazilian researcher to have articles published by the British magazine UFO Matrix. He is currently a columnist for UFO Truth Magazine.

Contacts:

E-mail: tlticchetti@yahoo.com.br

Revista UFO magazine: www.ufo.com.br

Facebook: www.facebook.com/Thiagoticchetti

Youtube: www.youtube.com/c/ThiagoLuizTicchetti

Twitter: https://twitter.com/TLTufologo

Instragram: @thiagoticchetti and @revistaufo

******

Un-X Media Publications

Haunted Independence Missouri by Margie Kay 2016, 2025
Gateway to the Dead: A Ghost Hunter's Field Guide by Margie Kay 2016
Family Secrets by Jean Walker 2017
The Kansas City UFO Flaps by Margie Kay 2017
A Sonoma County Phenomenon: Evidence for an Interdimensional Gateway by Margie Kay 2019
The Fast Movers: Evidence for High-Speed UFOs/UAPs
by Margie Kay, Bill Spicer, and Larry Tyree 2020
Journey to Spirit by Devin Listrom 2020
Winged Aliens by Margie Kay 2021
The Remote-Viewing Workbook by Margie Kay 2019 (on LULU)
The Master Dowsers Chart Book by Margie Kay 2021 (on LULU)
Rules for Goddesses by Margie Kay 1999
The Alien Colonization of Earth's Waterways by Debbie Ziegelmeyer 2021
50th Anniversary of the SE Missouri Ozarks UFO Flap
by Debbie Ziegelmeyer and Margie Kay 2022
Meeting Wallace by Larry Costa 2023
Poems by Pat Delap by Pat Delap 2025
Holiday Poems and Recipes by James Bair 2023
Adult Coloring Books for meditation by M.K. 2023 -2005
Earth's Unseen Inhabitants by Larry Tyree, Bill Spicer and Lily Nova 2025
Incident in Varginha: Space Creatures in the South of Minas by Vitório Pacaccini and Fernanda Pires 2025
How to Research a Haunted House by Margie Kay and Violet Wisdom 2025
All Monsters are Human by Derrick Smith 2025
We got it fromTHEM: by Dr. Gregory Rogers 2025
UFO ATTACKS IN BRAZIL: by Thiago Luiz Ticchetti 2025
Earth's Unseen Inhabitants by Bill Spicer, Larry Tyree, Lily Nova and Margie Kay 2025
Take a Haunted Road Trip on Route 66 by Margi e Kay 2025

Coming soon:
Missouri: UFO Hot Spot by Missouri MUFON 2025
THOR: The Extraterrestrial on Earth by Margie Kay 2025
Upgrade for Humons by Peleg Yagen 2025
How to Design a Medicinal Herb Garden by Faun Grey 2025

Magazines:

Un-X News Magazine 2011-2025 in print and digital
Wood-Fired Magazine 2007-2019
MCSC Magazine 2003-2019

Documentary Films:

PORTALS

The Cubes

Mysterious Missouri

Un-X Media is seeking authors who write fiction and non-fiction books about unexplained phenomena, paranormal subjects, alternative health, and esoteric knowledge. Contact us at www.unxmedia.com for more information.

Un-X Media and the Un-X Broadcasting Network are subsidiaries of G&M Enteprises, LLC.

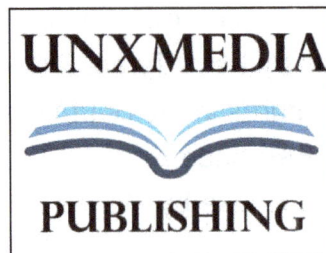
UNXMEDIA PUBLISHING

www.ingramcontent.com/pod-product-compliance
Lightning Source LLC
Chambersburg PA
CBHW081734270326
41932CB00020B/3273